Genetic Resources and Seed Enterprises

GENETIC RESOURCES AND SEED ENTERPRISES

Management and Policies

Part - 1

HARI HAR RAM
Professor and Head, Vegetable Science and Nodal Officer
Pantnagar Centre for Plant Genetic Resources
G. B. Pant University of Agriculture and Technology, Pantnagar (U.A.)

&

RAKESH YADAVA
Senior Research Officer (Cotton Breeding)
Emergent Genetics / Monsanto India, Hyderabad

New India Publishing Agency
Pitam Pura, New Delhi- 110 088

ISBN
Set : 978-81-89422-65-3
Part 1 : 978-81-89422-61-5
Part 2 : 978-81-89422-66-0

Typeset at:
Laxmi Art Creations
Delhi-110 034

Printed at :
Jai Bharat Printing Press
Delhi

Published by:
Sumit Pal Jain *for*
New India Publishing Agency
101, Vikas Surya Plaza, CU Block, L.S.C. Mkt.,
Pitam Pura, New Delhi- 110 088, (INDIA)
Phone: 011-27341717, Fax: 011-27341616
E-mail: newindiapublishingagency@gmail.com
Web: www.bookfactoryindia.com

Preface

Plant genetic resources (PGR) have a unique place within the overall ambit of biodiversity - and are of great value in providing food, fuel, clothing, medicine and shelter for the whole mankind. The significance of plant germplasm resources dates back to mid eighteenth century, when Sir Joseph Banks, the Director of Kew Botanical Gardens, England, accompanied Captain Cook on a plant collecting voyage. Plant genetic resources are, however, increasingly being threatened due to degradation of their habitats, changes in ecology, cropping systems, modernization of agriculture, rapid replacement of locally adapted indigenous cultivars by modern high yielding varieties and effect of urbanization. Plant genetic resources are going to be the back bone of plant breeding and seed industry enterprises which are occupying now the cenre stage in India. The success of seed industries in term of upgrading their products will greatly depend on the sustainable management of the plant germplasm. The scientific management of plant germplasm cannot be in isolation, any more. It has to be interwoven in the R and D programme of any successful plant breeding and seed production enterprise. These days, a lot of international agreements and national legislations on plant germplasm management and plant variety protection have been put in place and working knowledge of the same is essential for plant breeders, plant germplasm managers and seed industry personnel. The use of transgenics/GMOs has generated a lot of heat world-over. Regulatory procedures on their commercial release and management have been adopted and are under constant review and modifications. Almost all the CGIAR funded international agricultural research centres, have independent units/divisions on plant genetic resources management. IPGRI is coordinating and strengthening the PGR management programmes at the international level. National Bureau of Plant Genetic Resources, New Delhi is the nodal organization in India on PGR management. Crop based ICAR institutes are giving greater importance to PGR management. Even state agricultural universities are now initiating concerted efforts to strengthen teaching and research activities on PGR management. Indian Agricultural Research Institute, New Delhi has already started awarding Master level degree in PGR management. State Agricultural Universities and other organizations are giving greater coverage to this subject including seed technology, transgenics and intellectual property rights. Therefore, an attempt has been made in this book to put all these issues at one place in a simple and understandable manner without compromising on the core of the complex matter.

This exercise would not have been possible without the active support and opportunity provided by Dr. P. L. Gautam, Vice-Chancellor, GBPUAT and Dr. S. B. Singh, ex-Vice-Chancellor, GBPUAT and presently Vice-Chancellor, Dr. Ram Manohar Lohia Awadh University, Faizabad. We express our gratitude to them. Behind the scene players in this endeavour were Miss Gunjan Sharma, JRF and Dr. Kavita Sharma, SRF. They deserve our gratefulness and appreciations. We would like to single out Miss Gunjan Sharma who has just been superb in providing necessary assistance in this matter. She has put in genuinely sincere and exemplary efforts. It is just a privilege to thank her for ever. I profusely thank the entire staff of New India Publishing Agency for their promptness, support and co-operation.

November 13, 2006

Hari Har Ram
Rakesh Yadava

Contents

Section - II : Seed Enterprises

Part - 2

Section - II : Seed Enterprises (Contd...)

Section - III : IPR Issues

x-x-x

Section - I
Plant Genetic Resources

1

Plant Genetic Resources

Allard (1960) defines germplasm as "the sum total of the hereditary materials in species" The word germplasm in context of plants is now referred to as plant genetic resources i.e. PGR.PGR is a component of the world's biological diversity. The term biological diversity is a vast and often undervalued resource. Biological diversity includes every form of life, from the smallest microbe to the largest beast. Biodiversity is the variety and variability of all plants, animals and microorganisms and the ecological complexes of which they are part. The earth's biodiversity - its ecosystems, species and genes - are the product of over 3000 million years of evolution. Throughout this time, small changes have accumulated in populations, resulting in a multitude of living forms closely adapted to the physical conditions they face and to each other. They supply all our food, much of our raw materials and energy and many of our medicines. Intact eco-systems also play a central role in the functioning of biosphere. Plants are of crucial importance in stabilizing climate, protecting watersheds on soils and maintaining the chemical balance of the earth. When key species are lost, vital ecological services are disrupted (Zedan, 1995).

Importance of Plant Genetic Resources/Plant Diversity

Plant genetic diversity is the crucial component of any agricultural production system. In absence of plant genetic diversity, there will be no natural evolutionary adjustment of the system to changing environmental and biotic and abiotic conditions. As a matter of fact, the first step in any plant breeding programme is to identify plants that exhibit variation for the trait or plants of interest to the plant breeder. Plant germplasm is an essential resource for successful plant breeding. Plant biodiversity has been used as the foundation for modern plant breeding. 10,000 years ago, neolithic farmers in Middle East and elsewhere developed some of the most important crops and livestock from their wild relatives. It was soon recognized by them that certain species were more suitable than others for their needs and that within populations, some plants and animals were more suited to their specific requirements. They selected, bred and used these individuals in preference to others and this practice continues even today. This is precisely the art and science of plant breeding. Through selection and breeding using plant biodiversity as a base, human societies have developed thousands of local races of crops and livestock, each fitting a particular need in a specific physical environment and evolving in harmony with diverse systems of land and natural resources management of which they are integral parts.

These local land races are not only numerous and adapted to varying conditions, but are also genetically variable. This built-in variability ensures that even in bad years at least some individuals can survive and pass on adaptability to adverse climatic conditions, drought, water logging, low fertility, problem soils and diseases and pests. The maintenance of this diversity has until recently been ensured by traditional systems of agriculture and land use. In some cases, it depends on introgression with wild and weedy relatives growing nearby in both natural habitat and disturbed field margins.

However, with the advent of modern day scientific plant breeding new varieties began to be developed in response to the needs of modern day agriculture. In this process, breeders assembled desirable genetic traits from different varieties, and in some cases from different species leading to the development of high yielding genotypes to sustain modern societies. High performance in the intensive agriculture system requires genetic uniformity. This has been achieved by sacrificing diversity in favour of a limited number of characteristics that are needed at a particular time.

The development of high yielding varieties with focus on increasing productivities has helped in meeting the food requirement of a rapidly growing world population. However, this has created a number of problems also. The modern day, high yielding varieties are not well suited to the needs of marginal farmers who is dependent on low inputs. The high yielding varieties respond more favourably to substantial external inputs like water, fertilizers and pesticides which are often deleterious to the environment. The high potential of these varieties is not realized without much of costly inputs. These varieties are however more vulnerable to the diseases and insect pests. Large areas under genetically uniform seeds under one specific high yielding variety encourage the rapid spread of pests and pathogens. Plant breeders therefore look towards the availability of the pool of diverse genetic material (plant biodiversity) represented by local races and wild relatives so that they may remain one step ahead of tomorrow's unexpected calamities. Without a diverse genetic reservoir to draw from, further improvement in plant breeding may not be possible.

It is well understood that the value of biodiversity shall outweigh the conservation cost significantly. There are evidences on the economic benefits of using genetic diversity in conventional crop breeding programmes. Economic gains worth billions of dollars in higher crop yields have been obtained by using wild relatives in commercial crop improvement programmes. A few examples will illustrate the point. In Asia, by the mid 1970s genetic improvement had increased wheat production by US$2 billion and rice production by $1.5 billion a year by incorporating dwarfism into both crops. A 'useless' wild wheat plant from Turkey was used to give disease resistance to commercial wheat varieties worth $ 50 million annually to the USA alone. One gene from a single Ethiopian plant now protects some varieties of barley from yellow dwarf virus. An ancient wild relative of maize from Mexico-a perennial with resistance to seven major diseases and which can grow at high altitudes on marginal soils - can be crossed with modern varieties with potential savings to farmers estimated at $ 4.4 billion annually worldwide (Zedan, 1995).

The problem of crossability between wild and domesticated species is often a limiting factor for effective utilization of genes from wild relatives. The biotechnology revolution including recent developments in recombinant DNA technology which is often synonyms with genetic engineering/transgenics/GMOs raises the prospects of transferring the genes from species which were otherwise non-crossable. Moreover, plant biotechnology permits the better understanding of gene expression. This knowledge can be utilized to facilitate the use of plant germplasm in the development of superior crop varieties. Therefore, there is ever increasing need to maintain the richest possible plant genetic resource. The future of biotechnology, depends on conservation of wide range of genetic diversity. As the field of biotechnology develops, the range of plant genetic resource diversity is bound to increase. Thus, the stake which the plant biotechnology industry holds in the conservation of biodiversity should not be underestimated.

The role of plant germplasm diversity can be better understood if we realize that the genetic base of a staple food crops is generally very narrow. Half of the Canadian provinces are planted with just one variety of wheat. The USA's entire soybean industry originates from just 6 plants from one place in Asia. The number of different species on which mankind relies for food, fibre, timber, medicines and other natural products is extremely limited. Only small number of plant and animal species have yet been evaluated for their usefulness. Of an estimated 2,65,000 species of plants, only about 7000 have ever been cultivated for food. Of these, 20 species current supply 90 % of world's food and just three crops, namely, wheat, maize and rice supply more than half of the world's food. Many fruit crops of the regional or local importance have been rather neglected by the scientists. There is much potential to look beyond today's major crops at other species that may have value either directly or indirectly as sources of new genes. Therefore, the importance of plant biodiversity is obvious. Plant biodiversity is essential for:

- sustainable development,
- to meet the requirements for adaptation to a changing environment,
- continued functioning of the biosphere, and
- is essential for survival of mankind itself.

Sources of Plant Germplasm

Gene banks provide plant breeders with sources that may be of importance in plant breeding programme. The free flow of plant germplasm around the world is essential for continuous success in plant breeding and increasing food production. The seed and vegetatively propagated species from seed gene banks and field gene banks serve as basic genetic material to create variation following hybridization and selection. The common germplasm resources for plant breeders are as follows (Stoskopf, 1993) :

- Currently grown commercial cultivars
- Obsolete commercial cultivars
- Breeding lines and stocks
- Landrace cultivars
- Undomesticated forms of crop plants
- Plant introductions
- Related species
- Mutations
- Interspecific and intergeneric hybridization

Measures for Plant Biodiversity Conservation

There has been International consensus on the fact that intense pressure on biodiversity will continue to increase unless appropriate measures for conservation and sustainable use are initiated. This issue has gained momentum for more than two decades starting with the United Nations Conference on the Human Environment in 1972. Since then there has been the establishment of the International Board for Plant Genetic Resources (IBPGR) in 1974 which became the International Plant Genetic Resources Institute (IPGRI) in 1994. This was followed by the adoption, by the Food and Agriculture Organization (FAO) Conference in 1983 of the International Undertaking on Plant Genetic Resources and the establishment of the Inter-Governmental Commission on Plant Genetic Resources. Several documents as Environmental Perspective to the Year 2000 and Beyond, the Report of the World Commission on Environment and Development, Caring for the Earth : a Strategy for Sustainable Living, the Global Biodiversity Strategy and Agenda 21 of the United Nations Conference on Environment and Development also emphasize the importance of protecting biological diversity and provide general principles for action.

The complete action and policy resolves on plant conservation by government and developmental institutes need to be accelerated. The proactive action plan is hampered by :

- Lack of sound national biodiversity strategies
- Lack of action plans

- Lack of adequate trained manpower
- Lack of adequate funding

It is irony that the two thirds of all species occur in developing countries, particularly in the tropics which are not able to provide adequate finances for biodiversity conservation and sustainable use. On the other hand, the developed countries have the technology to exploit and benefit from plant biodiversity. It is understood that activities will be proposed within the framework of the Convention on Biological Diversity (CBD) which will no doubt add further to the financial burden of those countries that are rich in biodiversity or that host rare or endangered species. Therefore, ways of sharing the costs and benefits of conserving genetic resources will have to be devised. All the key players will be required to take social, political and economic decisions to derive benefits while using the available plant genetic resources and the technologies. The potential of plant genetic resources can best be utilized if they remain accessible to all users and if the information and technologies on how to use them is widely disseminated. For the welfare of mankind as a whole, there should be no exclusive rights or ownership to the valuable genetic resources and the technical capability to develop new products. No country is self sufficient in genetic resources. While possession and custody of a potential genetic resource might be limited to one nation, the benefits on this resource should flow to all the nations. There have to be a fair balance of benefit between custodian and consumer. The Convention on Biological Diversity (CBD) is engaged in addressing these important but difficult issues. Worldwide investment in the conservation of plant genetic resources has been very modest and is by no means secure. The integration of conservation and development plans is almost non-existent. National cross-sectoral cooperation is limited. The resources, power and capacity to promote conservation are unequally distributed. Local people are rarely brought into the process of planning the conservation and management of the genetic resources they shape and maintain. However, there is general agreement both in the scientific community and among the public in both industrialized and developing countries that there is a need for global effort to conserve as much of the gene pool crop plants as possible and to tap their reservoir for the benefit of all (Zedan, 1995).

References

Allard, R. W. 1960. Principles of Plant Breeding. John Wiley and Sons, Inc., New York and London. pp 485.

Stoskopf, N. C. 1993. Plant Breeding : Theory and Practice. West View Press, Inc. pp 531.

Zedan, H. 1995. Loss of plant diversity : a call for action. *In*: L. Guarino, V. Ramanatha Rao, R. Reid. (eds.). Collecting Plant Genetic Diversity : Technical Guidelines. CAB International. pp ix –xiv.

2

Plant Genetic Resources - National Perspective

The National Bureau of Plant Genetic Resources, commonly known as NBPGR was established by the Indian Council of Agricultural Research (ICAR) in 1976 with its main campus at New Delhi. Being the nodal organization in India it has been given the national mandate to plan, conduct, promote and coordinate all activities concerning plant exploration and collection and also for safe conservation and distribution of both indigenous and introduced genetic variability in crop plants and their wild relatives. The Bureau is also vested with the authority to issue Import Permit and Phytosanitary Certificate and conduct quarantine checks on all seed materials and plant propagules (including transgenic material) introduced from abroad or exported for research purposes.

Besides having a 40 ha experimental farm at Issapur village (about 45 km west of Delhi) the Bureau also has a network of 11 regional stations/base centers that provide access to representative agro-ecological situations in the country. It has strong linkages with leading crop- based Institutes, National Research Centers, All India Coordinated Crop Improvement Projects, State Agricultural Universities and other stake holders. NBPGR also works in close collaboration with several international institutes/organizations through memoranda/workplans developed under bilateral/multilateral agreements. The Bureau not only provides genetic resources to on-going crop improvement programmes to sustain continued advances in agricultural productivity and stabilize production, but also conserves them safely to meet needs of future generations.

Mandate

To act as the nodal institute at national level for acquisition and management of indigenous and exotic plant genetic resources (PGR) for agriculture, and to carry out related research and human resource development for sustainable growth of agriculture.

Objectives

- To plan, organize, conduct and coordinate exploration and collection of indigenous and exotic plant genetic resources
- To undertake introduction, exchange and quarantine of plant genetic resources
- To characterize, evaluate, document and conserve crop genetic resources and promote their use in collaboration with other national organizations
- To develop information network on plant genetic resources
- To conduct research, undertake teaching and training, develop guidelines and create public awareness on plant genetic resources

Organizational Set-Up

The Director, NBPGR is overall in-charge of administration, research management and coordination. The Institute Management Committee, Research Advisory Committee, Crop Germplasm Advisory Committees and the Staff Research Council play important roles. The Bureau functions through its four main Divisions, namely i) Plant Exploration and Collection, ii) Plant Quarantine, iii) Germplasm Evaluation and iv) Germplasm Conservation. The Bureau has units of Germplasm Exchange, Tissue Culture and Cryopreservation (TC&CP), Agricultural Research Information System (ARIS) and Policy Planning and Monitoring (PPM).

A principal scientist/senior scientist heads each Division/Unit. Other centralized services include units of Administration and Management, Purchase, Stores, Maintenance, Audit and Accounts, Security and Library. Regional Stations/Base Centers, headed by a principal scientist/senior scientist, are located at Shimla, Bhowali, Shillong, Jodhpur, Akola, Hyderabad, Amravati, Thrissur, Srinagar, Ranchi and Cuttack. It also houses NRC on DNA Fingerprinting, an All India Coordinated Research Project on Under-Utilized Plants, National Agricultural Technology Project on Plant Biodiversity (now phased out) and Household and Nutritional Food Security Project. The total sanctioned staff strength is 504 comprising 128 scientific, 127 technical, 78 administrative and 171 supporting staff.

National Genebank

The Indian National Genebank was established by the council at NBPGR to conserve national heritage of germplasm collections in the form of seeds, vegetative propagules, tissue/cell cultures, embryos, gametes, etc. Based on experiences gained from working with a built-in cold storage vault obtained from UK in 1983, four modules (two units of 100 m3 and two of 176 m3 capacity) were installed for long-term storage of seeds of orthodox species kept in laminated aluminium foils at -20°C after drying them to 5-7% moisture content. Stand-by diesel generator backs up the electricity supply. Vegetatively propagated clonal materials and recalcitrant seeds species are being maintained under field conditions backed up by tissue culture repositories. The Bureau has a strong programme on in vitro conservation and cryopreservation.

The new Genebank facility commissioned in 1997 has 13 modules, each with a storage capacity of 76,000 samples. One of these modules is used for medium term storage of active germplasm collections and the rest for base collections for long-term storage. Its cryopreservation facility contains six liquid nitrogen tanks (cryo-tanks), each containing 1000 litres of liquid nitrogen. These six cryo-tanks have a total capacity to store 0.25 million samples. Thus the new Genebank has a total capacity to store 1.25 million samples. This is one of the most modern Genebanks in the world.

Indian National Plant Genetic Resources System (INPGRS)

NBPGR is gradually developing and strengthening the national plant genetic resources system by linking up the National Base Collection (kept under long-term storage at NBPGR) with 57 National Active Germplasm Sites responsible for different crops where germplasm collections are evaluated and multiplied under field conditions, backed by medium-term storage facilities. The Research Advisory Committee and Crop Germplasm Advisory Committes for different crops advise the Bureau regarding improving the capability, efficiency and effectiveness of its services for management of PGR in the country.

International Collaboration

NBPGR implements work plans developed under MoU between ICAR and IPGRI. FAO and IPGRI also sponsor regional training courses on conservation and utilization of genetic resources of local crops of agricultural importance in South Asia and adjoining regions to be conducted by NBPGR.

Besides working closely with IPGRI, NBPGR also collaborates actively with the International Agricultural Research Centers (IARCs) like ICRISAT, IRRI, ICARDA and CIMMYT. It exchanges plant germplasm with more than 80 countries and implements work plans developed under bilateral, regional and international agreements.

Training Programmes and Information Services

The Bureau organizes advanced training programmes focusing on scientific procedures for collection, exchange, quarantine, biosafety, DNA fingerprinting, evaluation, documentation and conservation linked to use of plant genetic resources. Major accomplishments of its staff are published in Annual Reports. NBPGR Newsletter is brought out quarterly. Crop Catalogues, Plant Germplasm Reporter, Minimal Descriptors of Agri-horticultural Crops, Training Course Manuals, etc. based on computerized data are also developed and published.

Bureau's library at Headquarters specializes in information dealing with plant genetic resources and also subscribes to 50 foreign and 53 national journals.

Post-Graduate Teaching Programme

Since academic session 1997, Bureau is undertaking teaching in plant genetic resources leading to M.Sc. degree linked with Post Graduate School, IARI, New Delhi.

Extension Services

Bureau organizes Plant Biodiversity Fairs/field days for rabi and kharif crops and distributes seeds/planting material along with relevant literature on technical know-how for raising crops and management of PGR. Under the National Agricultural Technology Project- Plant Biodiversity (NATP-PB), special emphasis has been given to create PGR awareness among grass root level workers, tribal people, and farmers (particularly women) by organizing biodiversity fairs in villages. Students on educational tours from State Agricultural Universities are invited to visit the National Genebank, DNA fingerprinting, tissue culture and quarantine labs, plant quarantine glasshouses and containment facilities at New Delhi (NBPGR, 2003).

National Active Germplasm Sites Under Indian Plant Genetic Resources System

Recognising the need for conservation and management of Plant Genetic Resources (PGR) at national level, the ICAR entrusted the responsibilities for all activities related to PGR to National Bureau of Plant Genetic Resources (NBPGR) to function as a nodal institute and develop, operate and coordinate the Indian Plant Genetic Resources system. To fulfill these objectives a network approach has been followed, which comprises the National Genebank at NBPGR, New Delhi for conservation of base collection and crop based National Active Germplasm Sites (NAGS) for management and maintenance of active collections. In order to meet these obligations for sustainable conservation and use and biodiversity, and to discharge responsibilities under various national legislations and international agreements, it was decided to formally recognize the NAGS and charge them with specific responsibilities. The NBPGR shall facilitate need base strengthening of infrastructure and appropriate Human Resource Development at NAGS. The NAGSs are listed in Table 2.1.

Responsibilities of NAGS

- Conservation and maintenance of the active/working collections of germplasm of the mandated crop(s)/plants(s)
- Multiplication and regeneration of the germplasm as and when needed following the appropriate strategies

- Supply of the germplasm from the active collection for utilization in breeding programmes
- To assist in the exploration and collection of the mandated crop(s)
- Obtaining the National Accession Number from NBPGR for the germplasm and deposition of a set in the base collection as per the guidelines
- To document the information regarding passport and characterization data of the holdings and regularly update the information for the National Database maintained at NBPGR

In addition to have substantial germplasm of the mandated crops the NAGS must also ensure the following :

- A proper Genetic Resources Unit with at least one scientist (crop curator) having full-time responsibility for carrying out the responsibilities as mentioned above
- Infrastructure for proper maintenance, regeneration and storage of germplasm
- Facilities for documentation and effective communication

Table 2.1. Crop based national active germplasm sites recognized by NBPGR

Crops	Institute/AICRP/NRC	Location
Field Crops		
Cotton	Central Institute of Cotton Research	Nagpur
Crops of North East Region	ICAR, Research Complex NEH Region	Shillong
Fodder Crops	IGFRI	Jhansi
Groundnut	NRC For Groundnut	Junagarh
Jute and Allied Fibers	Central Research Institute of Jute and Allied Fibers	Barrackpore
Maize	Directorate of Maize Research	IARI, New Delhi
Oil Seeds	Directorate of Oil Seeds	Hyderabad
Pearl millet	AICRP Pearl millet	Mandore, Jodhpur
Pulses	IIPR	Kalyanpur, Kanpur
Rapeseed Mustard	NRC on Rapeseed Mustard	Sewar, Bharatpur
Rice	CRRI	Cuttak
Rice and Lathyrus	IGAU	Raipur
Small Millets	AICRP on small millets	UAS, Bangalore
Sorghum	NRC on Sorghum	Hyderabad
Soybean	NRC on Soybean	Indore

Contd...

Table 2.1. Contd...

Sugarcane	SBI	Coimbatore
Tobacco	CTRI	Rajamundry
Under -utilized crops	AICRP on UU and UEC	NBPGR, New Delhi
Wheat and Barley	Directorate of Wheat Research	Karnal
Horticultural/Agroforestry Crops		
Agroforestry spp.	NRC on Agroforestry	Jhansi
Arid Fruits	CIAH	Bikaner
Banana	NRC for Banana	Trichy
Cashew nut	NRC for Cashew nut	Puttur
Citrus species	NRC for Citrus	Nagpur
Grapes	NRC for Grapes	Pune
Litchi, Bael ,Aonla, Jackfruit	CHES	Ranchi
Medicinal and Aromatic Plants	NRC on Medicinal and Aromatic Plants	Anand
Mango	CISH	Lucknow
Mulberry	Silkworm and Mulberry Germplasm Station	Hosur
Oil Palm	NRC on Oil Palm	Iluru
Onion and Garlic	NRC for Onion and Garlic	Pune
Orchids	NRC for Orchids	Gangtok
Ornamentals and Non-Traditional Crops	NBRI	Lucknow
Plantation Crops	CPCRI	Kasaragod
Potato	CPRI	Shimla
Spices	IISR	Calicut
Temperate Horticultural Crops	IITH	Srinagar
Tropical Fruits	NBPGR RS, IIHR	Bangalore
Tuber Crops	CTCRI	Thiruvanthapuram
Vegetables	IIVR	Varanasi

Registration of Plant Germplasm

The need for recognition to the developers of new improved varieties as per release procedures available for this purpose has been served by the Indian Council of Agricultural Research (ICAR) through a Central Sub-Committee on Crop Standards, Notification and Release of Varieties (CSC on CSN & RV) of Agricultural Crops as part of the Indian National Agricultural Research System (NARS).

The ICAR operates the mechanism for identification and recommendation for release of crop varieties and the Ministry of Agriculture provides mechanism for notification of released varieties and production of quality seeds.

This duly meets the requirements of the intellectual property and related issues, to an extent keeping in view the developments concerning Plant Breeders' Rights (PBR), Farmers' Rights (FR) and other forms of Intellectual Property Rights (IPR).

It has been recognized that the recognition should also be accorded to those who are associated with the development and identification of improved and unique germplasm and/or genetic stocks.

Therefore, recognizing the need for an authentic national documentation system of valuable sovereign genetic resource with known characteristics and inheritance factors, a mechanism for "Registration of Plant Germplasm" has been provided at the National Bureau of Plant Genetic Resources (NBPGR), New Delhi by the Indian Council of Agricultural Research (ICAR).

The mechanism will finally develop into sole recognized institution for registration of "Plant Genetic Resources (PGR)" material at national level. Material registered with this institution will not be eligible for registration elsewhere without the permission of ICAR.

Guidelines For Registration of Plant Germplasm

Nature of Material to be Registered

Germplasm or genetic stock of agricultural, horticultural and other economic crops, including agro-forestry species, medicinal and aromatic plants, ornamental plants, which is unique and has potential attributes of academic, scientific or commercial value.

The following categories of materials shall not qualify for registration:

- Germplasm and genetic stock without accompanying documentary evidence for the claim made in the application
- Varieties/ hybrids and parents of hybrids released in the country, zone or state
- Materials for which any form of protection has been sought elsewhere

Eligibility Criteria for Registration of Plant Germplasm

All claims concerning the material submitted for registration should accompany scientific evidence for uniqueness and value in the form of:

- Publication in standard peer reviewed journal (a copy of reprint to be submitted) and/or

Evaluation data under All India Co-ordinated Research Project (AICRP) trial, nursery tests supported with relevant extracts of the document or verification by concerned Project Director (PD)/Project Co-ordinator (PC) and/or

- Publication of information on potential value of germplasm in institute annual report or any other such reports may be considered depending on importance and/or
- Certificate of the validation tests of the claimed attributes by any institution as per the advice of registration authority

Germplasm Registration Committee

A committee of the following constitution shall be the Authority for Registration of Plant Germplasm as per ICAR Order No. 24(2)/96-F.C. II/Seed dated 24.3.1996.

- Deputy Director General (Crop Science)- ICAR, Chairman
- Assistant Director General (Horticulture)- ICAR, Member
- Director- NBPGR, Member
- Assistant Director General (FC)- ICAR, Member
- One representative from SAUs,Member
- Representative of DoAC, Member
- Need based three or more specialists with reference to the material under consideration (To be identified by the Chairman), Member
- A senior level scientist of NBPGR, Member Secretary (To be identified by the Director, NBPGR)

Application Form

Application shall be made on the prescribed Proforma (Form A). The Germplasm Registration Committee shall meet at least twice a year, with the concurrence of the Chairman, for consideration of applications, related matters and decision thereupon.

Nodal Agency

- National Bureau of Plant Genetic Resources (NBPGR), New Delhi has been designated the nodal agency for registration of germplasm. The application should be addressed to the Director, NBPGR along with the seed sample or a certificate of submission of propagules with respective crop-based National Active Germplasm Site (NAGS) for conservation.
- The member secretary, Germplasm Registration Committee will duly acknowledge with date the receipt of the application and the seeds or genetic material(s).

- NBPGR will maintain a database and a permanent register for listing the germplasm materials approved by Registration Committee with user driven details.

Screening of Application(s)

The proposal(s) submitted as per the prescribed Proforma will be screened at NBPGR by the Member Secretary, Registration Committee, as per the Check - List.

Consideration of Proposals by the Germplasm Registration Committee

- After initial screening, the incomplete application(s) will be returned for revision.
- In the application in which the validation of the data is felt necessary by the Director, NBPGR, the applicant would be asked to produce a validation report from an appropriate institute. The revised application should accompany such report(s) duly endorsed by the competent authority of the institute, advised for the validation.
- The proposals complete in all respect will be put up to the Registration Committee for consideration.
- The Registration Committee will consider the proposal within a year. The decision of the Germplasm Registration Committee will be final.

Validity of Registration

The period for validity of registration shall be 18 years for germplasm of trees and vines and 15 years for other plant species.

Notification of Registered Materials

- All germplasm material approved for registration would be officially notified to the applicants along with registration number. A certificate will also be issued to this effect to the applicant. Official notification will be published along with salient description of not less than one page (on lines of Crop Science) in the subsequent issue of the following:

  Indian Journal of Plant Genetic Resources

  Indian Journal of Agricultural Sciences

  Indian Journal of Genetics and Plant Breeding

  NBPGR - Website (http ://nbpgr.delhi.nic.in)

  In addition, it may be published in :

NBPGR Newsletter, Published by the Director, NBPGR, New Delhi, ICAR News - Published by the Publication and Information Division, Krishi Anusandhan Bhavan, ICAR New Delhi

Concerned Crop Newsletter, if any

ICAR Annual Report

Conservation, Maintenance and Sustainable Utilization of Registered Germplasm

- Registered germplasm will be conserved either in National Genebank or at designated crop-based NAGS.
- Working stock for supply to users, would be maintained by the institution associated with the development of the germplasm.
- Active or working collection(s) will also be maintained with the NAGS with the responsibility for its multiplication and distribution to bonafide users.

Denotification

Denotification shall be done by the Registration Committee in case of false claim(s) or disputed IPR claim. Appeal for counter claim, if any, should reach the Registration Committee within a period of three months of the publication or notification in Indian Journal of Plant Genetic Resources - Published by the Indian Society of Plant Genetic Resources, New Delhi.

Procedure For Submission of Proposal/Germplasm Material

- All plant germplasm proposed to be registered should be submitted to the following address:

  The Director

  National Bureau of Plant Genetic Resources

  Pusa Campus, New Delhi- 110 012
- The material must be accompanied with properly filled in Form-A, duly signed by the applicant and the Head of the institution with rubber seal (10 copies are to be submitted).
- The Form-A must be accompanied by complete description of the germplasm material using standard descriptors (as per concerned crop AICRP or NBPGR requirements). It may include photograph(s) of plant/plant parts/crop and/or fingerprints, if available.
- A declaration to the effect that working stock for supply to users would be maintained by the institution associated with the development of the material.

FORM A

Application for Registration of Plant Germplasm
(To be submitted to the Director, NBPGR, New Delhi)
Please refer to guidelines for seed submission as given in the relevant chapter (For use of NBPGR)

1.	Status Code N/R	(i)	Application Number ....
2.	Crop Name	(ii)	Date of application ....
3.	Botanical Name	(iii)	Whether new or revised ....
4.	Crop group (code)	(iv)	If revised, date of .... Ist application
5.	Biological status of the .... material to be registered	(v)	If validation test suggested ....
	RE/MU/GP/GS		**whether report attached**
6.	Identity	(vi)	Action taken ....
7.	Criteria for registration (unique feature (s)		(a) Forwarded for registration (b) Sent for validation (c) Incomplete, sent for revision
8.	Nature of genetic material (Code) SD	(vii)	Whether registered or rejected
9	Quantity deposited actual	(viii)	Date of registration or rejection ....
10.	Value referred to (Code) SC/CM/AC	(ix)	Registration number ....
11.	Basis of eligibility (Code) **AR/PR/CT/OT**	(x)	Notified on : ....
12.	Validation test suggested Y/N N = New, R = Revised	(xi)	Remarks ....

13. Particular of the scientist(s)/Person(s) who developed germplasm/genetic stock

I. Name (Dr./Ms./Mr) ....
Designation ....
Address ....
Telephone ....
Fax ....
e-mail ....

II. Name (Dr./Ms./Mr.) ....
Designation ....
Address ....

III. Name (Dr./Ms./Mr.) ....
Designation ....
Address ....

IV. Name (Dr./Ms./Mr.) ....
Designation ....
Address ....

14. Passport information of germplasm

IC/EC No.	Other identity	Source	Place of origin	Remarks
....	....	....	....	....

15. Pedigree of the genetic stock

A. Pedigree .... B. Breeding Method (Code) ....

16. Salient characteristics/chief botanical and morpho-agronomic description (attach details)

Sl. No.	Trait description
1.	
2.	

** Provide salient description of the material as per All India Co-ordinated Research Programme (AICRP Proforma)

17. Year of seed production ....
18. Location of seed production ....
19. Quantity of seed available ....
20. Additional information/remarks (if any) ....
21. Recommendation of institute's Germplasm Identification Committee

Undertaking

1. I/We undertaken to ensure long-term conservation of the aforesaid germplasm/genetic stock at the National Gene Bank, NBPGR and also its sustainable use by maintaining appropriate quantity of active/working collection and providing access as appropriate on prior informed consent and on mutually agreed terms. I/We, also agree to provide any further information or data pertaining to the description and unique characteristics to the ICAR/NBPGR in a transparent manner.
2. I/We assure genetic purity and truthfulness of seed material supplied with the application.
3. That such germplasm does not contain any gene or gene sequence involving terminator technology.

Countersigned by Head of the Institute — Signature of the Depositor

Signature Seal — Name

Full name : — Designation
Designation — and address

Codes for Filling Information

Item I : Application status	
N	New
R	Revised

Item 4 : Crop groups	
CL	Cereals
FC	Fiber crops
FG	Forages
FT	Fruits
FR	Forestry
GL	Grain legumes
MT	Millets
MM	Minor millets
MP	Medicinal and Aromatic plants
NC	Narcotics
OS	Oilseeds
OR	Ornamentals
OT	Other (specify in Col. 19)
PC	Pseudocereals
SP	Spices
VG	Vegetable crops

Item 5 : Biological status of the material to be registered	
GP	Germplasm collection
GS	Genetic stock
MU	Mutant
RE	Recombinant

Item 7 : Nature of genetic material	
OT	Other (specify in Colums 20)
SD	Seed
TR	Tubers/Root/Bulbs
VP	Vegetative cuttings
WP	Whole plant

Item 8 : Quantity deposited with application
Actual

Item 9 : value referred to by applicant	
AC	Academic
CM	Commercial
SC	Scientific

Item 10 : Basis of eligibility for registration	
AR	Institute annual report
CT	All India co-ordinated trials data
OP	Any other report
PR	Published with peer review

Item II : Validation test suggested	
N	No
Y	Yes

Item 14 B : Breeding methods used	
BC	Backcross method
IN	Introduction or selection
MS	Mass selection
OT	Other (specify in Item 19)
PL	Pure line selection
PS	Pedigree selection
RS	Recurrent selection

Furnish the proposal with requisite quantity of seeds or propagules (propagules to the concerned NAGS) to the Director, NBPGR, Pusa Campus, New Delhi - 110 012

Reference

NBPGR, 2003. Annual Report 2003. NBPGR, New Delhi, India

3

Plant Genetic Resources International Perspective

Plants are vital part of the world's biological diversity and essential resource for human well being. Besides meeting our basic food and fibre requirement, many thousands of wild plants have immense economical and cultural importance and potential for providing food, medicine, fuel, clothing and shelter for vast number of people at the global level. Traditional Chinese medicines use around 5000 plant species and about 7000 different plants are used in traditional medicines in India. Plants play an important role in maintaining basic ecosystem functions essential for survival of animal life. Thus, it is of vital importance to collect and conserve the plant genetic resources on a global basis. It is all the more important as several cases of genetic vulnerability of plants that are grown for food and other uses have been recorded (National Research Council, 1972, Granett *et al.*, 1991). Genetic diversity in the germplasm of crop plants is an effective defense against genetic vulnerability. The genetic diversity is strategically used as modern plant breeding intervention to help increase stability and sustainability of food production using modern cultivars. Plant genetic diversity is not distributed randomly. The countries around the world are not equally rich in plant diversity. In the modern times there has been steady erosion in plant genetic resources especially in more productive environment because of the replacement of traditional cultivars with modern and uniform cultivars including the hybrids. Various natural and man made disasters have also contributed to the loss of plant genetic resources. Under such situation the role of international organizations dealing with crop plant research has increased tremendously towards conservation and utilization of plant genetic resources.

The Consultative Group on International Agricultural Research (CGIAR) is currently supporting a total of 16 international research centres out of which following 11 have been entrusted with the job of management of genetic resources of crop plants under their mandate. These are given in Table 3.1.

Table 3.1. Institutions supported by CGIAR and the mandate crops for research including PGR management.

	Name of centre	**Mandated crop (s)**
CIAT	Centro Internacional de Agricultura Tropical Apartado Aereo 6713, Cali, Colombia. Founded 1967	Beans, cassava
CIMMYT	Centro Internacional de Mejoramiento de Maizy Trigo Londres 40, Mexico. Founded 1966	Maize, wheat
CIP	Centro Internacional de Ia Papa. Apartado 5969, Lima, Peru. Founded 1970	Potato
IPGRI	International Plant Genetic Resources Institute, Food and Agriculture Organization of the United Nations Via delle Terme de Caracalla 00100. Rome, Italy. Founded 1974	Policy issues
ICARDA	International Center for Agricultural Research in the Dry Areas, Aleppo, Syria. Founded 1975	Broad bean, barley, lentil
ICRISAT	International Crops Research Institute for the Semi-Arid Tropics, Patancheru, P.O. Andhra Pradesh 502342, India. Founded 1972	Pigeon pea, chick pea, sorghum, pearl millet and ground nut
IITA	International Institute of Tropical Agriculture, P.O. Box 5320, Ibadan, Nigeria. Founded 1967	Cowpea
IRRI	International Rice Research Institute, P.O. Box 933, Manila, Philippines. Founded 1960	Rice
WARDA	West African Rice Development Association, EJ. Roye Memorial Building P.O. Box 1019, Monrovia, Liberia. Founded 1970	Rice
ICRAF	International Centre for Research in Agroforestry, P.O. Box 30677, Nairobi, Kenya. Founded 1977	Forestry plants
CIFOR	Centre for International Forestry Research P.O. Box 161, Bogor 16001, Indonesia. Founded 1993	Forestry plants
AVRDC	Asian Vegetable Research and Development Centre, Tainan, Taiwan	Sweet pepper, tomato, vegetable soybean

World Germplasm Holding

According to Plyckhett *et al.* (1987) there are about 2.5 million total plant accessions conserved in world gene banks. The number of germplasm holdings on global basis in certain crops is as follows :

❒ Wheat	1,10,000
❒ Barley	2,80,000
❒ Rice	2,15,000
❒ Maize	1,00,000
❒ Tomato	63,000
❒ Potato	59,000
❒ Brassicas	52,000
❒ Cucurbits	48,000
❒ Beans	36,000
❒ Sweet pepper	6,000

International Undertaking in Global Plan of Action

A major concern for IPGRI is the particular situation regarding agro-biodiversity in general and plant genetic resources for food and agriculture (PGRFA) in particular. When the CBD was adopted, the negotiating parties recognized the special status of PGRFA and FAO was asked to address the twin issues of Farmers' Rights and the status of collections assembled before the coming into force of the CBD. It was agreed to address these issues within the context of revising the International Undertaking (IU) on PGRFA (Rao, 2004).

The original IU, adopted in 1983, was superseded by the CBD and needed to be harmonized with it. The inter-governmental FAO Commission on Genetic Resources for Food and Agriculture (CGRFA) initiated formal negotiations to revise the IU in harmony with the CBD in 1994. Several countries participated in the negotiations. Since CGIAR is not a governmental organization, it has been participating as an 'observer' contributing through informal participation where possible. In several of the Working Groups associated with the negotiations, the CGIAR was asked to attend and contribute technical expertise. The primary articles of relevance to the CGIAR are: Article 15 (ex situ collections of PGRFA held by the IARC of the CGIAR and other International Institutions), Article 12 (on access), and Article 13 (on benefit-sharing).

Despite the complexity of the negotiations, significant progress has been made. While a great deal of IU text and provisions have been approved by The Sixth Extra-ordinary Session of the FAO Commission on Genetic Resources that took place at FAO Headquarters from 25 to 30 June 2001. Adoption of a complete IU is yet to be achieved. Once each paragraph has been successfully concluded, and formally adopted by the FAO Conference, the ratification process will begin in each country. This can be a lengthy process and may take years. A major feature of the new IU - which is expected to become a legal instrument in its own right (possibly a protocol to the CBD) - is the creation of a multi lateral system for access and benefit sharing with respect to PGRFA. It must be noted that, a major question facing both developed and developing countries has been whether to encourage the germplasm exchange and its use through a multi-lateral system, or it be handled through numerous bilateral agreements. Obviously, the later would restrain PGR exchange. The need for such a multi-lateral system is now well accepted by the negotiating parties, at least for major food crops for which interdependence among countries is very high and which often have wide and highly complex pedigrees. These factors, in particular, make the bilateral negotiation of terms of access and benefit sharing difficult, to achieve. The major stumbling blocks appear to be the definition of PGR and relationship with other treaties. Agreement of effective access is vital if scientists are to develop higher yielding and more biotic and abiotic stress resistant crop cultivars. These agreements are going to be quite significant in managing and utilizing the PGR held in public genebanks, particularly held in, trust by CGIAR centres.

A Global Plan of Action (GPA) for the conservation and sustainable utilization of plant genetic resources for food and agriculture was adopted during the International Technical Conference on Plant Genetic Resources, Leipzig, Germany,1996 (FAO, 1996a, FAO, 1996b). The GPA was developed based on FAO's Report on the State of the Worlds Plant Genetic Resources for Food and Agriculture and was prepared through a participatory country- driven process (Cooper *et al.*, 1998).

The GPA is part of the FAO global system for the conservation and sustainable utilization of PGRFA and an important element for the CGRFA in fulfilling its mandate. The Commission requested the development of a rolling GPA, with programmes and activities aimed at filling in gaps, overcoming constraints and facing emergency situations identified in the FAO Report on the State of the World. The periodically updated plan will permit the Commission to recommend priorities and to promote the rationalization and coordination of efforts. A discrete GPA for PGRFA is warranted because of their great importance to world food security and the unique features, as mentioned above, of this component of biodiversity (FAO, 1996a; FAO, 1996b). IPGRI was an active participant of the GPA development process.

The Plan comprises 20 priority activities based on the principle that only through the use of PGRFA can the social and economic benefits of the conservation of PGR be realized (Cooper et al., 1998). The 20 priority activities of global plan of action are listed at the end of the chapter. Accordingly, the GPA aims to strengthen the links between PGR conservation and use. National programmes are the basis for regional and global programmes and, hence, the Plan focusses attention at the national level, to assist countries and institutions responsible for conserving and using PGRFA.

Intellectual Property Protection

The protection of intellectual property related to germplasm - whether through the use of Plant Breeders' Rights or patents - is currently an issue that is being debated in many fora around the world. It is not possible to give a comprehensive overview of the debate here, but instead the current situation will be described with respect to the Future Harvest Centres supported by the CGIAR. These Centres collectively maintain the world's largest collection of PGR of major food and forage crops. More than 500,000 individual samples are maintained in the genebanks of 11 Future Harvest Centres, also known as CGIAR Centres. Under agreements signed in 1994 with FAO, the collections are recognized as being held by the Centres in trust for the world community.

The Centres are publically funded institutions working for the international public good, have adopted the position that the products of their research should, to the extent possible, be held in the public domain. Thus, full disclosure is the preferred course of action in most situations. It is recognized, however, that in certain circumstances it will be necessary to take out Intellectual Property Protection (IPP), for example, to facilitate the transfer of technology, or to prevent misappropriation by others.

The Centres have adopted a Set of Guiding Principles on Intellectual Property and Genetic Resources, which form the basis of individual Centre IPP policies. In summary, the Guiding Principles state that:

- Centres will not take IPP on the in-trust genetic resources and will require recipients to observe the same prescription.
- Centres and recipients may use the in-trust material as parents in plant genetic improvement. Plant Breeders' Rights can protect the varieties so produced provided they can be used without restriction as parental material in the development of further varieties (the breeders' exemption).
- Cells, organelles, genes or molecular constructs isolated from materials supplied by a Centre can only be patented with the Centre's permission. In granting such permission, the Centre will normally require that a share of any benefits derived from the use of the patent-protected material be provided for the international public good.

While the intention of the principles is clear, in practice there are a number of problems that still remain under debate (Hawtin and Reeves, 1998). For example, what is the minimum that a recipient of germplasm from a Centre must do before being allowed to file for IPP? Should patents (as opposed to PBR) be allowed at all on varieties bred using in-trust material in their parentage? Under what conditions, if any, should the Centres provide exclusive licenses and to whom? (Hawtin *et al.*, 2000).

These problems all reflect the fact that a large part of the germplasm collections maintained by the Future Harvest Centres comprises materials developed over millennia by farmers and farming communities, and , assembled through international public funding. This occurred predominantly at a time when such materials were widely considered to be 'the common heritage of humanity' and when IPP was an insignificant factor in their exploitation. Now the materials in the collections are increasingly being accessed for private gain and their use is being restricted through the application of IPP. Furthermore, such protection is often granted for what are in essence discoveries, with little if any intellectual input by the IPP claimant. It is widely argued that for most of the material in the in-trust collections, those farmers and, communities that bred and selected them over the course of history provided the major intellectual input. Yet today it is often the person or company that adds just one or a few traits, or makes some other relatively minor modification, that reaps the major benefit (Bragdon and Downes, 1998).

The nature of agricultural research worldwide is changing rapidly. Private companies are moving into areas that are traditionally the domain of the public sector and in many countries public support for agricultural research is on the retreat. The funds required for international agricultural research are becoming ever more difficult to obtain. If the in-trust germplasm collections are to survive in the international public domain, it is important that the necessary funding be sustained. It is unlikely that such would ever be maintained by the private sector.

With the new paradigm, countries are beginning to introduce access legislation that could hamper the flow of genetic materials across their borders. Such legislation might be important to ensure that benefits return to the country from the exploitation of its pharmaceutical and other industrially important biodiversity. However, unless the needs of agriculture are explicitly taken into account, there is a likelihood that such legislation could be to the detriment of genetic improvement of crops (Hawtin, 1999). The full effects of such legislation on the work of the Centres remain to be seen. In addition, the harmonization of IU with CBD (as discussed previously) is another most important factor that will contribute to the continued management, use and access of germplasrn maintained by the CG Centres.

Priority Activities of Global Plan of Action

The 20 priority activities are as follows:

i. In-situ Conservation and Development

- ❐ Surveying and inventoring plant genetic resources for food and agriculture
- ❐ Supporting on farm management and improvement of plant genetic resources for food and agriculture
- ❐ Assisting farmers in disaster situation to restore agricultural systems
- ❐ Promoting in-situ conservation of wild crops relatives and wild plants for food production

ii. Ex-Situ Conservation

- Sustaining existing ex-situ conservation
- Regenerating threatened ex-situ conservation
- Supporting planned and targeted collecting of plant genetic resources for food and agriculture
- Expanding ex-situ conservation activities

iii. Utilization of Plant Genetic Resources

- Expanding the characterization, evaluation and number of core collections to facilitate use
- Increasing genetic enhancement and base broadening efforts
- Promoting sustainable agriculture through diversification of crop production and broader diversity in crops
- Promoting development and commercialization of under-utilized crops and species
- Supporting seed production and distribution
- Developing new markets for local varieties and diversity rich products

iv. Institutions and Capacity Building

- Building strong national programmes
- Promoting networks for plant genetic resources for food and agriculture
- Constructing comprehensive information systems for plant genetic resources for food and agriculture
- Developing and monitoring early warning systems for loss of plant genetic resources for food and agriculture
- Expanding and improving education and training
- Promoting public awareness of the value of plant genetic resources for food and agriculture conservation and use

References

Cooper, H.D., Spillane, C., Kermali, I. and Anishetti, N.M. 1998. Harnessing plant genetic resources for sustainable agriculture. PGR Newsletter. 114 : 1- 8.

FAO, 1996a. Report on the state of the world's plant genetic resources – International Technical Conference on PGR, Leipzig, Germany. 17 – 23 June, 1996. FAO, Rome.

FAO, 1996b. Global plan of action for the conservation and sustainable utilization of plant genetic resources for food and agriculture, FAO, Rome, Italy.

Flower, C.M., Smale, M. and Gaiji, S. 2001. Unequal exchange? Recent transfers of agriculture resources and their implications for developing countries. Development Policy Review 19 : 181 – 204.

Garnett, J., DeBenedictus, J.A., Wolpert, J.A., Weber, E. and Goheen, A.C. 1991. Phylloxera on rice : deadly insect-pest poses increased risk to north coast vineyards. California Agriculture 45 : 30 – 32.

National Research Council. 1972. Genetic vulnerability of major crops. National Academy of Sciences, Washington D.C., USA.

Plucknett D., Smith, N.J.H., Williams, J.T. and Anishetty, N.M. 1987. Genebanks and the worlds food. Princetown University Press.

Rao, V.R. 2004. Management of plant genetic resources at international level for a food secure future. *In* : B.S. Dhillon, R.K. Tyagi, Arjun Lal and S. Saxena (eds.). Plant Genetic Resources Management, Narosa Publishing House, New Delhi.

4

National Bureau of Plant Genetic Resources (NBPGR)

Historically, Indian scientists have recognized the importance of conserving plant biodiversity with particular reference to indigenous and naturalized crops and their wild relatives. Indian scientists have recognized the importance of conserving plant biodiversity as early as 1935 by the 'Crops and Soil Wing' of the then Board of Agriculture and Animal Husbandry. Late Dr. B. P. Pal, in his classic paper 'Search for New genes' published in 1937 emphasized the importance of germplasm augmentation and utilization in crop improvement programmes. The need was reiterated in a meeting of the Indian Society of Genetics and Plant Breeding in 1941, which *inter alia* discussed the subject of economic crops. Dr. B. P. Pal, working at the then Imperial (now Indian) Agricultural Research Institute approached the then Imperial (now Indian) Council of Agricultural Research (ICAR) to set-up a unit for assembly of global germplasm under phytosanitary conditions in India. The ICAR scheme for plant introduction started functioning in 1946 under the leadership of late Dr. Harbhajan Singh as the first 'Operational Scientist'. The scheme was further expanded and strengthened as 'Plant Introduction and Exploration Organization' in the Botany Division in 1956 and later established as a separate Division of Plant Introduction in IARI in 1961 headed by Dr. Harbhajan Singh.

Subsequently, on the recommendations of the 'High Level Committee' constituted by the Government of India in 1970 and foresight of visionaries like Dr. B. P. Pal, Dr. M. S. Swaminathan and Dr. A. B. Joshi, the Division of Plant Introduction was developed into an independent institute 'National Bureau of Plant Introduction' in August 1976 which was rechristened as National Bureau of Plant Genetic Resources (NBPGR) in January 1977. Today NBPGR, is the nodal agency at the country level to plan, organize, conduct and coordinate plant exploration for collection of plant genetic resources in India and in specific cases abroad, to undertake introduction, distribution, exchange and plant quarantine, to characterize, evaluate and document germplasm collections of agrihorticultural crops and their wild relatives, to promote safe conservation of plant genetic resources on long term basis using *ex-situ* and *in-situ* on farm conservation of landrace diversity in collaboration with national/international organizations and to carry out other related activities (Gautam, 2003).

Mandate

- To plan, organize, conduct and coordinate plant explorations for collection of genetic diversity with particular reference to native and naturalized crops and their wild relatives in India and in specific cases abroad
- To undertake introduction, distribution, exchange and quarantine of plant germplasm for research purposes
- To characterize, evaluate and document germplasm collections of agrihorticultural crops and their wild relatives
- To develop and operate National Database for storage and retrieval of information on plant genetic resources, farmer's knowledge, innovations, practices and ethnobotanical knowledge
- To promote safe conservation of plant genetic resources on a long-term basis using *ex-situ* methods and *in-situ* on farm conservation of landrace diversity (agro-biodiversity) in collaboration with other national/international organizations.
- To conduct basic researches in the area of biosystematics, germplasm conservation and characterization for providing a sound scientific backup to Bureau's services and to develop the National Programme on plant genetic resources
- To undertake teaching and training on plant genetic resources for postgraduate students linked with PG School, IARI
- To organize suitable training programmes at the regional, national and international levels
- To act as sole repository of all notified/released varieties along with herbarium specimens with relevant information and original/parental lines (Dhillon *et al.*, 2001).

Organization

The NBPGR is headed by the Director who is overall in-charge of administration, research management and coordination. There are four main divisions under NBPGR. These are :

- Plant exploration and collection
- Germplasm evaluation
- Germplasm conservation
- Plant quarantine

In addition to four main divisions, there are two units, three cells and one national research centre as given below :

- Germplasm exchange unit
- Tissue culture and cryopreservation unit
- National research centre on DNA fingerprinting
- Plant genetic resources policy cell
- Agricultural research information system cell
- National agricultural technology project cell

Besides, the above divisions, units and cells, NBPGR also accommodates three all India coordinated research projects as listed below :

- All India coordinated research project on medicinal and aromatic plants
- All India coordinated research projects on underutilized crops
- All India coordinated research projects on guar/arid legumes

Regional Stations of NBPGR

NBPGR has 11 regional stations as given below :

1. National Bureau of Plant Genetic Resources, R. S. PDKV Campus, Akola -444 104 Maharashtra

❒ Year of establishment	1977
❒ Mandate crops	sesame, safflower, groundnut, niger, castor, linseed, soybean, pigeonpea, chickpea, horsegram, lentil, wingedbean, *lathyrus*, fingermillet, prosomillet, kodomillet,
❒ Germplasm holdings	31181

2. National Bureau of Plant Genetic Resources, Satellite Research Centre, Amrawati Campus, Amrawati – 444 602, Maharashtra

❒ Year of establishment	1976
❒ Mandate crops	chickpea, castor, pigeon pea
❒ Germplasm holdings	4839

3. National Bureau of Plant Genetic Resources R. S., Bhowali, Nainital – 263 132, Uttaranchal

❒ Year of establishment	1985
❒ Mandate crops	wheat, barley, lentil, chilli, Frenchbean and wild relatives
❒ Germplasm holdings	2665

4. National Bureau of Plant Genetic Resources Exploration Base Centre, Cuttack, CRRI Campus, Cuttack - 753 006, Orissa

- ❒ Year of establishment 1986
- ❒ Mandate crops rice, medicinal agricultural and horticultural crops
- ❒ Germplasm holdings 581

5. National Bureau of Plant Genetic Resources, R. S. Hyderabad 500 030, Andhra Pradesh

- ❒ Year of establishment 1985
- ❒ Mandate crops quarantine - pigeon pea, ground nut, sorghum, chickpea, pearl millet, small millets
  evaluation : chilli, tomato, brinjal, black gram
- ❒ Germplasm holdings 2903

6. National Bureau of Plant Genetic Resources, R. S. Jodhpur, Central Arid Zone Research Institute Campus, Jodhpur - 342 003, Rajasthan

- ❒ Year of establishment 1976
- ❒ Mandate crops pearl millet, cow pea, guar, moth bean, mung bean, castor, til, ber,, pome granate, acacia spp. jojoba, jatropha spp., khejri, tumba,
- ❒ Germplasm holdings 15169

7. National Bureau of Plant Genetic Resources, R. S. CHES Campus, Ranchi- 834 010 Jharkhand

- ❒ Year of establishment 1988
- ❒ Mandate crops behera, *Trminalia arjuna, Pongamia glabra,* mahua, bael, plash, jamun, barhal, jackfruit, aomla, samal,
- ❒ Germplasm holdings 226

8. National Bureau of Plant Genetic Resources, R. S. NEH Complex, Bara Pani Shillong - 793 013, Meghalaya

- ❒ Year of establishment 1978
- ❒ Mandate crops maize, paddy, ricebean, ginger, colocasia, turmeric, chilli, *Perilla,* buckwheat, adzuki, *Chenopodium*
- ❒ Germplasm holdings 1761

9. National Bureau of Plant Genetic Resources, R. S. Phagli, Shimla - 171 004, Himachal Pradesh

- Year of establishment: 1976
- Mandate crops: apple, peach, almond, pear, apricot, straw berry, feijoa, hazelnut, peconnut, plum, cherry, walnut, kiwi, minor fruits, Frenchbean, amaranth, buckwheat, chenopod, adzukibean, ricebean, meetha karela
- Germplasm holdings: 7322

10. National Bureau of Plant Genetic Resources Exploration Base Centre, Central Institute of Temperate Horticulture, Sri Nagar - Jammu and Kashmir

- Year of establishment: 1988
- Mandate crops: apple, almond, walnut, wheat, barley, brassica and other local crops,rice, medicinal aromatic plants
- Germplasm holdings: 800

11. National Bureau of Plant Genetic Resources, R. S. Vellanikkara, Thrissur - 680 654, Kerala

- Year of establishment: 1977
- Mandate crops: rice, horsegram, sesame and cow pea, okra
- Germplasm holdings: 10402

National Gene Bank

The Indian National Gene Bank has been established by the NBPGR to conserve national heritage of germplasm collections in the form of seeds, vegetative propagules, tissue/cell cultures, embryos, gametes, etc. Based on experience gained from working with a built-in cold storage vault obtained from UK in 1983, four modules (two units of 100 m3 and two of 176m3 capacity) were installed in 1986-1990 for long-term storage of seeds of orthodox species kept in laminated aluminium foils at -20°C after drying them to 5% moisture content. The genebank presently has 18 storage modules with a total capacity of about one million accessions. Currently it holds a total of 1.831ac accessions of different agri-horticultural crops and ranks fourth in the world. It also has a large cryo-bank with storage capacity of quarter million samples of small seeded crops. A total of over 6,000 accessions can be stored in the *in vitro* repository.

NBPGR at a Glance

The Bureau operates with a network of 11 regional stations/base centres spread in different phyto-geographical zones of the country. The Bureau also maintains links with more than 40 National Active Germplasm Sites (NAGS) and 131 other co-operators through which Indian National Plant Genetic Resources System (IN-PGRS) operates. Since its inception the Bureau has conducted 1,035 explorations, collecting 1,45,167 accessions. A total of 16.61ac samples were imported and 5.76 lac were exported through NBPGR. 21.8 lac samples have been processed at the quarantine facilities by which entry and spread of many important exotic pests have been prevented. Germplasm holdings of 2,00,665 accessions have been conserved in the national genebank, 3,076 in cryobank, and 1,138 in tissue culture repository. The characterization and evaluation data has been documented in 73 catalogues. Establishment of National Research Centre on DNA Fingerprinting and launch of National Agricultural Technology Project (NATP) on conservation of agrobiodiversity have significantly supplemented the ongoing efforts. In India, NBPGR is the only institution devoted to the human resource management in plant genetic resources. Commencement of post-graduate course in PGR and launch of team of excellence programme under NATP are significant developments towards this end.

References

Dhillon, B. S., Pareek, S. K., Singh, B. M. and Srinivasan, K. 2001. NBPGR : Past and Present. *In*: National Bureau of Plant Genetic Resources : A Compendium of Achievements. National Bureau of Plant Genetic Resources, New Delhi. pp 1-25.

Gautam, P. L. 2003. Inaugural Address *In* : R. V. Singh, Hari Har Ram and J. P. Tiwari (eds.). Proceedings of National Seminar on Role of Indigenous Germplasm in Improvement of Horticultural Crops. Pantnagar, 24-25 June, GBPUA&T, Pantnagar. pp vii-xiv.

5

International Plant Genetic Resources Institute (IPGRI)

IPGRI founded in 1974 is the world's largest international institute dedicated solely to the conservation and use of plant genetic resources. It is located at Rome in Italy. The institute is funded by the Consultative Group on International Agricultural Research (CGIR) which is an informal association of 45 public and private sector donors. CGIAR supports a network of 16 International Agricultural Research Centres. The Group was established in 1971. It is cosponsored by the Food and Agriculture Organization (FAO), the World Bank, United Nations Development Programme (UNDP), and the United Nations Environment Programme (UNEP).

IPGRI concentrates on supporting the work on Plant Genetic Resources conducted by national research and development systems in developing countries. It also has a special responsibility for bananas and plantains and for supporting the genetic resources work of the CGIAR system.

Mission

IPGRI's mission is to encourage, support and undertake activities to improve the management of plant genetic resources worldwide so as to help eradicate poverty, increase food security and protect the environment. IPGRI focuses on the conservation and use of plant genetic resources important to developing countries and has an explicit commitment to specific crops.

IPGRI's Vision

"Through the collective, concerted action of farmers, forest dwellers, pastoralists, scientists, development workers and political leaders, the full potential of the Earth's plant genetic diversity will be harnessed to eradicate poverty, achieve food security and protect the environment for the benefit of present and future generations".

Objectives

- To serve the plant genetic resources needs of countries, particularly developing countries
- To strengthen international collaboration in the conservation and use of genetic resources
- To develop and disseminate knowledge and technologies relevant to the improved conservation and use of plant genetic resources

Why Plant Genetic Resources Matter

Plants are vital for the development of human society. They are the key to food security, providing us with cereals and other foods in addition to feeding the livestock that produce milk, meat and eggs and provide farm power. The wise use of plant genetic resources can help eradicate poverty, both by lowering the price of food and other commodities and by raising and diversifying the incomes of producers and processors. They also protect and enhance the environment, for example by preventing erosion and desertification and by absorbing atmospheric carbon.

The multiple uses of plants depend on a crucial characteristic of plant life and its diversity. Diversity exists at three main levels, namely, the combination of species that make up different ecosystems, the number of different species, and the different combinations of genes in species. All three levels help sustain agricultural systems as well as ensure their productivity. Genetic diversity, in particular, provides species with the ability to adapt to changing stresses, such as pests and diseases or drought.

Today, the diversity of the Earth's plant life is under threat as never before. In agriculture, the widespread adoption of a few improved varieties has narrowed the genetic base of important food crops and led to the disappearance of hundreds of landraces. In the forests, up to 8% of all plant species are expected to disappear over the next 25 years as deforestation continues.

Conserving and using plant genetic diversity is vital to meeting the world's future development needs. The earth's population is predicted to double, perhaps even triple, before it stabilizes late in the 21st century. This will place enormous pressure on the environment, and will require a huge increase in the production of food and non-food commodities.

Strategic Choices

IPGRI has chosen to concentrate its efforts on eight areas of work. Each area addresses one or more of the institute's three objectives, is essential to the fulfillment of its mandate and reflects strong demand from its partners. All are areas in which IPGRI has a comparative advantage. The eight areas of work are as follows:

Strengthening National Systems

National institutions play the central role in promoting the conservation and use of plant genetic resources. IPGRI supports the development of national systems on plant genetic resource management and provides information, and advice on policy issues and training in key skills.

Working With Networks

Networks are important mechanisms for encouraging the collecting, conservation, exchange, evaluation, enhancement and use of genetic resources. IPGRI promotes the sharing of ideas, methodologies, technologies and results through networks. It develops and supports regional and crop-specific networks. It also contributes to global networks on themes of importance across regions

Improving Conservation Strategies and Technologies

Advances in conservation science can increase the effectiveness of efforts to conserve and use genetic resources. IPGRI investigates and develops new options for overcoming the problems of *ex situ* conservation, seeks to improve *in situ* conservation (including on-farm crop management), and explores the combined use of *ex situ* and *in situ* approaches.

Increasing the Use of Plant Genetic Resources

Plant genetic diversity can support development only if it is used. IPGRI works with genebanks and plant breeders to promote the use of genetic resources conserved *ex situ*. It also conducts research to understand and enhance the roles played by diversity in production systems. Options for broadening the genetic base of important food crops receive special attention.

Managing and Communicating Information

To promote the use of plant genetic resources, IPGRI helps national programmes to analyze and document their own collections and to gain access to information about the collections of others. In addition, the Institute provides information services and has extensive publishing and public awareness activities.

Addressing Socioeconomic and Policy Issues

Understanding the 'human dimension' is vital to the successful conservation and use of plant genetic resources. IPGRI investigates the socioeconomic factors influencing farmers' management of diversity, including the different roles played by men and women. It also contributes to the international debate on key policy issues affecting the management, availability and use of plant genetic resources worldwide.

Conserving and Using Specific Crops

IPGRI takes special responsibility for the conservation and use of certain important and potentially important crop species, especially when these are not included in the mandates of other international agricultural research institutes. A major focus of IPGRI's work is on bananas and plantains. In addition, coconut and cacao also receive attention at present. Neglected or underused species that are regionally or locally significant and that are threatened by genetic erosion also receive support, as do wild or semi-domesticated species, especially the wild relatives of major food crops.

Conserving and Using Forest Genetic Resources

The conservation of forests and the diversity they contain is one of the most urgent tasks facing humanity. IPGRI's work on forestry deals mainly with the *in situ* conservation of intraspecific diversity in those key species that are at the greatest risk of overexploitation. The institute also aims to improve *ex situ* conservation methods for forest genetic resources.

International Agreements

- IPGRI Establishment Agreement
- 1992 Convention on Biological Diversity
- 1996 FAO Global Plan of Action for the Conservation and Utilization of Plant Genetic Resources
- 1983 International Undertaking on Plant Genetic Resources for Food and Agriculture
- Agreement on Trade-related Aspects of Intellectual Property Rights
- The International Convention for the Protection of New Varieties of Plants

Beneficiaries

Beyond the immediate circle of IPGRI's partners lies the broader circle of people who benefit from the Institute's work. IPGRI's ultimate beneficiaries are the farmers, forest dwellers and other members of the rural community in developing countries. The people who support these primary producers and processors form a group of intermediate beneficiaries. This group includes plant breeders and other scientists in the public and private formal research sectors, and development workers in both the government and non-government sectors. Lastly, urban consumers benefit when IPGRI's work leads to the increased production and diversity of food and other commodities or to lower prices.

Because IPGRI has a global mandate, its beneficiaries know no geographical limits. However, the Institute's ultimate goal is to contribute to improving the lives of the poorest producers and consumers in developing countries.

Partnerships

Many different kinds of institutions are concerned with plant genetic resources. IPGRI seeks to achieve its objectives by working in partnership with these institutions. Such collaboration increases the effectiveness of its work. The Institute works particularly closely with National Genebanks and Research Institutions These institutions are at the core of national systems. They bear the primary responsibility for conserving and using plant genetic resources at the country level. IPGRI works in partnership with them on all aspects of plant genetic resources. Strengthening their activities, particularly through capacity-building, is central to IPGRI'swork.

Other National Public-Sector Institutions

Government ministries, seed multiplication and extension services, universities, botanical gardens and policy-making bodies have activities relevant to national genetic resources work. IPGRI works to involve them in the conservation effort, as well as forging and sustaining links among them.

Advanced Research Institutes

These institutes, in both developing and developed countries, have special expertise and IPGRI works with them on particular research topics, with the aim of transferring the results to other partners, especially those in developing countries.

Farmers and Forest Dwellers

These groups are important custodians and users of plant genetic resources. They are IPGRI's partners in the development of management systems that maintain or enhance diversity while meeting production or other objectives.

Non-Government Organizations

Many NGOs are active locally and internationally in the conservation and use of plant genetic resources. IPGRI works with them to conduct locally based research and to raise awareness and build the necessary consensus for action.

Private-Sector Companies

Many private-sector companies are expanding their activities in such areas as biotechnology research, varietal development and the supply of planting materials to farmers. IPGRI is strengthening its links with the private sector in areas of mutual interest.

The CGIAR

IPGRI's sibling centres in the CGIAR system hold important collections of genetic resources in trust for the world community, and work on their conservation and use. IPGRI collaborates in this research and also in the centres' training activities. In addition, it provides advice and information on the policy and legal aspects of genetic resources, both to individual centres and to the system as a whole. Through its role in SGRP, IPGRI has a special responsibility for promoting and coordinating collaborative activities in the CGIAR.

Other International Institutions

The Food and Agriculture Organization of the United Nations (FAO) is a major partner for IPGRI in many aspects of the Institute's work. IPGRI also collaborates with several other UN bodies.

Donors

The donor and development agencies that support IPGRI are important in shaping the Institute's programme. IPGRI works with its donors to identify priorities and to develop and implement projects.

Programmes

The Plant Genetic Resources Programme, which supports the efforts of national, regional and international systems to conserve and use plant genetic resources is IPGRI's largest programme. It consists of regional offices for the Americas, Europe, sub-Saharan Africa, Central and West Asia and North Africa, and Asia, Pacific and Oceania, and two thematic groups: Genetic Resources Science and Technology, and Documentation, Information and Training.

The International Network for the Improvement of Banana and Plantain (INIBAP), works to increase and sustain the productivity of bananas and plantains grown on smallholdings.

The CGIAR Genetic Resources Support Programme, provides advice and services to the CGIAR in the area of genetic resources policy, and enhances the system's work on genetic resources through the CGIAR System-wide Genetic Resources Programme (SGRP).

Projects

IPGRI's activities are implemented through 20 multidisciplinary projects that are regional or global in scope.

1. Support to Plant Genetic Resources Programmes and Regional Networks in the Americas

This project aims to assist countries in Latin America and the Caribbean in building their capacities to conserve and use plant genetic resources. It also establishes and maintains links with institutions in North America.

2. Support to Plant Genetic Resources Programmes and Regional Networks in Asia, the Pacific and Oceania

This project aims to assist countries in Asia, the Pacific and Oceania in building their capacities to conserve and use plant genetic resources. It also promotes the adoption of integrated approaches to conservation in the region.

3. Support to Plant Genetic Resources Programmes and Regional Networks in Europe

This project aims to assist countries in Europe in building their capacities to conserve and use plant genetic resources. It emphasizes the security of crop collections under threat in Eastern European countries and encourages technology and information exchange between Western and Eastern Europe.

4. Support to Plant Genetic Resources Programmes and Regional Networks in Sub-Saharan Africa

This project aims to assist countries in sub-Saharan Africa in building their capacities to conserve and use plant genetic resources. It also aims to raise the awareness of policy-makers and the public of issues relating to plant genetic resources.

5. Support to Plant Genetic Resources Programmes and Regional Networks in Central & West Asia and North Africa

This project aims to assist countries in West Asia-North Africa in building their capacities to conserve and use plant genetic resources. It also covers Central Asia. A further aim is to raise the awareness of policy-makers and the public of issues relating to plant genetic resources.

6. Global Capacity-Building and Institutional Support

This project aims to enable countries to build effective national plant genetic resources conservation and use programmes. It trains scientists and trainers and develops training tools.

7. Global Forest Genetic Resources Strategies

This project aims to support strategic research on the conservation and use of intraspecific diversity of useful forest tree species. It also aims to develop an information system on forest genetic resources.

8. Promoting Sustainable Conservation and Use of Coconut Genetic Resources

Through COGENT, this project aims to promote national, regional and global collaboration among coconut-producing countries and partner institutions in the conservation and use of coconut genetic resources. It seeks to strengthen national research on coconut.

9. Locating and Monitoring Genetic Diversity

This project aims to develop methods for locating and measuring genetic diversity in cultivated and wild species, combining ethnobotanical with genetic and agro-ecological approaches. It also develops methods for monitoring genetic erosion.

10. Ex Situ *Conservation Technologies and Strategies*

This project aims to develop improved and, whenever possible, low-input technologies for the *ex situ* conservation of plant genetic resources. It also investigates *ex situ* conservation strategies.

11. In Situ *Conservation of Crop Plants and their Wild Relative*

This project aims to develop a scientific basis for effective on-farm conservation in ways that meet farmer and community needs and that provide for the effective maintenance and evolution of diversity. It also assists national systems in locating, monitoring and maintaining viable *in situ* populations of the wild relatives of crops.

12. Linking Conservation and Use

This project aims to strengthen the links between plant genetic resources conservation and use, with a view to enhancing both. It focuses on both *ex situ* and *in situ* approaches, and on the development of complementary conservation strategies. It emphasizes neglected and underused crops and supports the use of cacao genetic resources.

13. Livelihoods and Institutions: Social, Cultural and Economic Aspects of Agrobiodiversity

The project aims to strengthen the links between plant genetic resources conservation and the well-being of people, particularly poor rural people. It emphasizes such factors as gender, nutrition, incomes, indigenous knowledge, traditional resource rights, participatory approaches, access and benefit-sharing policies and intellectual property rights legislation. It also informs policy-makers on these subjects.

14. Information Management and Services

This project aims to provide information services and to build capacity in information management so as to meet national, regional and international responsibilities for the conservation and use of genetic resources. It also provides publications and information to support the research activities of IPGRI staff and their partners.

15. Public Awareness and Impact Assessment

This project aims to build and sustain financial and institutional support for plant genetic resources activities worldwide by raising awareness among key target audiences of the role of these resources in sustainable development and food security. It also aims to assess IPGRI's impact on the conservation and use of plant genetic resources.

16. Musa Genetic Resources Management

This project aims to collect the germplasm of *Musa* and its wild relatives and to promote its safe storage and movement, and its use. It also develops standardized tools for retrieving and exchanging information on *Musa* germplasm.

17. Musa Germplasm Improvement

This project aims to identify disease- and pest-resistant *Musa* genotypes, to carry out strategic research on *Musa* pathogen diversity, screening methods and molecular genetics, and to develop improved *Musa* genotypes. In addition to providing *Musa* germplasm, the project also distributes beneficial root microflora such as mycorrhizae.

18. Musa Information and Communication

This project aims to coordinate, facilitate and support the production, collecting and exchange of information on banana and plantain. It publicizes *Musa* issues and the work of INIBAP to both scientific and non-technical audiences.

19. Support to Regional Musa *Programmes*

This project aims to support INIBAP's global regional and national networks and other partnerships in Latin America and the Caribbean, in Asia, the Pacific and Oceania, and in sub-Saharan Africa.

20. CGIAR Genetic Resources Support Programme

This project aims to provide support to the CGIAR system in two areas: genetic resources policy, and genetic resources conservation and use. Support in the second area is provided in IPGRI's capacity as convening centre of the CGIAR's System-wide Genetic Resources Programme (SGRP).

Reference

http://www.ipgri.cgiar.org/system/page.asp?theme=2

6

Centres of Origin of Cultivated Plants

The place of origin of the crop plant can be found out on the basis of genetic affinities. This would mean to find out the geographical distribution of the wild progenitors of the cultivated plants. If one can locate the distribution of a given progenitor, one can delimit the area in which domestication could have occurred. Vavilov based his judgment on variation patterns in cultivated plants. Information on wild plants in those days was much more fragmentary than what it is today. Even today, genetic and ecological information on wild relatives of several cultivated plants is incomplete and in such cases assessment of a place of origin is difficult to make. However, in most cases required information is available on genetic affinities between wild and tame, on distribution areas and on ecological affinities of the wild progenitors. On this basis it is possible to determine the initial place of origin.

Vavilov proposed that the centres for diversity are places of origin or centers of origin of cultivated plants. In fact in most of the publications, centers of diversity are referred to as centers of origin. Vavilov's concepts in a nut shell are "The place of origin of a species of a cultivated plant is to be found in the area which contains the largest number of genetic varieties of this plant". According to Vavilov (1951) most of the varietal wealth in our crop plants was concentrated in 8 great centers of diversity. The main world centres as proposed by Vavilov (1951) are enumerated in Table 6.1. each with a representative list of crops (Zohary, 1970). The Abyssinian Centre is now known as Ethiopia.

Table 6.1. World centres of diversity of cultivated plants

1. The Chinese Centre	2. The Indian Centre
Avena nuda, Naked oat (secondary center of origin). *Glycine hispida*, Soybean	*Oryza sativa* (Rice) *Eleusine coracana*, African millet *Cicer arientinum*, Chick pea

Contd...

Table 6.1 Contd...

Phaseolus angularis, Adzuki bean *Phaseolus vulgaris,* Bean (recessive form; secondary center) *Phyllostachys* spp, Small bamboos *Brassica juncea,* Leaf mustard (secondary center of origin) *Prunus armeniaca,* Apricot *Prunus persica,* Peach *Citrus sinensis,* Orange *Sesamum indicum,* Sasame (endemic group of dwarf varieties, secondary centre) *Camellia (Thea) sinensis,* China tea	*Phaseolus aconitifolius,* Moth bean *Phaseolus calcaratus,* Rice bean *Dolichus bioflorus,* Horse gram *Vigna sinensis,* Asparagus bean *Solanum melongena,* Egg plant *Raphanus caudatus,* Rat's tail radish *Colocasia antiquorum,* Taro yam *Cucumis sativus,* Cucumber *Gossypium arboretum,* Tree cotton, 2x *Corchorus olitorius,* Jute *Piper nigrum,* Pepper *Indigofera tinctoria,Indigo*
2a. The Indo-Malayan Centre	**3. The Central Asiatic Center**
Dioscorea spp., Yam *Citrus maxima,* Pomelo *Musa* spp., Banana *Cocos nucifera,* Coconut	*Triticum aestivum,* Bread wheat *Triticum compactum,* Club wheat *Triticum sphaerococcum,* Shot wheat *Secale cereale,* Rye (secondary center) *Pisum sativum,* Pea *Lens esculentum,* Lentil *Cicer arietinum,* Chick pea *Sesamum indicum,* Sesame (one of the centres of origin) *Linum usitatissimum,* Flax (one of the centres of origin) *Carthamus tinctorius,* Safflower (one of the centres of origin). *Dacus carota,* Carrot (Basic centre of Asiatic varieties). *Raphanus sativus,* Radish (one of the centres of origin). *Pyrus communis,* Pear *Pyrus malus,* Apple *Juglans regia,* Walnut
4. The Near Eastern Centre	**5. The Mediterranean Centre**
Triticum monococcum, Einkorn wheat *Triticum durum,* Durum wheat *Triticum turgidum,* Poulard wheat	*Triticum durum,* Durum wheat *Avena strigosa,* Hulled oats *Vicia faba,* Broad bean

Contd...

Table 6.1 Contd...

Triticum aestivum, Bread wheat (endemic awnless group, one of the centres of origin) *Hordeum vulgare*, endemic group of cultivated two - rowed barleys *Secale cereale*, Rye *Avena byzantina*, Red oat *Cicer arietinum*, Chick pea (secondary centre) *Lens esculenta*, Lentil (a large endemic group of varieties) *Pisum sativum*, Pea (a large endemic group, secondary center) *Medicago sativa*, Blue alfalfa *Sesamum indicum*, Sesame (a separate geographic group) *Linum usitatissimum*, Flax (many endemic varieties) *Cucumis melo*, Melon *Amygdalus communis*, Almond *Ficus carica*, Fig *Punica granatum*, Pomegranate *Vitis vinifera*, Grape *Prunus armeniaca*, Apricot (one of centres of origin) *Pistacia vera*, Pistachio (one of the centres)	*Brassica oleracea*, Cabbage *Olea europaea*, Olive *Lactuca sativa*, Lettuce
6. The Abyssinian Centre American Centre *Triticum durum*, Durum wheat (an amazing wealth of forms) *Triticum turgidum*, Poulard wheat (an exceptional wealth of forms) *Triticum dicoccum*, Emmer wheat *Hordeum vulgare*, Barley (an exceptional diversity of forms) *Cicer arietinum*, Chick pea (a centre) *Lens esculenta*, Lentil (a centre) *Eragrostis abyssinia*, Teff *Eleusine coracana*, African millet	**7. The South Mexican and Central** *Zea mays*, Corn *Phaseolus vulgaris*, Common bean *Capsicum annuum*, Pepper *Gossypium hirustum*, Upland cotton *Agave sisalana*, Sisal hemp *Cucurbita* spp., Squash, Pumpkin, Gourd

Contd...

Table 6.1 Contd...

Pisum sativum, Pea (one of the centres) *Linum usitatissimum*, Flax (a centre) *Sesamum indicum*, Sesame (basic centre) *Ricinus communis*, Caster bean (a center) *Coffea arabica*, Coffee	
8a. South American (Peruvian-Ecua-Dorean-Bolivian Centre	**8a. The Chiloe Center**
Ipomea batatas, Sweet potato *Solanum tuberosum*, Potato *Phaseolus lunatus*, Lima bean *Lycopersicum esculentum*, Tomato *Gossypium barbandense*, Sea Island Cotton (4 x) *Carica papaya*, Papaya *Nicotiana tabacum*, Tobacco	*Solanum tuberosum*, Potato
8b. Brazilian Paraguayan Centre	
Manihot utilissima Manioc *Arachis hypogea*, Peanut *Theobroma cacao*, Cacao (secondary center) *Hevea brasiliensis*, Rubber tree *Ananas comosa*, Pineapple *Passiflora edulis*, Purple granadilla	

Now nine centres of diversity of crops are recognized. These are as given in Table 6.2 (Kloppenburg and Kleiman, 1987; Harlan, 1971).

Table 6.2. Nine centres of diversity with leading food and industrial crops

Centre of diversity	Leading crops
Sino-Japanese	Soybean, orange, rice, tea
Indo-Chinese	Banana, coconut, yam, rice, sugarcane
Hindustani	Rice, jute
West Central Asiatic	Wheat, barley, grape, apple, flax (oil and fibre), sesame
Mediterranean	Cabbage, sugar beet, olive, rapeseed
African	Sorghum, millet, oil palm, coffee
Euro-Siberian	Oat, rye
Latin American	Maize, potato, sweet potato, cassava, tomato, cotton, rubber, cocoa
North American	Sunflower

Centres of Diversity

We frequently confuse the concept of centres of diversity with another Vavilovian concept namely centres of origin. When distribution patterns of variation in crop plants became obvious to Vavilov, he also proposed that areas of maximum diversity are places of origin. In fact, in his publications, Vavilov used the term centres of origin instead of centres of diversity and frequently this tradition is in use. It must be understood that while centres of diversity are a biological fact, the term centres of origin is only an interpretation.

Centres of diversity were recognized and mapped by Vavilov and his group on the basis of the tremendous amount of material which they had collected and brought back to Russia in 1920s and early 1930s. In crop after crop, Vavilov found that over very extensive territories cultivated plants had relatively few varieties while in some restricted areas there were multitudes of varieties and aggregates of species. Wheat is a classical example to substantiate this point.

In most of Europe as well Siberia, wheat cultivation is based on a rather restricted number of varieties consisting mostly hexaploid bread wheats. However, when Vavilov and his colleagues explored the Middle East, they encountered the multitude of forms and types of wheat grown in these territories. They found here a wealth of species, subspecies and varieties - diploid tetraploid and hexaploid. In addition to this Near Asiatic centre of diversity, Vavilov found that in wheat several other variation centres existed as well. In the relatively small and geographically isolated area of Ethiopian plateau, Vavilov noted hundreds of endemic varieties and subspecies of tetraploid wheats. Thus, Ethiopia is indeed an excellent illustration of variation centre. Other concentrations of wheat variability were found by Vavilov in Afghanistan and in the Mediterranean basin. Wheat serves as an example for the patterns of variation and distribution found in scores of cultivated crops. Each agricultural crop shows concentration of genetic variation in one (or relatively few) geographic areas.

Primary Centres vs Secondary Centres

Primary centers are the places were domestication has occurred. In contrast, secondary centres are those places which are centres of diversity in terms of tremendous variability for a particular crop but these centres are far away from the areas in which their wild relatives occur and where domestication is supposed to have taken place. A typical example of this situation is the Ethiopian centre. In this centre there is rich diversity for wheats, barleys, peas, flax and lentils. There is widest possible variation for tetraploid wheats in this centre. However, none of these crops has a single wild relative in Ethiopia suggesting that these crops could not have possibly been domesticated in Ethiopia. The wild relatives of these crops and particularly wheats are found only in Middle East which is located thousands of kilometers away from Ethiopia. This example illustrate that the domestication of wheat took place in Middle East but surprisingly Ethiopia represents a rich collection of varieties. In order to meet this anomalous situation, Vavilov had to reconsider his concept of centres of origin and to distinguish between primary centres (places where domestication originated) and secondary centres which are only explicable in terms of what happened after domestication was achieved (Zohary, 1970).

Plant Germplasm Resources

The plant breeders can look towards following categories of plant materials as germplasm resources (Stoskopf, 1993).

- Currently grown commercial cultivars
- Obsolete commercial cultivars
- Breeding lines and stocks
- Landrace cultivars
- Undomesticated forms of crop plants
- Plant introductions
- Related species
- Mutations
- Interspecific and intergeneric hybridization

Realizing the fact that plant genetic resources may be the most precious asset of the world, several gene banks have been established for conservation of plant genetic resources. Exploration and conservation programmes are initiated with the following objectives:

- To meet the needs of the plant breeders for desirable genetic features like, pest resistance or adaptation, etc. for continued progress in plant breeding
- To reduce the threat or actual loss of genetic diversity in the field in form of landraces/primitive forms of crop species which are being replaced by high yielding modern varieties

It has been realized by all concerned that several of early collections/plant introductions have been lost because of poor seed storage facilites. Concern about this situation led to the concept of national gene storage units. In 1958, a milestone in US was achieved in the conservation of plant germplasm when the United States opened the National Seed Storage Laboratory at Colorado State University, Fort Collins, Colorado. Other nations followed this example and constructed controlled storage rooms designed to preserve the viability of seeds during intervals between periodic regeneration in the field. In India, National Bureau of Plant Genetic Resources (NBPGR), New Delhi has controlled storage gene bank facilities.

References

Harlan, J. R. 1971. Agricultural origins: Centers and non-centers. Science 174: 468-474.

Kloppenburg, J., Jr., Kleiman , D. L. 1987. The plant germplasm controversy. BioScience 37: 190-198.

Stoskopf, N. C. 1993. Plant Breeding : Theory and Practice. Westview Press, Inc. pp 531.

Vavilov, N. I. 1951. Phytogeographic basis of plant breeding. The origin, variation immunity and breeding of cultivated plants. Chronica Bot. 13, I - 366.

Zohary, D. 1970. Centres of diversity and centres of origin. *In* : O. H. Frankel and E. Bennett (eds.). Genetic Resources in Plants - Their Exploration and Conservation. Blackwell Scientific Publications, Oxford and Edinburgh. pp 33 - 42.

7

Plant Diversity and Genetic Erosion

Biodiversity is the foundation stone of human survival on the earth. Very appropriately, the theme of the World Food Day celebration, 16 October, 2004 was "Biodiversity for Food Security" Biological diversity is fundamental to agriculture and food production. Scientists have identified so far about 1.4 million of the plant and animal species that exist on earth. A rich variety of cultivated plants and animals serves as the foundation for agricultural biodiversity. Yet people depend on just 14 mammal and bird species for 90 % of their food supply from animals. Just four plant species, namely, wheat, maize, rice and potato provide half of our energy from plants (FAO, 2004).

Plant Biodiversity in India

India is one of the 12 mega-diversity centres of the world. India is rich in plant resources and its richness of plant biodiversity can be assessed from the fact that India accounts for about 10.5 %, of species on only 2.4 % land area of the world. India holds the 10th position among the major plant rich countries of the world (Groombridge and Jenkins, 2002). The rich biodiversity in India is on account of a variety of climatic and topographic conditions ranging from equatorial to temperate latitudes, from warm tropical sea shores like the Coromandel and Malabar coasts to the alpine and ice-capped mountains in the Himalayas, from humid tropical forests in Western Ghats and North East India to the hot desserts in Rajasthan and cold desserts of Laddakh, from the highest rainfall area in Cheerapunji to the rainless Thar dessert. Thus, India is remarkably diverse in ecosystems, species and their genetic assemblages. Rich biodiversity is matched by an equally rich and dynamic indigenous technical know how (ITK) also known as dynamic traditional knowledge systems which is being continuously involved and improved or modified by innovative members of successive generations of the local and indigenous communities. Indian gene centre (Vavilovian Hindustani Centre) is considered primary and secondary centres of origin and diversification of as many as 356 major and minor crop plans and 320 wild relatives. The India includes hot spots of biodiversity namely Eastern Himalayas and Western Ghats, out of total 25 hot spots identified in the world. The total number of angiosperms in the world is reported to be 27,000 out of which India accounts for 17,500. There are about 4,25,000 species of living plants in the plant kingdom from unicellular alga to the highly evolved flowering plants. Out of 17,500 reported number of angiosperms in India, 5,725 species belonging to 140 genera are endemic to India. These 5,725 endemic taxa which represent about one third of the Indian flora (about 45,500 plant species) are distributed in 3 major phytogeographical areas having endemic and floristic diversity. These are :

- The Himalayas (3471 species)
- The peninsular India (2015 species) and
- The Andman and Nicobar Islands (239 species)

The details can be seen in Tables 7.1 and 7.2 (Groombridge and Jenkins, 2002). India is homeland of 167 cultivated species and 329 wild relatives of crop plants (Gautam, 2004).

The Indian sub-continent is a centre of domestication and diversification of several economically useful wild plant species comprising about 3,000 plants of edible value, 4,000 species having known reputed medicinal value, 700 plants of traditional and social significance, 500 fibre yielding species, 400 fodder plants, 40 plants having insectivorous uses, 300 gum and dye yielding plants, 100 aromatic and essential oil yielding species (MoEF, 1994; Chowdhery and Murti, 2000).

Table 7.1. Major plant diversity countries of the world (after Groombridge and Jenkins, 2002)

Country	Maga diversity country ranking	No. of centres of plant diversity	No. of vascular plant taxa	Biodiversity hotspot with No. of endemic plant species
Brazil	1	12	56000	Brazilian Cerrado, 6000; Atlantic forest, 4400
Colombia	2	8	51000	Tropical Andes, 20,000 (in part); Choro, 2500 (in part)
China	3	8	27100	Indo-Burma, 7000 (in part); Eastern Himalayas, 5000 (in part)
Mexico	4	12	20000	Mesoamerican Forest (in part)
Australia	5	10	15600	South Western Australia, 3724
Indonesia	6	18	12000	Wallacea, 1500; Sundaland, 5000 (in part)
Peru	7	8	18200	Tropical Andes, 20,000 (in part)

Contd...

Table 7.1. Contd...

Ecuador	8	6	20000	Choco, 2500 (in part)
Malaysia	9	13	14000	Sundaland, 5000 (in part); Indo-Burma, 7000 (in part)
India	10	6	17000*	Western Ghats, 2182 (in part); Eastern Himalayas, 5000 (in part)' Indo-Burma, 7000 (in part)
Zaire	11	12	11000	–
Madagascar	12	1	10000	Madagascar, 9700 (in part)
United States	n/a	8	20000	California, 2125 (in part); Caribbean Islands, 7000 (in part)
Philippines	n/a	6	8931	Philippines, 5832
South Africa	n/a	6	23420	Karoo, 1940; Cape, 5682
Turkey	n/a	6	8650	Mediterranean basin, 13000 (in part)

* 17500 species according to BSI (2005)

Table 7.2. Phytodiversity of India (including fungi, but excluding virus/bacteria)

Taxonomic group	No. of species			% of world flora
	India	(endemic)	World	% of world
Angiosperms	17500	(5725)	270000	6.48
Gymnosperm	48	(10)	836	5.74
Pteridophytes	1200	(193)	13125	9.14
Bryophytes	2850	(938)	16000	17.81
Algae	6500	(1924)	40000**	16.25
Fungi	14500	(3500)	79000	50.40
Lichen	2075	(518)	13500	15.37
Total	44673	(12808)	423461	10.54

* Additionally there are 28 exotic species

Hot Spot Concept

Myers (1988) proposed the concept of hot spots which he further elaborated (Myers, 1990; Myers *et al.*, 2000). The hot spots are species rich tropical forests with high degree of species endemism and they are characterized by exceptionally high concentration of species and these species face high degree of threat in the tropical forests ecosystems. Initially, 10 hot spots were identified in the tropical forests on the basis of species and endemism analysis. Later on, these were expanded to 18 hot spot areas and Myres *et al.* (2000) have now proposed 25 hot spots on the basis of analysis of plant and vertebrate species and their endemism. These 25 hot spots are :

Tropical Andes, Mesoamerica, Caribbean, Brazil's Atlantic Forest, Choc/ Darien/Western Ecuador, Brazil's Cerrado, Central Chile, California Floristic Province, Madagascar, Eastern Arc and Coastal Forests of Tanzania/Kenya, Western African Forests, Cape Floristic Province, Succulent Karoo, Mediterranean Basin, Caucasus, Sunland, Wallacea, Philppines, Indo-Burma, South-Central China, Western Ghats/Sri Lanka, South-West Australia, New Caledonia, New Zealand and Polynesia/Micronesia. Of the 25 hotspots, the tropical forests support 15 hotspots, while Mediterranean types support 5 hot spots and other are islands, mostly tropical. In a revised estimate these 25 hotspots support 133,149 plant species in an area of 1.4 % of the earth's land surface (Myres *et al.*, 2000).

Threatened Species

World Conservation Monitoring Centre (WCMC) provides a conservation status of many of the threatened plant species. According to this about 3000 - 4000 higher plants may be under a degree of threat in India. The project on study, survey and conservation of endangered species of flora (POSSCEP) has partially documented these plants and published its finding in Red Data Books (Nayar and Sastri, 1987). Table 7.3 provides summary statistics for this information.

Table 7.3. Summary of plant conservation status information in India based on WCMC

IUCN (International Union for Conservation of Nature and Natural Resources) of Threat Category	Number of species
Extinct	19
Extinct/Endangered	43
Endangered	149
Endangered/Vulnerable	2
Vulnerable	108
Rare	256
Indeterminate	719
Insufficiently known	9
No information	1441
Not threatened	374
Total	3120

Assessment of Threat of Genetic Erosion

Agenda 21 states that "the current decline in biodiversity is largely the result of human activity and represents a serious threat to human development". According to World Conservation Monitoring Centre (WCMC, 1992), the following factors are considered responsible for endangering biodiversity.

- Habitat loss or modification, often associated with habitat fragmentation
- Over-exploitation for commercial or subsistence reasons
- Introduction of exotic species which may compete with, prey on or hybridize with native species
- Disturbance and uprooting
- Incidental take
- Disease
- Limited distribution.

Gomez-Campo *et al.* (1992) present a very comprehensive check list of specific factors potentially affecting the persistence and genetic diversity of individual populations of wild plants as given in Table 7.4.

Table 7.4. Check list of specific factors affecting the genetic diversity of wild plants

- Drainage works that destroy humid habitats and lower the water table in adjacent areas
- Dam building and the resultant flooding
- Clearing of land for agriculture
- Change in agricultural techniques, particularly increasing use of chemicals and heavy machinery
- Forestry plantation
- Decrease in pollinator populations due to increasing insecticide use
- Overgrazing by domestic livestock or wild herbivores
- Scrub regeneration as a result of lessened grazing pressure on pasture
- Increased or decreased frequency of forest fires
- Water pollution
- Air pollution
- Contamination of the soil
- Industrialization and urbanization
- Tourism and touristic development
- Road construction
- Mining and quarrying
- Intensification of traditional exploitation
- Horticultural collecting
- Competition with introduced plants
- Genetic contamination by hybridization with other species
- Introduced pests and diseases
- Small population size

Dahl and Nanbhan (1992) have discussed the threats endangering the genetic diversity of cultivated plants in context of global environmental change and international economic pressures to crop specific problem. These are listed in Table 7.5.

Table 7.5. The factors endangering the genetic diversity of cultivated plants

- Introduction of modern varieties and exotic crops
- Loss of seed-saving and vegetative propagation skills
- Acculturation of traditional caretakers (or their death)
- Change in economic base
- Land conversion to industrial agriculture
- Destruction (urbanization) of habitat and farmland
- Herbicide and pesticide impact
- Environmental contamination
- Introduction of exotic pests
- Loss of seeds and pests
- Net reduction in the number of farmers
- Inadvertent crossing of varieties.

Measuring the Risk of Genetic Erosion

Quantitative approaches have been adopted in developing models which can be used to estimate the threat of genetic erosion to a particular crop plant. The models are based on scoring a variety of factors like biological, environmental and socio-economic and summing the factor scores to give a total which is proportional to level of increasing threat of genetic erosion. A model proposed by Guarino (1995) is given in Table 7.6.

Table 7.6. A model for quantifying the threat of genetic erosion

FACTOR	
1. General	**Score**
1.1 Taxon distribution	
❒ Rare	10
❒ Locally common	5
❒ Widespread or abundant	0
1.2 Drought	
❒ Known to have occurred in two or more consecutive years	10
❒ Occurring on average one or more times every ten years, but not in consecutive years.	5
❒ Occurring less than once every ten years on average.	0

1.3. Flooding

❐ Area known to be very flood prone	10
❐ Area not known to be flood prone	0

1.4. Accidental fires

❐ Area known to be very prone to fires	10
❐ Area not known to be prone to fires	0

1.5 Potential risk from global warming

❐ Summit areas or low-lying coastal areas	0

2. Crop species

2.1. Area under the crop

❐ Declining rapidly	10
❐ Increasing or static	0

2.2. Modern cultivars of the crop

❐ Available and used by more than 70 % of farmers	15
❐ Available and used by 50 – 70 % of farmers	10
❐ Available and used by less than 50 % of farmers	5
❐ Not yet available, but introduction planned	2
❐ Not available	0

2.3. Performance of agricultural services

❐ Very strong, and biased towards modern varieties	10
❐ No agricultural services	0

2.4. Mechanization

❐ Tractors used by > 30 % of farmers	10
❐ Animal traction used by > 50 % of farmers	5
❐ Manual labour used by > 50 % of farmers	0

2.5 Herbicide and fertilizer use

❐ > 50 % of farmers	10
❐ 25 % of farmers	5
❐ None	0

2.6 Farming population

❐ Declining rapidly	10
❐ Increasing or static	0

3. Wild species

3.1. Extent of wild habitat of target species within study area

❐ Very restricted (< 5 %)	15
❐ Restricted (5 – 15 %)	10
❐ 15 – 50 %	5
❐ Extensive (> 50 %)	0

3.2. Conservation status of target species

❐ Species not known to occur in any protected area	10

- Species known to occur within a protected area, but protection status poor or unknown. 5
- Species known to occur within a protected area, and protection status good 0

3.3. Extent of use of wild habitat of target species

- Industrial exploitation 15
- Exploitation by surrounding populations (e.g. fuelwood gathering from nearby towns) 10
- Hunting and gathering by small local communities 2
- Completely protected 0

3.4. Extent of use of target species

- Industrial exploitation 15
- Exploitation by surrounding populations 10
- Local exploitation 5
- Protected or not used 0

3.5. Agricultural pressure on wild habitat

- Large scale cultivation within habitat margins 15
- Subsistence cultivation areas within habitat margins 12
- Land suitable for cultivation, cultivated areas within 3 km of habitat margins 10
- Land suitable for cultivations, cultivated areas within 3 – 10 km of habitat margins 5
- Land unsuitable for cultivation 0

3.6. Human population growth rate per year

- > 3 % 10
- 1 – 3 % 5
- < 1 % 0

3.7. Availability of agricultural land

- > 70 ha km^{-2} cultivated 10
- 30 – 70 ha km^{-2} cultivated 5
- < 30 ha km^{-2} cultivated 0

3.8. Species palatability

- High 10
- Medium 5
- Low 0

3.9. Ratio of present livestock density to estimated carrying capacity

- > 1.0 10
- 0.5 – 1.0 5
- < 0.5 0

3.10. Average proximity to borehole or other all-year round water supply

- < 10 km 10
- 10 – 20 km 5
- > 20 km 0

3.11. Distance to major population centre

❒ < 20 km	10
❒ 20 – 50 km	5
❒ > 50 km	0

3.12. Distance to major road

❒ < 10 km	10
❒ 10 – 30 km	5
❒ < 30 km	0

3.13. Distance to development projects (irrigation scheme, tourism complex, mining site, hydroelectric power scheme, land reclamation scheme)

❒ < 20 km	10
❒ 20 – 50 km	5
❒ > 50 km	0

References

Chaudhary, H.U. and Murti, S.K. 2000. Plant diversity and conservation in India – An overview. Bishen Singh Mahendra Pal Singh, Dehradun, India.

Dahl, K. and Nabhan, G.P. 1992. Conservation of Plant Genetic Resoruces. Grassroots Efforts in North America. ACTS Press, Nairobi.

FAO, 2004. Biodiversity for Food Security. World Food Day, 16 October, 2004.

Gautam, P.L. 2004. Trends in plant genetic resources management. *In* B.S. Dhillon, R.K. Tyagi, Arjun Lal and S. Saxena (eds.). Plant Genetic Resource Management, Narosa Publishing House Pvt. Ltd., New Delhi. pp 18 – 29.

Gomez-Campo, C. and Collaborators 1992. Libro Rojo de Especies Vegetales Amenazadas de Espana Peninsular e Islas Baneares. Ministry de Agricultura y Alimentacion, Madrid.

Groombridge, B. and Jenkins, M.D. 2002. World Atlas of Biodiversity : Earth's Living Resources in 21st Century. Univ. of California Press, Berkeley, Los Angeles, London.

Guarino, L. 1995. Assessing the threat of genetic erosion. *In* Guarino, V. Ramnatha Rao and R. Reied (eds.). Collecting Plant Genetic Diversity – Technical Guidelines. pp 67 – 74. CAB International.

Ministry of Environmental and Forests, Govt. of India. 2005. Zero order draft of third national report, Workshop for preparation of India's third national report to the convention on biological diversity, May, 20 – 21.

MOEF 1994. Ethanobotany in India – A status report. All India Coordinated Research Project on Ethanobotany. Ministry of environment and Forest, Govt. of India. New Delhi.

Myers, N. 1990. The biodiversity challenge : Expanded hot spots analysis. The Environmentalist 10 : 243 – 543.

Myers, N., Mittermeler, R.A., Mittermeier, C.G., da Fonseca, G.A.B. and Kent, J. 2000. Biodiversity hot spots for conservations priorities. Nature 403 : 853 – 858.

Myers, N.1988. Threatened biotas: "Hot spots" in tropical forests. The Environmentalist 88 : 187 – 208.

Nayar, M.P. and Sastri, A.K. (eds.). 1987. Red Data Book of Indian Plants Vol. I. Botanical Survey of India. Calcutta. pp 367.

WCMC 1992. Global Diversity : Status of the Earth's Living Resources. Chapman and Hall, London.

8

Plant Diversity in Indian Gene Centre

India is located between 8°- 38° N and 68°- 97°E and exhibits extreme variation in altitude — from sea level to heights above vegetational limits in the Himalayas (ca 4500 m), and climate — from monsoon- tropical in south to temperate and alpine in the north-western Himalayas and extremely arid to semi-arid in the north-western plains. It is floristically extremely rich with about 33 percent of its botanical wealth (over 15,000 species of higher plants) being endemic. There are about 141 endemic genera distributed over 47 families. Further, of the 4,900 endemic species, larger percentage is localised in the Himalayas (about 2,532 species) than in other regions, namely, the peninsular tract (1,788 species), and the Andaman and Nicobar Islands (185 species). It is also estimated that floristic richness is maximum in the north-eastern region, which holds about 50 percent of India's total species diversity, i.e., more than 7,000 species, and is considered as the cradle of flowering plants. Of 990 species of orchids, 700 species occur in this region (Nayar, 1989). This, thus, makes the Indian region botanically unique and interesting (Arora, 1991).

Indian Subcontinent as Gene Centre

India is one of the twelve centres/regions of diversity of crop plants in the world (Zeven and de Wet, 1982). As is evident from the above brief account, antiquity of agriculture as also the ethnic diversity in the sub- continent have played a major role in the diversification of crop resources in this region. Thus, rich genetic diversity occurs in several crop plants and their wild progenitors. As it is, the diverse agro-eco climate of the 'Hindustani Centre', apart from possessing about 166 species of agri-horticultural crop plants, also has rich diversity in wild relatives of crop plants numbering about 320 species. In addition, as already reflected above, the Indian agriculture has been enriched by a continuous stream of introductions of new crops and their cultivars by man since the ancient times. The Mughals, Spaniards and the Portuguese brought several new crops to this country. Prominent amongst these were pear, grape, nut fruits, datepalm, garlic, opium, maize, potato, sweet potato, tomato, chillis, French bean, peas, etc. The Arabs possibly brought clove, coriander, cumin, ammi or lavage (ajwain) and fennel. This process continued during the British period which saw the introduction of coffee, cocoa, cashew nut, strawberry, litchi, Cinchona, etc., besides collection of tea in Assam and adjoining region. The current matrix of diversity consists of the gene pool of indigenous crop plants, their wild and/or weedy relatives and the well adapted introductions from practically all over the globe. The introduced types, depending on the time of introduction, extent of material introduced and the areas of introduction, exhibit enormous build-up of diversity within the Indian subcontinent. Some good examples are : cereals — wheat, barley, oats, maize, pulses — gram, French bean (rajmah) and peas; vegetables — potato, onion, cauliflower, cabbage, carrot and tomato, fruits — apple, pear, grapes, cherry, peach and apricot, oilseeds — soybean, sunflower and groundnut, fibre crop - cotton, to a very limited extent also in medicinal plants—mint, liquorice, Digitalis (foxglove), Cinchona (quinine), Hyoscyamus (henbane) and others such as *Humulus lupulus* (hops).

Prominent examples of crop species that originated in this region are given in Table 8.1.

Table 8.1. Prominent crops having Indian continent as origin

Cereals and millets	:	Rice (*Oryza sativa*), little millet (*Panicum sumatrense*), kodo millet (*Paspalum scrobiculatum*).
Legumes/Pulses, etc.	:	Black gram/urd (*Vigna mungo*), moth bean/moth (*V. aconitifolia*), pigeonpea or arhar (*Cajanus cajan*), horsegram/kulthi (*Macrotyloma uniflorum*), *Mucuna utilis*.
Fruits	:	Mango (*Mangifera indica*), banana (*Musa* spp.), jamun (*Syzygium cumini*), jackfruit/kathal (*Artocarpus heterophyllus*), Citrus group - lime and others, karonda (*Carissa congesta*), khirni (*Manilkara hexandra*), phalsa (*Grewia subinaequalis/G. asiatica*), bael (*Aegle marmelos*), wood apple (*Feronia limonia*), kokam (*Garcinia indica*).
Vegetables	:	Eggplant/brinjal (*Solanum melongena*), ridge gourd and smooth gourd/tori (*Luffa* spp.), round gourd/tinda (*Citrullus lanatus* var. *fistulosus*), pointed gourd/parval (*Trichosanthes dioica*), taros/ arbi (*Colocasia esculenta*), yams (*Dioscorea* spp.), jimikand (*Amorphophallus campanulatus*), kundru (*Coccinea indica*), cucumber (*Cucumis sativus*), *Raphanus caudatus* — mungra type.
Oil seeds	:	*Brassica* spp. (rai, sarson and toria types).
Spices and condiments	:	Turmeric (*Curcuma domestica*), ginger (*Zingiber officinale*), cardamom (*Elettaria cardamom*), Bengal cardamom

Contd...

Table 8.1. Contd...

		(*Amomum aromaticum*), long pepper, black pepper (*Piper nigrum*), betel leaf and cinnamon/ dalchini (*Cinnamomum zeylanicum*).
Other crops	:	Titapatljute (*Corchoruscapsularis*), tree cotton (*Gossypium arboreum*), sugarcane (*Saccharum officinarum*), sunnhemp (*Crotalaria juncea*), bamboos, dhancha (*Sesbania sesban*) and tea (*Camellia sinensis*).

Plant Diversity Rich Regions in Indian Sub-Continent

Arora (1988) has listed 6 regions which exhibit richness in crop plant diversity. These are as follows :

1. *Western Himalayas including cold arid tract:* In Kinnaur, Lahaul and Spiti, Pangi, Ladakh and adjoining areas, rich genetic diversity occurs in wheat, barley (particularly hull-less types), buckwheat, prosomillet, amaranth, chenopods, field peas, lentil and several other crops, possessing adaptability to cold and tolerance to drought conditions. In barley, bluish/black grain types occur at high altitudes (above 3300 m), exhibiting akinness to local Tibetan types. In wheat, both awned and awnless, tall types occur and collections from Lahaul and Spiti have exhibited resistance/tolerance to rust. Further, rich diversity in rice, maize, French bean, peas, buckwheat, amaranth and other crops - radish, leafy Brassicas, pome, stone and nut fruits, forages and medicinal plants, occurs in the sub-humid hill regions of Jammu and Kashmir, Himachal Pradesh and Uttar Pradesh (now the hilly region is the state of Uttaranchal).
2. *North-eastern region and the eastern Himalayas :* The tribal dominated belts of Mizoram, Manipur, Meghalaya, Tripura, Sikkim, North Bengal and parts of Nagaland and Arunachal Pradesh, are rich in local variability of cereals-rice, maize (including the primitive pop-corn), barley, wheat, pseudocereals - buckwheat, Chenopodium, amaranth and soft-shelled forms in Coix; millets - finger millet, foxtail millet; legumes - rice bean, winged bean, adzuki bean, black gram, sem /Dolichos, soybean, sword bean, peas, oilseeds - Brassica spp., Perilla, sesame, niger; vegetables - cucurbits (*Cyclanthera, Cucurbita, Momordica, Cucumis, Luffa, Lagenaria, Benincasa*); fibres - *Corchorus* spp., tree cotton, kenaf and mesta; taro/yam and several other tuberous/ rhizomatous/ bulbous types, and Citrus, Musa and bamboos.
3. *Eastern peninsular region :*The tribal dominated tract of Orissa and the Chotanagpur plateau may serve as good examples of areas that hold rich crop diversity *viz*, rice, sorghum, finger millet, fox-tail millet, prosomillet and other minor millets, Dolichos, rice bean, chickpea, pigeonpea, horsegram/ kulthi, sesame, niger, Brassicae, linseed, castor, several cucurbits, brinjal, chillis, mango and wild types such as in rice (*Oryza nivara*).

4. *Western arid/semi-arid region :* In Rajasthan and Gujarat (including Saurashtra), rich diversity occurs in sorghum, pearl millet, moth bean, cluster bean guar, cowpea, black gram, green gram, Brassicae, sesame, cucurbits, forage legumes and grasses.

5. *Central tribal region :* Madhya Pradesh and adjoining tract of Maharashtra possess rich diversity in wheat, rice, sorghum, minor millets, grain legumes - particularly cowpea, chickpea, pigeonpea, black gram and green gram, oilseeds - niger, sesame, Brassicae; and cucurbits.

6. *Western peninsular region :* The Western Ghats tract has promising diversity in tuber crops - Colocasia, Dioscorea and Coleus; rhizomatous types like Curcuma species and ginger, and other crops like rice, finger millet, several grain legumes, and much variation in cucurbits, forage legumes and grasses; several fruit trees such as jackfruit and spices like cardamom and black pepper.

Crops and Areas Rich in Diversity for Landraces and Primitive Cultivars in Indian Sub-Continent

This information is summarized in Table 8.2.

Table 8.2. Crops and areas where rich diversity in landraces and primitive cultivars occurs in Indian subcontinent (Mehra and Arora, 1982 with additions by the author)

Cereals and pseudo-cereals		
Rice	:	North-eastern region in Arunachal Pradesh, Garo and Khasi hills of Meghalaya; Mizoram, Manipur and Nagaland; north Bengal and Sikkim; and the mountain tracts of upper Assam including Mikir hills and Cachar (more attention required in these areas)
	:	Tribal dominated tracts of eastern and central India, including Jeypore tract of Orissa, Koraput, Keonjhar, Kalahandi, Santhal Parganas and Chotanagpur hills of Bihar (now in Jharkhand), Bastar, Rewa, Bilaspur, Sarguja, Raigarh, Raipur and adjoining tract in M. P, and hilly areas of east Godavari and adjoining region in Andhra Pradesh
	:	Chikmagalur, Shimoga and adjoining tracts in Karnataka extending to mountains of Tamil Nadu and Kerala in the Western Ghats
	:	Western Himalayan region including H. P., J & K, and Uttaranchal hills
Maize	:	North-eastern region, Cachar and Mikirhills, Arunachal Pradesh and north Bengal.

Contd...

Table 8.2. Contd...

	:	Tribal areas of peninsular region in A. P., Orissa, Bihar and northern M. P.
	:	Mountain belt of Nilgiris and Pulney hills of Karnataka adjoining Chikmagalur
	:	Western Himalayas, Chamba and adjoining area in H. P., and Uttaranchal hills
Wheat	:	Particularly in northern India, hilly tracts in Western Himalayas, Kinnaur, Lahaul and Spiti, Pangi valley in H.P., extending to Ladakh in Kashmir; also Baramula, Usmaro, Gurez valley in J & K
	:	Eastern region in north Bengal, Sikkim and Arunachal Pradesh
	:	Peninsular north-western, central plains; drier tracts of Maharashtra, Karnataka and A.P.
Barley	:	Peninsular tract, western and northern M.P., and western India in Rajasthan, eastern Maharashtra and parts of Karnataka
	:	Western and Eastern Himalayas in areas indicated above for wheat (naked and hooded types), in north Bihar and in other areas of peninsular India in wheat belt (husked types)
Secale	:	Dry cold arid areas of Ladakh, Lahaul and Spiti, Pangi valley and Kinnaur in particular, sporadic distribution in higher elevation, above 2100 m
Buckwheat	:	Hilly tract of north-eastern region, particularly in Arunachal Pradesh and Sikkim
		Western Himalayas - in Pangi valley, Lahaul and Spiti, Kinnaur and adjoining region in H. P, Uttaranchal hills and in Kashmir valley including Ladakh
Amaranth	:	North-eastern region in north Bengal, Sikkim and in particular Arunachal Pradesh
		Western Himalayas in H. P., Uttaranchal and Kashmir
		North-western and central plains; eastern Maharashtra
Millets		
Sorghum	:	Tribal tracts of M. P. in Bastar, and adjoining region in A.P. - Cuddapah, east- Godavari, Srikakulum and Vishakapatnam and Khammam; Santhal Parganas and Chotanagpur in Bihar, Chandrapur and adjoining tract in Maharashtra; and Gunjam, Kalahandi and adjoining hilly tract in Orissa

Contd...

Table 8.2. Contd...

		Western part of Tamil Nadu in Coimbatore and Salem, and also hills of Karnataka
		The foot hill region of the Himalayas
		North-eastern hills, bordering Burma
Pearl millet	:	Particularly tribal belt of Gujarat and Western Rajasthan, also in adjacent M.P., and Karnataka
Lesser millets		
Foxtail millet (*Setaria italica*)	:	Western and north-eastern region including the Himalayan tract and sub-tropical Assam, hilly region and adjoining areas in particulars, Uttaranchal hills
	:	Tribal belt of peninsular India, particularly in M. P., A.P., Orissa and adjoining Bihar
	:	In Western Ghats, eastern Maharashtra and southwards, more in drier tract
Finger millet (*Eleusine coracana*)	:	Tribal belt of peninsular India, particularly in north of M.P. and Tamil Nadu, A. P. and Orissa; and also lower hills- up to 2,100 m in the Himalayas
Kodo millet (*Paspalum scrobiculatum*)	:	Central peninsular belt in M. P Bihar and Orissa in particular
Sawan (*Echinochloa colonum*)	:	The above areas (for kodo) and the hills of eastern region
Cheena or Prosomillet (*Panicum miliaceum*)	:	Hills of H. P., Uttaranchal particularly higher elevation, also in parts of Gujarat and Maharashtra and in Bihar and Bengal, extending to Arunachal Pradesh
Panicum miliare	:	Tribal belt of A.P., Orissa, Bihar, M. P. and Maharashtra particularly in Orissa - Kalahandi, Keonjhar and Phulbani, adjacent parts of M. P. and Bihar; and in W. Bengal
Coix lacryma – jabi (soft-shelled forms)	:	North-eastern part in Meghalaya, Manipur, Mizoram and Nagaland, particularly in Garo and Khasi hills of Meghalaya
Grain and other legumes		
Green gram (*Vigna radiata*)	:	Tribal agency areas of east Godavari and adjoining tract in A. P., also in Warangal and Khammam and Orissa - Phulbani, Kalahandi and other tribal tracts extending to M.P. southwards, east/westwards
		Northern India, in comparatively drier areas of Hisar (Haryana) and Ferozepur (Punjab)

Contd...

Table 8.2. Contd...

	:	Western India, tribal tracts in Dhulia, Parbhani, Aurangabad in Maharashtra; Panchmahal and Kaira tract in Gujarat; also Dharwar and neighbouring tract in Karnataka
Black gram (*Vigna mungo*)	:	Hill tracts of H. P., valley and other areas Uttaranchal hills Nainital, Rampur and adjoining tracts; in north Bengal and upper Assam
	:	Tribal belt of peninsular India as indicated above for green gram, areas of Godavari and adjoining tract, Guntur and Krishna in A.P; Raipur, Durg and Bilaspur in M.P.; and Santhal Parganas, Ranchi, Bhagalpur, Singhbhum and Purnea in Bihar/Jharkhand
	:	In western India, Kutch, Surat and Panchmahal in Gujarat and eastern Maharashtra
	:	Northern plains, in U.P. particularly Sitapur, Barabanki, Hardoi and Bareilly, Ambala in Haryana and Ferozepur in Punjab
Rice bean (*Vigna umbellata*)	:	North-eastern region including Assam plains, adjoining hill regions, north Bengal and Sikkim and particularly Garo and Khasi hills ofMeghalaya,Maniptr and Mizoram tracts), Uttaranchal hills
	:	Tribal belts of Orissa and Bihar and the areas of east Godavari in A. P., particularly Kalahandi and Phulbani in Orissa and Santhal Parganas in Bihar
Moth bean (*Vigna aconitifolia*)	:	Kutch, Kaira and other dry areas in Gujarat, west Rajasthan; and eastern Maharashtra (mainly north west plains, Uttaranchal hills
Pilipasara (*Vigna trilobata*)	:	Tribal areas of A. P., east Godavari; Krishna and adjoining tract of Tamil Nadu
Red gram	:	Tribal belt of peninsular India, particularly in Kalahandi, Phulbani, Keonjhar and other tracts of Orissa and adjoining U. P., and Bihar (largely eastern peninsular tract and adjoining region)
		Hill region of north-east, mainly in Mizoram
Bengal gram (*Cicer arietinuin*)		Bundelkhand area in UP., Banda, Jhansi; Raipur, Chattarpur, and Bilaspur in M.P., and other areas of eastern U.P.
Lentil (*Lens culinaris*)		Western Himalayan region of H.P., and Uttaranchal hills in particular and parts of M. P., central peninsular tract

Contd...

Table 8.2. Contd...

	:	Eastern Himalayas in north Bengal and adjoining areas (sporadic)
Cowpea (*Vigna unguiculata*)	:	Hill regions of northern and eastern India, including north-eastern hilly tribal tract
	:	Tribal tracts of A. P., Orissa, Bihar and M. P.
	:	Partly in Western Ghats, and Maharashtra southward and particularly Karnataka hills, Chikmagalur and other areas
Vegetables		
Eggplant	:	Eastern and western peninsular region, north-eastern region
Okra	:	Gangetic and western arid plains, north-eastern region
Cucurbits		
Pointed gourd	:	Gangetic plains and the north-eastern region
Round gourd	:	Parts of Haryana mainly Gangetic plains, northern part, region of UP, Aligarh
Bitter gourd	:	Gangetic plains; central highlands and north-eastern region
Ridge & smooth gourd	:	Gangetic plains mainly
Snake gourd	:	Eastern and western peninsular region, north-eastern region
Cucumber	:	Sporadic diversity available all through, including the Himalayan region
Leafy types	:	Gangetic plains and north-eastern region. (Amaranth, Brassicae); Himalayan region
Fruits/Nuts		
Banana/Musa spp	:	Eastern peninsular and north-eastern region, north-eastern region
Jackfruit	:	Western peninsular region, Western Ghats, central highlands and adjoining Gangetic plains
Jamun	:	Gangetic plains, western peninsular region, northern part in Maharashtra, western plains
Indian gooseberry/ *Emblica officinalis*	:	Gangetic plains mainly
Pyrus/Prunus Rubus	:	Western and Eastern Himalayas

Contd...

Table 8.2. Contd...

Walnut	:	Western Himalayas, cold arid and temperate (Ca 3000 m) belt
Miscellaneous		
Jute and tree cotton	:	North-eastern region
Ginger	:	North-eastern region, parts of H.P., eastern and western peninsular region
Turmeric	:	Eastern and western peninsular region
Black pepper	:	Western Ghats - mainly southern part
Black caraway	:	Cold arid belt - Western Himalayas
Sugarcane	:	Eastern and western peninsular region mainly, sporadic variability in north-eastern region

References

Arora, R.K. 1988. The Indian gene centre – Priorities and prospects for collection. pp 66 – 75. *In* R. S. Paroda, R. K. Arora and K. P. S. Chandel (eds.). Plant Genetic Resources : Indian Perspective. NBPGR Publ., New Delhi 545 p.

Arora, R.K. 1991. Plant diversity in the Indian gene centre. *In* : R.S. Paroda and R.K. Arora (eds.). Plant Genetic Resources : Conservation and Management. NBPGR, New Delhi. pp 26 – 49.

Mehra, K.L. and Arora, R.K. 1982. Plant genetic resources of India : Their diversity and conservation. NBPGR Sci. Monogr. No. 4. 60 pp.

9

Plant Germplasm Sampling Strategy

Marshall and Brown (1975) have considered the subject of optimal sampling strategies for use in the genetic conservation of crop plants. The theme of this paper has been relatively narrow. These authors have defined a strategy for populations under imminent threat of extinction applicable to particularly the traditional landraces of some major crops such as wheat, rice and barley on account of danger of replacement of the landraces on a very broad scale due to the spread of modern cultivars. According to Brown and Marshall (1995) the emphasis in collecting of crop genetic resources has changed radically over the years. Instead of crisis-driven broad scale crop specific programmes, the collection of plant genetic resources now involves 3 important elements as given below :

- Greater emphasis on the collecting and conservation of germplasm of wild and weedy relatives of the major crops
- Increased focus on the nationally or locally important crops used for food, fibre, medicine or fuel
- Sampling and resampling of major crop germplasm to fill specific gaps to replace lost or poorly representative samples or to meet current specific needs to breeders

Sampling Strategy

The basic questions which a plant explorer faces are the genetic variation that are aimed to be sampled and the quantification of genetic variation present in a set of population. The variation can be described by several parameters like allele and genotype frequencies, gene diversities, heterozygosity levels and disequilibrium coefficients.

Marshall and Brown recognized four types of alleles, namely (i) common, widely distributed, (ii) common, locally distributed, (iii) rare, widely distributed, and (iv) rare, locally distributed. The first category alleles are included in small samples from a few populations.

For the last two classes adequate ex-situ conservation will be limited by the population sizes which the plant explorers are prepared to collect and conserve. According to Marshall and Brown (1975) the key indicator for optimum sampling is the number of such alleles that each population possesses i.e. the number of alleles that attain appreciable frequencies in only one population or in a few adjacent populations. The conservation of such populations will largely depends on the identification of populations closely adapted to specific environment and agricultural practices. Other things being equal populations with a higher number of common alleles deserve priority.

Based on these data, a plant explorer is required to roughly divide a region into a limited number of areas clearly distinct because of physiographic, edaphic, climatic, botanical, agricultural or cultural differences. Total time available is then divided among the areas according to travelling convenience and differences in genetic diversity among areas. More sites could be samples in areas where the target species is more common. Generally, a set of about 50 sites comprises a justifiable sample. The sample size at a site should be about 50 individuals. This figure could be increased based on any splitting of the samples, possibility of non-viable seeds from individuals and possible loss of some individuals in the sample in transport and quarantine.

Brown and Marshall (1995) have outlined the following 4 elements of a basic sampling strategy.

- Sample about 50 populations in an ecogeographic area or mission
- Sample about 50 individual plants in each population
- In general, sample individuals at random at each site, but sample separately within distinct local microenvironments if the habitat at the site is heterogeneous
- Sample sufficient seeds or vegetative material per plant to assure representation of each original plant in all duplicates

Factors Affecting Basic Sampling Strategy

Brown and Marshall (1995) have listed the following attributes which may affect sampling in practice.

Geographic Distribution

A species found only in a narrow geographic range would merit sampling from fewer sites but with an increased number of individuals at each site and increased number of propagules per individual.

Local Abundance

For the species particularly the wild ones which are locally rare, it may be difficult to include a target of 50 individuals per population. This limitation can be compensated by increasing the number of sites and by increasing the number of propagules per plant. Brown and Briggs (1991) gave guidelines for the collecting of endangered species of 10 individuals at each of up to 5 sites.

Inter Population Migration

When migration rates appear to be high between populations, sampling of fewer but more widely spread populations is recommended.

Habitat Diversity

Species that grow in a wide range of ecological situations are more likely to have diverged genetically among their different habitat. In such cases an increased number of populations or distinct subpopulations are sampled at the expense of the number of individuals per population.

Life History

Life history is taken into account as normally perennial fruit trees are outbreeding and annual crops are self and cross-pollinated both. Because of this, outbreeding annuals and outbreeding perennials tend to have similar levels of genetic divergence among populations.

Duration of Life Cycle

Collection of perennial species is normally not season bound as vegetative material is available throughout the year. There can be a return visit to the site if collected material proves to be insufficient or does not survive. Therefore, number of propagules collected from perennials may not be as high as for the annuals.

Population Age Structure

If a population of perennials consists of individuals of various ages, random sampling is done irrespective of size or age to maximize genetic diversity. In case of distinct age groups, random sampling can be done in a stratified manner.

Vegetative Reproduction

For vegetatively propagated species, it is advisable to include samples of seeds as well as vegetative material wherever possible.

Fecundity

Fecundity has direct bearing on the number of propagules/plant to be collected while germplasm collecting. Sampling from more individuals/site is done for species that produce less seeds.

Determinacy of Flowering and Seed Maturation

Well timed exploration is needed for highly synchronized flowering species. For indeterminate flowering species, it is important to sample plants with different maturities.

Mating System

For outbreeding species, the number of populations in an eco-geographic region can be reduced and number of individuals per site increased. For inbreeding species, the reverse could be adopted.

Mode of Pollination

For wind pollinated species, the plant population at a particular site tends to contain half-sibs as they are the progenies/products having common mother and many sources of pollen. In such populations, there is less divergence. In contrast, in animal pollinated outbreeding species, seeds from a single fruit may have largely the same male parent and such species tend to show more population divergence than wind pollinated outbreeders. More the population divergence, more are the populations during sampling.

Sampling for Specific Goals

Sometimes plant germplasm exploration is organized to collect plants/genes for resistance to a particular disease or certain abiotic stresses like drought or waterlogging. The desired rare allele may be the basis of the desired phenotype or one could look for a specific site where the target gene may be available. If the gene is rare, more individuals per site or more sites are samples. The size of sample in terms of the number of random gametes(s) required to be 95 % certain of including at least one copy of the target allele or genotype is as follows (Brown and Marshall, 1995).

$S = -3 \log e (1-p)$ where

p = frequency of allele in the population

This formula gives the sample sizes required for an allele or character that is increasingly rare ($p = 0.05, 0.03, 0.01, 0.001$) are 59, 99, 299 and 2999. The same calculations apply for the number of sites also if p shows the frequency of the desired site and S the required number of sites to sample to be 95 % certain of a successful mission.

Thus, the germplasm explorer or the collector may well decide to increase substantially the sample size at a site that evidently meets a specific objective of the mission particularly when the desired variants are likely to be rare and restricted to that site.

Sedcole (1977) has computed the sample size (s) required to be 95 % certain of recovering a minimum number (r) of plants with a trait that occurs in a population with frequency p. Table 9.1 lists a selection of these values of S. An approximate formula for S for relatively infrequent traits ($p < 0.20$) is :

$$S = \{r + 1.645 + \sqrt{r}\ 0.5\}/p$$

The values in Table 9.1 show that the required sample size increases with increasing required number of copies, as would be expected. However, the increase is less than proportionate. Thus, increasing the sample size fivefold (for 0.05, from 63 to 314) increase the assured number of copies tenfold.

A relatively large sample is thus justified in missions that are aimed at finding several examples of specific rare and valuable variants. The final figure will depend on the practical limits on handling larger samples weighed against the number of copies that are actually needed.

Table 9.1. The sample size (s) required to be 95 % certain to recover a minimum number (r) of plants with a trait that occurs in a population with frequency p (from Sedcole, 1977)

P	Number of copies to be recovered (r)								
	1	2	3	4	5	6	8	10	15
0.25	11	18	23	29	34	40	50	60	84
0.125	23	37	49	60	71	82	103	123	172
0.0625	47	75	99	122	144	166	208	248	347
0.03125	95	150	200	246	291	334	418	500	697
0.015625	191	302	401	494	584	671	839	1002	1397
0.05	63	97	127	156	184	211	264	314	438

Re-sampling Region

The re-sampling of a region is required to replace the samples in a gene bank that have become inviable. Re-sampling is also needed in populations that are undergoing co-evolutionary changes in response to changes in cropping system, predators, pathogens and weeds, etc. Jain (1975) argued a two step sequential method in which the first season collection is done on a coarse grid and the second round on a finer scale in areas of interest. Brown and Marshall (1995) have suggested the following general strategic questions facing the plant collector in the case of re-sampling.

- Should the collector give particular emphasis to uncollected or largely unstudied areas within the region? The risk of doing this is that the species may be rare or absent from such areas. The benefit is that any new samples could add new ecogeographic representation to the collection.
- Should the collector emphasize areas where the previous data and samples indicate high genetic diversity? The risk here is that the new samples may

contain much that is redundant with the accessions already in gene banks. The benefit is that the collector will be assured of getting a very rich collection.

- Should the collector aim for specific ecotypes or landraces reported to exist but no longer available from gene banks? The disadvantage of this strategy is that such specific localities are often very scattered and the material may have subsequently disappeared from the site. For crop populations in particular, the field situation is likely to be relatively labile. The advantage is that the specific-target strategy has a good chance of locating at least some of the targets.

References

Brown, A.H.D. and Briggs, J.D. 1991. Sampling strategies for genetic variation in *ex-situ* collections of endangered pant species. *In* : D.A. Falk and K.E. Holsinger (eds.). Genetics and Conservation of Rare Plants. Oxford University Press, New York. pp 99 - 119.

Brown, A.H.D. and Marshall, D.R. 1995. A basic sampling strategy : theory and practice. *In* : L. Guarino, V. Ramantha Rao and R. Reid (eds.). Collecting Plant Genetic Diversity. CAB International. pp 75 - 91.

Jain, S.K. 1975. Population structure and the effects of breeding system. *In* : Frankel, O.H. and J.G. Hawkes (eds.). Crop Genetic Resources for Today and Tomorrow. Cambridge Univ. Press. Cambridge. pp 15 - 36.

Marshall, D.R. and Brown, A.H.D. 1975. Optimum sampling strategies in genetic conservation. *In* O.H.Frankel and J.G. Hawkes (eds.). Crop Genetic Resources for Today and Tomorrow. Cambridge University Press, Cambridge. pp 53 - 80.

Sedcole, J.R. 1977. Number of plants necessary to recover a trait. Crop Science. 17 : 667 - 668.

10

Plant Germplasm Exploration

The diversity collection and making it available for future use is one of the most important activities related to management to plant germplasm resources. The collection of plant germplasm from their natural habitat is broadly known as plant germplasm exploration. The main reasons for collecting germplasm of given gene pool in a given geographical location are as follows (Engels *et al.*, 1995):

- It is in danger of genetic erosion or even extinction
- A clear need for it has been expressed by the users, at national level or internationally
- The diversity it represents is missing from, or insufficiently represented in, existing *ex situ* germplasm collections
- More needs to be known about it

These are not mutually exclusive. Germplasm may be both threatened and useful, and there may be gaps both in collections of a gene pool and in what is known about it.

Rescue Collection

This is adopted when the genetic diversity is imminently threatened in an area and *in situ* conservation methods are not feasible or insufficient. Small populations and genotypes with restricted distributions are most at risk and qualify for rescue collecting. Replacement of diverse traditional crop types by less number of uniform modern varieties has become a global threat to traditional landraces of major crops specially since late sixties in India. In many areas, locally important traditional food crops are being replaced by exotic staples or by cash crops. Modernization of agriculture including shift on use of hybrid cultivars has added additional factors leading to the loss of biodiversity.

The irreversible loss of genetic diversity are also caused by over grazing in case of forages or by uncontrolled harvesting from the natural habitat in case of medicinal plants, firewood species, timber trees, etc. Natural disasters like disease epidemics, prolonged drought, flood, tsunami, etc. can also result in loss of genetic diversity. Under such situations, the rescue collecting is used.

Collecting for Immediate Use

Plant breeders usually maintain their own active collections which include selected genotypes for use in their breeding programme. However, there is always a continuous need for new, specific traits and combinations of traits in view of new problems which keep on arising. Therefore, a close linkage between germplasm collection and its use has often proved to be the most effective way of its conservation.

In the new era of modern biotechnology, there is always the need of transferring genes from one species to another. This would imply that the breeders will be interested even in the species which are genetically less closely related to the species with which they are working. This will necessitate collecting new germplasm lines. Such germplasm lines could be collected from existing *ex-situ* collections which have been characterized/evaluated or which may include material from geographical locations or habitats likely to harbour the desired traits, for example, the genes conferring cold resistance could be perhaps looked into the landraces or wild relatives which shall be collected from high altitude areas and thus such regions shall be high priority for germplasm collection. Such collections are also required if the material in gene banks is unavailable to the prospective user or available only in inadequate quantities. Repeat collections may also be required if it is wanted to estimate genetic erosion in an area.

Gap Filling for Future Use

Immediate user need is an important reason for germplasm collection but material not considered important today may turn out to be of vital importance tomorrow and therefore, germplasm collecting for future use is also essential. These germplasms shall be of high value in view of changing priorities, eco-systems and food habits. Potential future use is a good enough legitimate justification for germplasm collection. There are cases of random distribution of genes over the geographical areas. It is quite possible that cold tolerance gene might be available in the germplasm lines collected from tropical regions. Likewise, there could be disease resistance in the germplasm from outside the geographic range of the pathogen. Therefore, collections under such situations are also required.

Germplasm Collection for Research

Germplasm collection are also required for resolving particular research problems. Such problems could include understanding the mating system of the species, their taxonomic boundaries, evolutionary relationships and demystification.

Types of Germplasm Collection Missions

The collecting missions are of the following types (Engels *et al.*, 1995) :

Multi-species vs Species Specific Collecting

In multi-species collecting mission, a region is targeted and efforts are made to sample as mush as possible of the diversity of as many species as possible. Such missions are area driven and area selected is the one which is difficult to reach and from where no systematic collections have been made. The multi-species collection missions have more to do with conservation than immediate use.

Species specific (gene pool specific) missions are done by the actual users of the germplasm, e.g. typical plant breeders and plant introduction scientists. The targets of such missions may be broad or very specific. In specific target missions, the specialists like entomologists, plant pathologists, microbiologists and biotechnologists, etc. may be involved.

Wild Species vs Crop Collection

For collecting wild species the collections are restricted to a narrower range of time because the wild species are prone to shattering. In case of crop plant collections, the seeds usually stay longer on the plant and even collections can be done by visiting harvested stacks, farmers' stores, shops, family seed companies and markets. Since wild species are usually difficult to find, much more time will be required to be spent for such collections.

Single Visit vs Multiple Visit Collection

Usually there are single and short visits to the target areas for collection of germplasm. However, because of variation within a region in the timing of fruiting due to latitudinal, altitudinal and climatic differences, a single visit may not be sufficient. In a single short visit, early and late maturing materials within a population as well as among populations may be missed. Similarly if there are two or more growing seasons for a crop in the particular region, the germplasm lines specifically adapted to a particular season may be missed. Year to year variation may also be overlooked by single visits. Therefore, multiple visit collections are usually required. Repeat visits are also necessary to monitor genetic erosion at a site or in a region.

Institutional (Formal Sector) vs Community (Informal Sector) Germplasm Collection

The formal sector germplasm collections are carried out by national gene banks and the institutions engaged in agricultural research. In Indian context the formal sector institutions could be National Bureau of Plant Genetic Resources and its regional stations, ICAR crop institutes and the State Agricultural Universities, etc. Lately, the non-governmental organization (NGOs), community groups and other grass-root organizations are showing interest in the germplasm collection which falls under informal sector. According to Mooney (1992) the institutional gemrplasm collection has largely been one of *ex-situ* germplasm storage in gene banks while the community germplasm collection strategy is generally one of *in situ* conservation. Mooney (1992) desired the community strategy as follows :

Cultivars are kept as part of the farming system or, where possible, in small plots for endangered cultivars and/or seed samples are cleaned, dried, stored under cool/dry conditions within the community and monitored by local people knowledgeable about the species.

Technical Planning Before Initiating Plant Germplasm Exploration

The following points need to be considered for plant germplasm exploration and collection :

- Transport (four wheel vehicle/motor cycle)
- Provisional itinerary drawn up at an early stage of the planning process showing the target areas and precise locations
- Intimation to State Biosafety Board established under section 22 of the Biological Diversity Act-2002 of Government of India
- Camping equipment if needed
- Checking of seed and vegetative sample on a regular basis for insect and fungal attack and discarding or treating infected samples
- Checking for rotting of samples of fleshy fruits in plastic bags
- Sending samples back to the gene bank at various stages during the course of the mission rather than waiting until the end
- Sorting and preparing germplasm and any reference ancillary specimens
- Collating, completing and editing the collecting data
- Distributing the germplasm samples/safe storage

The Ingredients of Success of Plant Germplasm Exploration and Collection

Engels *et al.* (1995) has listed the following basic points :

- Planning well ahead
- Involving local people
- Incorporating flexibility
- Developing a search image including experienced person in the exploration team

- Choosing collecting and processing technique with care
- Documentation of the collection scrupulously
- Taking trouble with the samples
- Thinking about safety
- Implementation of follow up programme which shall include among other things proper storage in duplicate base collections, deposition of voucher specimens in herbaria and identification as necessary, characterization, evaluation, other experimental work and writing a report for wider circulation

References

Engels, J.M.M., Arora, R.K. and Guarino, L. 1995. An introduction to plant germplasm exploration and collecting : planning, methods and procedures follow-up. *In* L. Guarino, V. Ramantha Rao and R. Reid (eds.). Collecting Plant Genetic Diversity. CAB International. pp 31 - 63.

Mooney, P.R. 1992. Towards a folk revolution. *In* D. Cooper, R. Vellve and H. Hobbelink (eds.). Growing Diversity. Intermediate Technology Publications, London. pp 125 - 138.

11

Collection of Wild Plant Species

Active plant breeding programmes call for continuous search for new genes to be incorporated into advanced breeding lines to face the new problems and the challenges. The rapid technological developments in gene transfer technology in recent years have opened new avenues for introducing useful genes from wild and weedy relatives of crop plants. It is also a fact that due to genetic erosion, wild species, are getting lost. Therefore, systematic collection of germplasm of wild and weedy species have become more relevant today then ever before for the following reasons (Bothmer and Seberg, 1995).

- Collection for taxonomic, phylogenetic and biosystematics research
- Collection for genetic diversity study and conservation
- Collection for immediate use in a breeding programme

The Gene Pool Concept

For understanding of genetic diversity in wild crop relatives, the concept of gene pool has been advocated by Harlan and deWet (1971). The primary gene pool consists of taxa including cultivated, weedy and wild forms of a crop among which there are no sterility barriers and gene transfer is therefore simple and straightforward. The cultivated and the wild forms of the crop plants among which crossing is possible but gene transfer is difficult are included in the secondary gene pool. The tertiary gene pool includes taxa from which gene transfer is very difficult due to strong sterlity barriers.

The examples of primary gene pool could be listed as cultivated barley, *Hordium vulgare*, spp. *vulgare* and spp. *spontaneous*. The cultivated soybean, *Glycine max* and the wild soybean, *Glycine formosana* constitute another primary gene pool. In such cases there are no problems in gene transfer from wild to the cultivated types. However, the disadvantage here is that several undesirable genes of the wild form (e.g. shattering, small seed size and shrunken seed, etc.) are transferred along with the desired characters. Under such circumstances, there should be a prebreeding programme including some generations of backcross to the cultivated types and repeated selection of desired genotypes.

The secondary gene pool of barley includes only one species namely *Hordium bulbosum*, a perennial Mediterranean type from which gene transfer to barley through sexual hybridization is putatively possible. This species is of importance for the production of barley haploids through selective elimination of the *H. bulbosum* chromosomes in the interspecific hybrids. The tertiary gene pool of barley includes the remaining species of the genus *Hordium*.

Collection of Wild Species

The first step in planning and collecting programme for wild species is basically a background study of taxonomy and distribution of the target species which is called as eco-geographic survey. The floras and the monograph are studied as the primary sources of such information. Flora is an account of the plants found in a specified region usually providing means of identifying the different taxa. On the other hand, a monograph is a comprehensive treatment of the available taxonomic data pertaining to all the taxa in a specified group. Understanding of the passport data along with past germplasm collecting and other information and distribution gives an indication of geographical gaps in the existing collections. Naturally such areas become of high priority for conservation collecting. However, it is always safe that even areas from where the material has already been collected for the gene bank, revisit and recollection usually gives some new genotypes.

Centres of diversity are the priority geographic areas for collecting the germplasm to elucidate taxonomic and genetic relationships. The isolated populations are also given special consideration. Populations in marginal areas or unusual habitats are likely to have germplasm with unique traits. The populations or species which face the danger of severe genetic erosion or extinction should be adopted for such collecting.

The samples collected for study of genetic diversity are subjected to evaluation by morpho-metric, biochemical and molecular methods. Isozyme electrophoresis has been widely used in understanding of genetic diversity in plant species (May, 1992). Variation in the seed storage proteins has also been used extensively through seed protein electrophoresis to study genetic diversity in the plant species. The seed protein electrophoresis has the advantage that the protein profiles are stable across the environments and are reproducible. Advances in molecular marker technique (RAPD, RFLP, AFLP, SSR, etc.) have revolutionized the diversity analysis in plant germplasm. The collection for plant breeding programmes are usually for a specific trait for e.g. source of resistance to an abiotic stress (salinity, drought, frost, etc.) or to a pest. The desired trait could be located through systematic screening of gene bank holdings of the primary and secondary gene pools. If the traits are not available in the existing collections, new collecting missions are organized in appropriate areas. Such collecting missions are usually carried out in the areas where conducive environmental conditions for the abiotic stress or the pathogen are available. Thus, the germplasm search having tolerance to abiotic stress should be done in areas where the species has been exposed to the stress factor for a considerable period. Collection for pest resistance is likely to be successful in the areas where there is a long documented history of co-existence of the host and the pathogen. But it should also be kept in mind that resistance may occur anywhere (Harlan, 1978).

In summary, when collecting for specific traits, the target area should be chosen after studying the available ecogeographic and pest information (including information on the distribution of alternate hosts), analyzing existing characterization and evaluation data and/or screening gene bank accessions for the desired (or correlated) trait. Relatively small areas should be selected for careful, intensive sampling. Comparatively large random samples should be collected from populations likely to harbour the desired trait. In addition to a population sample, a selective sample (e.g. of healthy individuals in an infested field) can be collected if the expression of the target trait is obvious and stable (Bothmer and Seberg, 1995).

References

Bothmer, R.V. and Seberg, O. 1995. Strategies for collecting of wild species. . *In* L. Guarino, V. Ramantha Rao and R. Reid (eds.). Collecting Plant Genetic Diversity. CAB International. pp 93 - 111.

Harlan, J.R. 1978. Sources of genetic defense. Annals of the New York Academy of Science. 287 : 345 - 356.

Harlan, J.R. and J.H.J. de Wet 1971. Towards a rational classification of cultivated plants. Taxon. 20 : 509 - 517.

May, B. 1992. Starch gel electrophoresis of allozymes. *In* A. R. Hoelzel (ed.). Molecular Genetic Analysis of Populations. pp 5 - 27. IRL Press, Oxford.

12

Collection of Seed in the Field

For seed storage in the gene banks, the seed stored should have high quality which is greatly affected by seed handling during collection. The details on these issues have been given by Smith (1995). In terms of seed storage characteristics seeds are of following 3 steps.

- Desiccation intolerant (recalcitrant)
- Partially desiccation tolerant (intermediate)
- Fully desiccation tolerant (orthodox)

In general, smaller seeds are desiccation tolerant as compared to larger seeds which tend to be desiccation intolerant. For seed storage behaviour prediction, habitat of species can also be taken into consideration. The seeds from better habitats are more likely to be desiccation tolerant. Plant habit can also be added to the list of factors associated with seed storage behaviour. Generally, seed weights of herbaceous species are lower than those of trees. This suggests that herbaceous species may have mostly desiccation tolerant seeds. Aquatic species are usually desiccation intolerant.

Desiccation Intolerant (Recalcitrant Seeds)

These seeds are unable to withstand drying. Aquatic tropical grasses and temperate trees seeds lose viability when they are dried to water potential of – 5 to – 3 MPa (Pritchard, 1991). These water potentials are in equilibrium with RH of > 95 %. Such RHs are difficult to be available in nature and hence recalcitrant seeds dry to lethal water potentials even under ambient conditions in which the species occurs in the wild. In fleshy fruits such as mango, the flesh acts as a reservoir of moisture. In such fruits, high moisture in seed is maintained by allowing seeds in fleshy fruits while collecting these. Loss of moisture can be minimized by keeping seeds in impermeable plastic bags. Temperature also affects the longevity of recalcitrant seeds. For temperate species of oak, reducing the temperature as low as -1.5°C increases seed storage life but further reduction to -

5°C is lethal. Harvesting desiccation intolerant seeds just before natural dispersal is better than collecting fallen fruits.

In practice, there should be minimum possible time gap between harvesting and dispatch of desiccation intolerant seeds. The gap should not exceed 30 days between harvesting and receipt of seed in the seed bank.

Intermediate Seeds

Coffee and oil palm are common crops under this category. Under field conditions, these seeds usually behave like desiccation tolerant species. Drying of such seeds under the sun or using silica gel should be done with care.

Orthodox Seeds

Fully desiccation tolerant seeds are known as orthodox seeds. Tolerance to drying in dry fruited cereals, pulses and crucifers is obtained about halfway through seed development. In cereals, seed storage potential continues to increase until well after maximum seed weight is reached and seed moisture has fallen to equilibrium with the surrounding air. Now onwards there is loss of seed storage potential with advancement of seed aging in the field. In fleshy fruited tomato, tolerance to rapid desiccation does not occur until just before or just after maximum seed weight has been reached. Seed storage potential then continues to increase for the next 10 days and remains constant for a further period of 40 days (Demir and Ellis, 1992).

Viability Loss in Desiccation Tolerant Seeds at High Moisture Level

Smith (1995) has outlined following 3 important conclusions of practical importance in this regard.

- Great care must be taken to keep seeds aerated when they are held at high moisture contents, for example those at which many non-domesticated species are shed. Holding the seeds in sealed conditions, such as in a plastic bag, or packing the seeds tightly into limited space will result in an anaerobic environment rapidly developing around the seeds, due to their respiration. This will dramatically increase the rate of loss of viability.
- The switch to air-dry storage behaviour appears to occur at relative humidity close to 90 %.
- The benefits of drying will only be realized if ambient relative humidity is < 65 %.

Practical Advice to Seed Collectors

According to UN Conference on Environment and Development (UNCED) and as described by Smith (1995), the advice is as follows :

All Seeds

- Attempt to collect equal numbers of seeds from each plant sampled, and at the same maturity, ideally when seed storage potential/desiccation tolerance is highest. Avoid damaged seeds (mechanical damage, pest attack).
- If seeds must be cleaned during the trip, do so by hand to minimize the chance of mechanical damage.
- If it is possible to avoid quarantine seed treatments without breaking quarantine regulations, for example through post-entry quarantine, do so.
- Personally ensure that seed arrives at the seed bank without undue delay.

Desiccation-Intolerant Seeds

For desiccation-intolerant seeds, which are more likely to be large seeds from dominant trees growing under wet conditions.

- Collect close to fruit fall. Do not collect from the ground unless you can be sure seeds are only recently dispersed.
- Keep seeds aerated and moist in inflated polythene bags, changing the air at least weekly by deflation and reinflation.
- Do not allow such seeds collected in the tropics to either cool below 20°C or heat up above ambient shade temperatures in the field or during transport.
- Plan your activities so that no more than one month elapses between collecting and reception by the seed bank.

Desiccation-Tolerant Seeds or Their Fruits

For desiccation-tolerant seeds or their fruits, which are more likely to be small seeds from herbs growing under dry conditions.

- If meteorological data suggest that more than 0.1 probity/month will be lost during the collecting trip, either modify your itinerary or prepare to dry actively with silica gel.
- For fleshy-fruited species, if logistically possible keep the seeds in the fruits and the fruits aerated and at ambient temperatures.
- If the above is not logistically possible, and for fruits that are dry dehiscent or indehiscent, air-dry the hand-cleaned seeds in a thin layer (to ensure aeration) under shade for three days or more (larger seeds need longer) to reduce the seed moisture content towards equilibrium with ambient relative humidity before packing, to use space more efficiently.

References

Demir, I. and Ellis, R.H. 1992. Changes in seed quality during seed development and maturation in tomato. Seed Science Research 2 : 81 – 87.

Pritchard, H.W. 1991. Water potential and embryonic axis viability in recalcitrant seeds of *Quercus rubra*. Annals of Botany 67 : 43- 49.

Smith, R.D. 1995. Collecting and handling seeds in the field. *In* L. Guarino, V. Ramantha Rao and R. Reid (eds.). Collecting Plant Genetic Diversity. CAB International. pp 121 – 129.

13

Collecting Vegetatively Propagated Plants

The most important root and tuber crops cultivated on a worldwide scale are potato (*Solanum tuberosum*), cassava (*Manihot esculenta*), sweet potato (*Ipomoea batatas*), yams (*Dioscorea* spp.) and taro (*Colocasia esculenta*), but there are many others that are of regional, national or local importance, in a total of over a dozen dicot and monocot families. Most of these originated in tropical or subtropical areas and are mainly used as sources of carbohydrates. Many minor root and tuber crops, such as turmeric (*Curcuma longa*) and arrowroot (*Maranta arundinacea*, *Tacca leontopedaloides*), are also used in folk medicine and as spices (Sastrapradja *et al.*, 1981). All these crops are vegetatively propagated. There are also vegetatively (or clonally) propagated crops that are not roots or tubers, e.g. bananas and sugarcane. A selection of the more important crops is given in Table 13.1 (Huaman *et al.*, 1995).

Changes in the environment and in agricultural practices can cause particularly serious genetic losses in vegetatively propagated crops. This is because many set seeds only sparingly or not at all. Their capacity for adaptation is therefore lower than that of sexually propagated species, especially outcrossers, in which genetic recombination is much faster. Clonally propagated crops are also threatened by the accumulation of viruses in propagules (Lebot, 1992). It is therefore all the more important to conserve such species ex situ under controlled conditions. Among the Consultative Group on International Agricultural Research (CGIAR) commodity centres, the Centro Internacional de Agricultura Tropical (CIAT) maintains collections of cassava, the Centro Internacional de la Papa (CIP) potato and sweet potato, the International Institute of Tropical Agriculture (IITA) cassava and yams and the International Network for the Improvement of Banana and Plantain (INIBAP) bananas and plantains. Collections of vegetatively propagated crops, whether in international centres or national programmes, are commonly stored in field gene banks or in vitro.

It is usually necessary to collect vegetative material of clonally propagated crops, normally the farmer's planting material. Even when possible, seed collecting

Table 13.1. Selected vegetatively propagated herbaceous crops and the plant parts (–) suitable for germplasm collecting (Smith *et al.*, 1993).

	Stools	Rhizomes	Bulbs	Offsets	Shoots	Suckers	Tubers cuttings	Stem	Bulbs	Stolons	Storage roots
Acorus calamus	•	•	-	-	-	-	-	-	-	-	-
Allium sativum	-	-	•	-	-	-	-	-	•	-	-
Ananas comosus	-	-	-	-	•	•	-	-	-	-	-
Agaves pp.	-	-	-	-	-	•	-	-	•	-	-
Arracacia xanthorrhiza	-	-	-	•	•	-	-	-	-	-	-
Asparagus officinalis	-	•	-	-	-	-	-	-	-	-	-
Canna edulis	-	•	-	-	-	-	•	-	-	-	-
Colocasia esculenta	-	-	-	-	-	•	•	-	-	-	-
C. gigantean	-	-	-	-	-	•	-	-	-	-	-
Curcuma longa	-	•	-	-	-	-	-	-	-	-	-
Dioscorea spp.	-	-	-	-	-	-	•	-	-	-	-
D. bulbigera	-	-	-	-	-	-	-	-	•	-	-
D. esculenta	-	-	-	-	-	-	-	•	-	-	-
Eleocharis dulcis	-	-	-	-	-	-	-	-	-	•	-
Fragaria vesca	-	-	-	-	-	-	-	-	-	•	-
Hedychium coronarium	-	•	-	-	-	-	-	-	-	-	-
Ipomoea batalas	-	-	-	-	-	-	-	•	-	-	•
Lepidium meyenii	-	-	-	-	•	-	-	-	-	-	-
Manihot esculenta	-	-	-	-	-	-	-	•	-	-	-
Maranta arundinacea	-	-	-	-	-	•	-	-	-	-	-
Mirabilis expansa	-	-	-	•	-	-	-	•	-	-	•
Musa spp.	-	-	-	-	-	•	-	-	-	-	-
Oxalis tuberose	-	-	-	-	-	-	•	•	-	-	-
Pachyrrhizus ahipa	-	-	-	-	-	-	•	-	-	-	-
Polymnia sonchifolia	-	-	-	•	-	-	-	•	-	-	-
Saccharum spp.	-	-	-	-	-	-	-	•	-	-	-
Sansevieria spp.	-	•	-	-	-	-	-	-	-	-	-
Solanum spp.	-	-	-	-	-	-	•	•	-	-	-
Tacca leontopedaloides	-	•	-	-	-	-	-	-	-	-	-
Tropaeolum tuberosum	-	-	-	-	-	-	•	-	-	-	-
Ullucus tuberosus	-	-	-	-	-	-	•	•	-	-	-
Xanthosoma nigrum	-	-	-	-	-	•	-	-	-	-	-
Vanilla planifolia	-	-	-	-	-	-	-	•	-	-	-
Zingiber officinale	-	•	-	-	-	-	-	-	-	-	-

is not necessarily desirable, as it may result in the loss of the specific gene combinations that make a clonally propagated variety distinctive. This presents several difficulties at different stages of collecting (Hawkes, 1975, 1980,Ford-Lloyd and Jackson, 1986), but all can be overcome by careful planning and appropriate techniques.

Planning an Exploration

Normally only those vegetative parts are collected that are used by the farmers to propagate the crop. Thus appropriate time of germplasm collection will be the harvesting season of the target crop. Collecting immature storage organs or the sprouted ones is not advised. Local information is of importance to decide the appropriate time of collection. Potatoes are collected during dormancy stage when the above ground parts are completely dry. There are many kinds of vegetative plant materials that are used for vegetative propagation. These are defined as follows (Bell, 1991 Huaman *et al.* 1995).

Some Definitions of Vegetative Propagating Structures

Bulb : A short, fleshy, underground shoot, mostly made up of colourless, swollen leaf bases and scale leaves enclosing next year's bud. A bulbel or bulbet is a small bulb arising from a parent bulb.

Bulbil : A small bulb arising in an axil on the aerial part of a plant.

Corm (e.g. Crocus, Gladiolus) : A short, swollen, annual underground stem of several nodes and internodes, with that of the next year forming above it at the base of each flowering stem. A cormel is a corm arising vegetatively from a parent corm.

Knob (e.g. *Amorphophallus campanulatus*) : Sprout from the main tuber.

Offset (e.g. *Sempervivum tectorum*) : Small plantlet taken from the base of the above ground part of the main stem.

Rhizome (e.g. Acorus, Zingiber) : An underground stem, usually horizontal and often elongated, that is distinguished from a root by the presence of nodes and internodes and/or the presence of buds in the axils of reduced, scale-like leaves.

Shoot (e.g. Ananas) : The apical or lateral meristematic dome together with the leaf primordia, emerging leaves and subadjacent stem tissue.

Stem cutting (e.g. *Lpomoea batatas*) : Section of a mature stem with several leaves and their axillary buds.

Stolon or runner (e.g. *Fragaria vesca*) : A trailing stem, capable of forming roots, shoots or both (and hence a new plant) from its axillary buds.

Stool (e.g. Acorus) : A clump of roots or rootstocks (short, erect underground stems, like vertical rhizomes).

Storage root or root tuber (e.g. *Mirabilis expansa*) : A fleshy, secondarily thickened root, without buds (except sometimes adventitious buds giving rise to suckers).

Sucker (e.g. Agave) : A shoot arising adventitiously from a root, often at some distance from the main plant.

Tuber (e.g. Colocasia, Solanum) : A swollen, subterranean stem, with numerous buds or eyes

Sampling Procedure

According to Hawkes (1980), it is appropriate to collect a bulk sample of 2 – 4 vegetative propagules from each of 10 – 15 randomly chosen individual clones. Individual clones may be identified by morphological features. Duplication should be avoided by taking samples from sites not too close. Wherever possible, a bulk seed sample should also be collected and given a separate collector's number. While collecting germplasm from farmers' field, the recommended procedure is selective sampling to avoid over representation of abundant clones at the expense of rare ones. In the processing of selective sampling, all morphotypes in a field or market are identified and two to four popagules of each morphotype are collected. This process is repeated at each sampling site. Duplicate samples between sites are identified later on when the material is grown for characterization and/or evaluation. A morphotype is defined as being composed of plants which are apparently phenotypically identical, without implying genetic identity (Ford-Lloyd and Jackson 1986).

In case seed collection is possible, the numbering of samples should be as follows (Hawkes, 1980).

- Allotment of same collecting number if seeds are taken from the same plant(s) from which a vegetative sample has been taken.
- Allotment of separate collecting number if the seeds are taken from more than one randomly chosen morphotype and bulked.

It should be kept in mind that morphologically similar cultivars may have different isozyme alleles at some loci or may show different reactions to biotic and abiotic stresses. In such cases more samples need to be collected combining selective and random sampling techniques.

Handling Vegetative Material During Collection in the Field

Vegetative materials are prone to rapid deterioration after harvest. They are easily damaged during the transport. Therefore, the collected samples should be free of insect damage and disease symptoms. The underground organs should be free of soil. The detailed guidelines for the field handling of material of several vegetatively propagated crop like *Colocasia, Dioscorea, Heavea, Ipomea, Manihot, Musa, Saccharum, Vanilla* and *Zingiber* have been given by Leon and Withers (1986).

Mechanical damage during digging up of the underground parts should be avoided. Bruises can be cut out and treated with a mild alkali such as chalk hydrated lime or wood ash. In case of large tuber crops like yam, head of proximal end could be collected. The material can be kept for gentle drying in the sun followed by cleaning of excess soil and fungicidal dusting or dip. Afterwards these could be wrapped in semi-permeable material and kept under cool conditions. It is advisable to have specially made crates allowing the storage of single layers of samples in trays or shelves. There should be adequate ventilation and extremes of temperatures should be avoided. Growing tissues require more care. In general, stem cuttings should be taken from the middle portion of the stem and these cuttings should have several nodes. The cut ends should be dipped in disinfectant to avoid rotting. Shoots, suckers and cuttings from freshly harvested lots keep reasonably well with the proximal end wrapped in slightly wet paper towels or newspapers. After this these are stored in insulated cool boxes. Sample should be examined regularly to keep the wrapping paper moist and the developing roots should be excised. In case of sweet potato usually 30 cm long stem cuttings with at least 6 nodes are retained. Cassava stem cuttings called stakes are usually 40 – 50 cm in length and can be simply wrapped in paper bags. In sugarcane stem cutting ends are sealed by dipping in hot paraffin after a short hot water treatment (50°C for 30 min).

References

Bell, A.D. 1991. Plant Form. Oxford University Press, Oxford.

Ford, L. B. and Jackson, M. 1986. Plant Genetic Resources : An Introduction to Their Conservation and Use. Edward Arnold, London.

Hawkes, J.G. 1980. Crop Genetic Resources Field Collection Manual. IBPGR and EUCARPIA, Rome.

Huaman, Z. Puente, F. de la and Arbizu, C. 1995. Collecting vegetatively propagated crops (especially roots and tubers). *In* L. Guarino, V. Ramantha Rao and R. Reid (eds.). Collecting Plant Genetic Diversity. CAB International. pp 457 – 466.

Leon, J. and Withers, L.A. 1986. Guidelines for seed exchange and plant introduction in tropical crops. FAO Plant Production and Protection Paper 76. FAO, Rome.

14

Molecular Characterization of Plant Genetic Resources

Molecular characterization involves the use of techniques that are available to analyze variation in plants and animals at the DNA level. Differences in gene sequences can be directly observed and described with a precision previously impossible to achieve. Many of the techniques that have been developed have already been used in evolutionary and taxonomic studies. They have immense value in studies of accessions identity and for the detection of novel useful variation in the plant germplasm. Molecular markers provide genetic information of direct value in key areas of conservation both ex situ and in situ. For ex situ conservation the key issues are:

- *Acquisition:* Data on the diversity of existing collections can be used to plan collection and exchange strategies. In particular, calculations of genetic distances based on molecular data can be used to identify particular divergent sub- populations that might harbour valuable genetic variation that is under-represented in current holdings.
- *Maintenance:* Genetic data are essential to identify duplicate accessions in order to ensure best use of available resources. Genetic markers are also needed to monitor changes in genetic structure as accessions are generated. Molecular markers provide markers suitable for both of these.
- *Characterization:* The genetic diversity within collections must be assessed in the context of the total available genetic diversity for each species. Existing passport data document the geographic location where each accession was acquired. However, passport records are often missing or incorrect. Molecular markers may extend and complement characterization based on morphological or biochemical descriptions, providing more accurate and detailed information than classical phenotypic data.
- *Distribution to Users:* Users of collections benefit from genetic information that allows them to identify valuable traits and types quickly. On a more

fundamental level, molecular marker information may lead to the further identification of useful genes contained in collections. Molecular data on diversity may provide essential information to develop core collections (Hodgkin *et al.*, 1995) that accurately represent the entire collection.

Basic Molecular Techniques

The techniques of most relevance in plant genetic resources work have been described by Hoelzel and Green (1994), Weising *et al.* (1995), Anolles and Gresshoff (1997) and Karpt *et al.* (1997). A list of common techniques is given in Table 14.1.

Table 14.1. DNA marker techniques

AFLP	Amplified fragment length polymorphism
ALP	Amplicon length polymorphism
AP-PCR	Arbitarily primed PCR
AS-PCR	Allele-specific PCR
CAPS	Cleaved amplified polymorphic sequences
DAF	DNA amplification fingerprinting
ISA	Inter-SSR amplification
RAPD	Random-amplified polymorphic DNA
RFLP	Restriction fragment length polymorphism
SAP	Specific amplicon polymorphism
SCAR	Sequence characterized amplified region
SSCP	Single strand conformation polymorphism
SSLP	Microsatellite simple sequence length polymorphism
SSLP	Minisatellite simple sequence length polymorphism
SSR	Simple sequence repeats
STS	Sequence tagged sites

The most commonly used techniques have been RAPD, RFLP and SSR and these are described as follows.

Random Amplified Polymorphic DNA (RAPD)

RAPD (random amplified polymorphic DNA) analysis is a PCR (polymerase chain reaction) based molecular marker technique by amplification of genomic DNA using at least one short oligonucleotide primer in low stringency conditions, results in multiple amplification products from loci distributed throughout the genome (Wiliams *et al.*, 1990, Welsh and McClelland, 1990). Depending on the specific conditions of amplification or product separation and detection, the arbitrary primer amplification methods were termed RAPD. AP-PCR (arbitrarily primer PCR; Welsh and McClelland, 1990) or DAF (DNA amplification fingerprinting, Caetano-Anolles *et al.* 1991) is generated from a region which is flanked by a part of 10 bp priming sites in the appropriate orientation. Genomic DNA from two different individuals often produces different amplification patterns. A particular fragment generated for one individual but not for other represents DNA polymorphism and can be used as a genetic marker. RAPD amplification is performed in conditions resembling those of polymerase chain reaction using genomic DNA from the species of interest and a single short oligonucleotide (a10 base primer).

Polymerase Chain Reaction

The polymerase chain reaction is used to amplify a segment of DNA that lies between the two regions of known sequence where two oligonucleotides as primers can bind the opposite strands of DNA due to complementary nature of base sequence. It is based on the features of semi-conservative DNA replication carried out by DNA polymerases in prokaryotic and eukaryotic cells. The only requirement is that the sequences at the borders of the selected DNA region must be known so that two short oligonucleotides can anneal to the target DNA molecule for amplification. It involves three steps:

1. Denaturation 94° C
2. Annealing 35°-65° C
3. Extension 72° C

The amplification reaction is catalyzed by Taq DNA polymerase. The template DNA is first denatured by heating at 94^0 C. The reaction mixture is then cooled to a temperature that allows the primers to anneal to their target sequences. These annealed primers are then extended (i.e. synthesis of DNA) with Taq DNA polymerase. The cycle of denaturation, annealing and DNA synthesis is repeated many times, because the product of one round of amplification serves as template for the next, each successive cycle essentially doubles the amount of the desired DNA product.

Components of PCR

1. Oligomer primer

Primer should have high GC content. Primers are used at a concentration of 1 μM which is sufficient for at least 30 cycles of amplifications.

2. Amplification buffer

This buffer contains KCl, Tris Cl and 1.5 mM magnesium chloride.

3. Deoxyribonucleoside triphosphates

dNTP's are used at a saturating concentration of 200μM for each dNTP. It consists of 4 dNTP's such as ATP, CTP, GTP and TTP.

4. Target sequence/genomic DNA

The concentration of the target sequence is 5-100 ng.

5. Taq DNA polymerase

This enzyme is purified from *Thermus aquaticus*. It consists of 5′-3′ polymerase dependent exonuclease activity. Addition of excess enzyme may lead to amplification of nontarget sequences.

Equipment

- Pestle mortar
- Eppendorf tubes
- Micropipettes
- Tips
- Waterbath
- Refrigerated centrifuge (10,000 – 15,000 rpm)
- Electrophoresis system (horizontal)
- Biophotometer for the determination of DNA concentration
- PCR machine (thermal cycler)
- Gel documentation system

Reagents

1 M Tris pH (8.0)

Tris	12.11 g
Distilled water	80 ml

After adjusting the pH (8.0), volume is made upto 100 ml.

10% Sodium dodecyl sulphate (SDS)

10 g SDS is dissolved in 100 ml distilled water.

0.5 M EDTA (Ethylene diamine tetra acetic acid) - pH 8.0

EDTA	18.62 g
Distilled water	80 ml

After adjusting the pH (8.0), volume is made upto 100 ml.

5 M Potassium acetate	(pH 5.8)
Potasium acetate	49.12 g
Distilled water	80 ml

After adjusting the pH (5.8), volume is made upto 100 ml.

3 M Sodium acetate	(pH 5.8)
Ready made solution available	
RNase	10 mg/ml
Proteinase K	20 mg/ml

Chloroform : isoamylalcohol	(24 : 1)
Isopropanol	
70% Ethanol	

SDS extraction buffer

1 M Tris (pH 8.0)	100 ml
0.5 M EDTA (pH 8.0)	167 ml
Polyvinyl pyrolidone(PVP)	10 g
SDS (10%)	170 ml

The volume is made upto 1000 ml.

Tris-EDTA buffer (TE)

1 M Tris (pH 8.0)	1 ml
0.5 M EDTA (pH 8.0)	0.2 ml

The volume is made upto 100 ml

Tris-acetate (TAE) 50 X (pH 8.0)

Tris base	24.2 g
Glacial acetic acid	5.71 ml
0.5 M EDTA	10.0 ml

The pH is adjusted to 8.0 with concentrated HCl and the volume is made upto 100 ml. This stock solution is diluted 50 times for 1 X working solution.

Ethidium bromide	(10 mg/ml)

100 mg ethidium bromide is dissolved in 10 ml distilled water. The bottle is wrapped in aluminum foil paper.

DNA loading dye

Bromophenol blue (0.25%)	0.025 g
Glycerol (30%)	2 ml

The volume is made upto 10 ml with distilled water.

Agarose

Primers (usually 10 base nucleotide sequence to be obtained from standard supplier based on specified sequence)

PCR amplification kit

λ Hind-III digested marker DNA

Protocol

DNA extraction and amplification

Normally CTAB (cetyl trimethyl ammonium bromide) method (Murray and Thompson, 1980) is used. However, SDS method (Dellaporta *et al.*,1983) is also common and is described as follows :

1. The leaves are washed with sterilized distilled water and blot dried and cut into small pieces with a sterilized blade.
2. 1 g of leaf material is placed into a pre-cooled mortar.
3. Liquid nitrogen is added to it and a fine powder is made and transferred to oakridge tube/eppendorf tube (eppendorf tubes are appropriate for 1 g leaf powder).
4. After making a fine powder 800 μl extraction buffer sodium dodecyl sulphate (SDS) is added.
5. The sample is homogenized and incubated in water bath at 65° C for 1 hr.
6. 300 μl potassium acetate and equal volume of chloroform : isoamylalcohol (24 : 1) are added.
7. The sample is centrifuged at 4° C and 12,000 rpm for 30 min.
8. The supernatant (200-300μl) is taken into another eppendorf tube of 1.5 ml vol. and equal volume of isopropanol is added.
9. The sample is centrifuged at 4°C and 12,000 rpm for 20 min. and the supernatant is discarded.
10. The pellet is washed with 70% ethanol (approximately 200μl).
11. After washing the eppendorf tube is kept in inverted position over paper towel/tissue paper for removal of ethanol leading to drying of DNA.
12. The pellet is dissolved in 250 μl TE (tris EDTA) buffer and sample is kept at -20° C over night/till needed for PCR.
13. After storing overnight at -20° C 4 μl RNase and 4 μl proteinase K are added to each sample.
14. The sample is shaken gently through tapping by finger and incubated in water bath at 65°C for 1 hr.
15. Equal volume (approximately 250 μl) of chloroform : isomylalcohol (24 : 1) is added and the sample is centrifuged at 4°C and 12,000 rpm for 35 min. and supernatant is transferred by micropipette to another eppendorf tube.
16. Step 15 is repeated on freshly collected supernatant 2-3 times in order to remove phenolic compounds.
17. The supernatant is taken and 3 M sodium acetate and ethanol are added.
18. The sample is centrifuged at 4° C and 12,000 rpm for 35 min.
19. Pellet is dried with paper towel and dissolved in TE (tris EDTA) buffer. The sample is now ready for PCR.
20. The conditions vary according to crop and primer and the system standardized and usually giving satisfactory results across a range of crops at Pantnagar is as follows:

Denaturation : 94° C (5 min), annealing : 35° C (1 min), extension : 72° C (5 min) and number of cycles : 40.

21. The amount of template and primers used for 25 µl of PCR reaction mixture are 20 ng DNA (2µl) and 1 µl of primer.
22. The concentration of Taq DNA polymerase used in PCR is 0.5 units/25 µl.

PCR Reaction

Reagent	Stock solution	Working concentration	Amount used per reaction mixture (25 µl)
Taq polymerase buffer	10 x	1x	2.5 µl*
dATP	10 mM	200 µM	0.5µl**
dCTP	10 mM	200 µM	0.5µl
dGTP	10 mM	200 µM	0.5µl
dTTP	10 mM	200 µM	0.5µl
Taq polymerase	3 units/µl	1.5units/µl	0.5 µl
Random primer	20 ng/µl	20 ng/µl	1µl

Reaction volume 25 µl

Taq polymerase buffer*

$N_1V_1 = N_2V_2$
$10 \times V_1 = 1 \times 25$
$V_1 = 25/10$
$= 2.5$ µl

dNTP**

$N_1V_1 = N_2V_2$
$10 \times 1000 \times V_1 = 200 \times 25$
$V_1 = 200 \times 25 / 10 \times 1000$
$= 0.5$ µl

For 10 reactions

Total reaction volume		25 µl × 10	= 250 µ l
Taq buffer (10 X)		2.5µl × 10	= 25 µl
dNTP's	ATP	0.5 µl × 10	= 5 µl
	CTP	0.5µl×10	= 5 µl
	GTP	0.5 µl × 10	= 5 µl
	TTP	0.5 µl × 10	= 5 µl
Taq polymerase		0.5 µl × 10	= 5 µl
Random primer		1µl × 10	= 10µl
Total			**60 µl**

Template(genomic DNA)	2 µl × 10	= 20 µl
Sterile distilled H_2O		= 250 µl-20 µl
		= 230 µl-60 µl
-		= 170 µl

230 µl is divided into 10 PCR @23 µl/tube and the PCR is performed in thermal cycler (Eppendorf make or equivalent).

Agarose Gel Electrophoresis

To know about quality and quantity of genomic DNA, agarose gel electrohphoresis is performed on genomic and amplified DNA. Larger molecules migrate more slowly because of greater friction drag and they form their way through the pores of gel, less efficiently than smaller molecules. Range of separation in gel containing different amount of agarose is as indicated below.

Amount of agarose in gel (% w/v)	Efficient range of separation of linear DNA molecules (kb)
0.3	5-60
0.6	1-20
0.7	0.8-10
0.9	0.5-7
1.2	0.4-6
1.5	0.2-3
2.0	0.1-2

The steps are as follows :

1. The open ends of the clean, dry plastic tray, supplied with the electrophoresis apparatus are sealed to form a mold. The mold is set on a plain horizontal table.
2. 0.8% agarose solution is prepared for genomic electrophoresis and 1.8% for that of amplified DNA . 0.5 µg/ml ethidium bromide of 1 X TAE is added into the agarose after boiling. The boiling can be done either in microwave oven (about 10 min.) or water bath for about 30 min. till the solution becomes transparent. A comb with adequate number of wells (13 or 24, etc.) is inserted vertically into the gel before polymerization to make the wells. The thickness of gel is about 1 mm.
3. After the gel is completely set, the comb is removed carefully and the gel is mounted over the electrophoretic tank containing 1X TAE buffer in amount sufficient to immerse the gel in the buffer.
4. The λ Hind III is used as marker DNA.

5. The DNA loading dye is mixed with DNA samples in ratio of 1 : 5.
6. The sample is loaded in wells of gel with micropipette.
7. Electrophoresis is done at 100 volt till the dye front reaches the end of the gel.
8. The gel is documented through gel documentation system (e.g. Syngene-Gene Geneus, etc).

Advantages

- Non radioactive detection
- No prior DNA sequence information for genome required
- Universal primers work in any genome
- Very small amounts of genomic DNA are sufficient
- Multiplex detection of polymorphism
- Experimental simplicity
- No need for expensive equipment beyond a thermal cycler and a transilluminator

Limitations

- RAPD polymorphisms are inherited as dominant-recessive characters. This causes a loss of information relative to markers which show codominance.
- RAPD primers are relatively short, a mismatch of even a single nucleotide can often prevent the primer from annealing, hence there is a loss of band.
- RAPD is sensitive to changes in PCR conditions resulting in changes to some of the amplified fragments.

Applications

- Generating genetic maps
- Mapping traits in segregating populations
- Mapping traits in near –isogenic lines
- Mapping traits using bulk segregant analysis
- Saturating regions of a genome with markers
- Fingerprinting individuals
- Germplasm analysis
- Measurement of genetic distance between individuals
- Estimating relative parental contributions in backcross (Rafalski, 1997)

Restriction Fragment Length Polymorphism (RFLP)

Restriction enzymes are site-specific DNases that cleave a DNA molecule whenever the recognition sequence, which is usually a 4-6 base palindrome, is present. Because of the enzyme's sequence specificity, digestion of a particular DNA results in a reproducible array of fragments. RFLP, or length differences in homologous fragments between different DNA, are caused by changes in the primary sequence of the DNA. These length differences can be the result of:

- A point mutation resulting in the loss or gain of restriction enzyme cut site
- An insertion or deletion of DNA between two restriction enzyme cut sites
- A deletion which overlaps a restriction enzyme site
- A DNA rearrangement where one end of the rearranged segment resides between two restriction enzyme sites

When such differences occur, they can be detected by DNA hybridization and used as molecular markers in fingerprinting or genetic studies. The steps are as follows:

Isolate DNA to be Analyzed

Choose a DNA isolation procedure that works for the species being analyzed. The main requirements are that the DNA be of high average molecular weight and that it be pure enough to be cleaved with restriction enzymes.

DNA with Restriction Enzyme

Restriction enzymes are site-specific nucleases that reproducibly cleave DNA. Over 100 restriction enzymes are currently available with many different sequence specificities.

Perform Agarose Gel Electrophoresis

Agarose is a long chain polysaccharide that forms a mesh of molecules when it is melted and allowed to solidify. During electrophoresis this mesh acts as a molecular sieve for the restriction fragments, which are separated by size with the smallest fragments moving fastest through the gel.

Transfer DNA from Gel to Membrane

Transfer of the DNA from the gel to a membrane in a form suitable for hybridization analysis is known as Southern transfer or blotting. This step consists of:

- Denaturation of the DNA to single strands by treatment with alkali
- Transfer of the DNA to a membrane by capillary action and permanently fixing the DNA to the membrane by UV treatment and/or drying

During this step water and DNA move together from the gel to the stack of dry paper towels. However, since the DNA can't pass through the membrane, at the end of the transfer process the membrane contains all of the DNA originally in the gel in the exact spatial orientation that it had in the gel.

Hybridize with Labeled Probe

DNA hybridization takes advantage of the fact that DNA is normally a double stranded molecule with the two strands having complementary base sequences and held together by relatively weak hydrogen bonds between the base pairs. Since the DNA on the membrane is single stranded, the parts of the DNA molecules that are involved in interstrand base pairing are exposed. When the membrane is incubated with probe under conditions that promote the formation of hydrogen bonds between complementary base pairs, the probe will hybridize wherever it finds a DNA sequence on the membrane that is complementary to it.

Wash Membrane to Remove Unhybridized Probe

After hybridization there will be probes stuck nonspecifically to the membrane as well as weakly hybridized to DNA fragments where short segments of homology occurred by chance. Both of these events cause unwanted background and they can be removed by high temperature washes in a low salt solution. Conditions are chosen so that true probe will target hybrids are stable but the nonspecifically bound probes are not.

Autoradiography

At this point, the membrane has tiny amounts of radioactivity bound to it. This can be detected by placing a piece of photographic film in contact with the membrane, a process called autoradiography. The film will be exposed at the regions that are close to radioactivity. This exposed sheet of film, with its variously placed bands, is the raw data of an RFLP experiment.

Reuse of membrane

Membranes can usually be reused up to ten times. Before reuse any previously hybridized probe should be removed by denaturing the probe : target DNA hybrid. After prehybridization, the membrane is then ready for another round of hybridization.

Advantages of RFLP

- Present everywhere
- Mendelian inheritance
- Codominant expression
- No pleiotropic effects
- Independent of the environment
- Present at each developmental stage
- Long stability of cDNA probes
- Different loci may be identified by one probe
- Heterologous genes may be used as probes

- ❐ Any number of probes can be produced
- ❐ Probes are producible for coding and silent sequences
- ❐ Probes show the variability of flanking sequences
- ❐ Several characters can be screened in the same sample

Simple Sequence Repeats (SSRs)

Brown and Kressowich (1996) have described various technique of molecular characterization for plant genetic resources conservation. SSRs (also known as Microsatellites) occur quite frequently in the genome Allelic variation occurs at an SSR locus as a result of changes in the number of repeated sequences altering the length of the region. Differences in length at an SSR locus are detected by DNA amplification via PCR using a pair of oligo-nucleotide primers that complement unique sequences flanking the SSR. The sizes of the amplified products are determined by acrylamide gel electrophoresis. This type of assay generates single locus data that provide alleles and genotypes. Evolutionary distances can be calculated between the various alleles at a locus based on the exemption of a stepwise mutation process that results in the alleles closet in size being most closely related by descent from a common progenitor.

SSRs exhibit high levels of polymorphism in most plant species including closely related genotypes. These allelic variations usually demonstrate co-dominant Mendelian inheritance. Since SSR can be assayed by PCR, small quantities of sample are required. Further the reactions are inexpensive and can be completed from sample preparation to genotype analysis just in 2 days. The primary drawback of SSR based marker is higher effort and cost required to develop specific primer pairs for each polymorphic locus. This involves the cloning and sequencing of large number of genomic DNA fragments. The labs using RFLP and RAPD markers and agarose gels provide a familiar method to visualize SSR length polymorphisms which does not require the use of radiolabelled compounds or sophisticated or expensive equipment. The materials and reagents required are agarose, standard PCR reagents, including 25 M $MgCl_2$, PCR primers-forward and reverse both, dNTPs, 10 X PCR buffer, 500 mM KCl, 100 mM Tris HCl, (pH 8.3), and Taq DNA polymerase. There are several applications of microsatellite markers in the plant genetic search. These include:

- ❐ Exact allele sizing
- ❐ Genetic mapping
- ❐ Markers assisted selection (MAS)

Application of Molecular Marker Data in PGR Management

The most common approach to use molecular data is to produce a dendrogram or tree diagram linking together in cluster samples that are more genetically similar to each other than two samples in other clusters. Clusters are linked to each other at progressively lower levels of similarity until all the samples being analysed are included in a single cluster. The cluster analysis may proceed according to a range of different algorithms but common ones include unweighted paired group method with arithmetic averages (UPGMA), Neighbour joining method and Ward's method. Brown and Kresovich (1996) have assigned the following tasks to the plant curators using data from molecular markers.

1. *Acquisition:* Data on the diversity of existing collections can be used to plan acquisition strategies. In particular, calculations of genetic distances can be used to identify particularly divergent subpopulations that might harbour valuable genetic variation that is under- represented in current holdings.

2. *Maintenance:* Molecular markers can be used to confirm duplicate accessions. Markers can be applied to monitor changes in heterogeneity and heterozygosity (genetic drift) as accessions are regenerated. Molecular data on diversity may provide essential information for the development of core collections that accurately represent the entire collection.

3. *Characterization:* The genetic diversity within collections must be assessed in the context of the total available genetic diversity for each species. When available, existing passport data documents the geographic location where each accession was acquired. However, many records are missing or incorrect. Molecular markers may allow for characterization based on gene, genotype, and genome, which provide more accurate and detailed information than classical phenotypic or passport data.

4. *Utilization:* Users of collections benefit from genetic information that allows them to quickly identify valuable traits. On a more fundamental level, molecular marker information may lead to the identification of useful genes contained in collections and transfer of these genes into well-adapted cultivars.

References

Anolles, G.C. and Gresshoff, P.M. (eds.). 1997. DNA Markers : Protocols, Applications and Overviews. Wiley Liss, Inc.

Brown, M.S. and Kresovich, S. 1996. Molecular characterization. *In* A.J. Paterson (ed.). Genome Mapping in Plants. R.G. Lands Company pp 85 – 93.

Hodgkin, T., Brown, A.H.D., Van Hinthum, T. H. J.L. and Morales, E.A.V. 1995. Core Collections of Plant Genetic Resources, John Wiley and Sons, UK.

Hoelzel, A.R. and Green, A. 1994. Analysis of population level variation by sequencing PCR amplified DNA. Pp. 159 – 187. *In* A. R. Hoezel (ed.). Molecular Genetic Analysis of Populaitons. IRL Press, UK.

Karp, A., Kresovich, S., Bhat, K.V., Ayad, W.G. and Hodgkin, T. 1997. Molecular Tools in Plant Genetic Resources Conservation : A Guide to The Technologies. IPGRI Technical Bulletin No. 2. International Plant Genetic Resources Institute, Rome, Italy.

Kresovich, S., William, J.G. K., Sachaal, B.A., McFerson, J J.R. and Routman, E. 1992. Characterization of genetic identities and relationships of *Brassica oleracea* L. via random polymorphic DNA assay. Theor. App. Genet. 85 : 190 – 196.

15

In Vitro Method of Germplasm Conservation

Aplant collector and conservationist faces many problems. These include adverse weather, limited availability of personnel, insufficient material because of poor growing season, general scarcity of the plant in the target region, grazing of the plant by animals, damage by pests or diseases or immaturity of the material. These problems are common to both seeds and vegetative propagules. The vegetatively propagated plant material face another unique problem related to excessive weight and bulk. The cost and inconvenience of transporting large amounts of material can be a limitation on the scope of a plant exploration mission. There is additional need to maintain selected clones in a disease-free condition for distribution to national programmes.

In 1982, the International Board for Plant Genetic Resources (IBPGR) constituted an advisory committee on *in vitro* storage to examine the status of *in vitro* conservation technology and to highlight opportunities for developing and applying new approaches to overcome some of the bottleneck problems in the conservation of plant genetic resources. Two major bottlenecks recognized were those associated with recalcitrant seeds and vegetatively propagated plants. Withers (1980, 1982) recognized that *in vitro* techniques could be helpful in the safe conservation of problem crops. On the basis of IBPGR (1983), the following issues emerged regarding *in vitro* conservation.

- The basic conceptual framework for *in vitro* conservation was clarified.
- Species requiring priority attention for *in vitro* conservation were identified.
- The following critical areas of research in *in vitro* conservation were identified:
- The problem of genetic stability of cultures
- The need for parameters for characterization of accessions entering *in vitro* storage
- The need for disease indexing techniques at the in vitro level

- Development of techniques for in vitro germplasm collection and exchange
- Development of an efficient documentation and data management system for *in vitro* conservation

In vitro conservation has to be considered part of the overall conservation strategy for a particular plant species, as a valuable adjunct to genetic resource conservation. In some cases, in vitro storage would be the only strategy for a given species, for example, for some tropical fruit trees such as cacao (*Theobroma*). Tuber and root crops like potatoes, cassava, and sweet potatoes (*Ipomoea*) would probably be stored mainly as seed, in vitro methods being supplementary to the conservation of particular genotypes (cultivars, hybrids, elite clones, etc.). For other tropical tuber and root crop genotypes and fruit species (e.g., Musa) that seldom produce seed and are totally sterile, in vitro storage and ex situ field gene banks would probably be complementary. In the case of tropical species with recalcitrant seed *in vitro* collection and ex situ conservation should play a basic role (Roca *et al.*, 1989).

Approaches to *In Vitro* Conservation

The range of crop species whose conservation is potentially achievable through in vitro techniques is wide. In addition to the species listed above, several genera of asexually propagated plants can be included such as *Colocasia*, *Xanthosoma*, *Vitis*, and *Saccharum*. Furthermore, genera traditionally propagated by sexual means but whose seed is recalcitrant or presents sterility problems are also potential candidates for the in vitro approach. This range of species includes *Theobroma*, *Citrus*, *Coffee*, *Cocos*, and *Elaeis*. As previously mentioned, however, careful scientific assessment of the advantages and disadvantages of the in vitro approach to genetic resource conservation should be carried out for every species.

Conservation of plant genetic resources by means of in vitro culture methods can be achieved by changing the culture environment to slow down or totally suppress the growth of cells and tissues, the objective being to either increase the culture transfer period as much as possible or extend it indefinitely.

Two types of *in vitro* gene-bank conservation have been proposed (Withers and Williams, 1985):

- The in vitro active gene bank (IVAG), where cultures are maintained under slow growth.
- in vitro base gene bank (IVBG), where culture are cryo-preserved.

The IVAG is being developed, to large extent, for cassava, potatoes, sweet potatoes, banana, and sugar cane and constitutes a working collection. Its counterparts would be field collection and sexual seed collection for short-term storage. The IVBG constitutes a base collection; cryo-preservation is still not fully developed for a given crop. The latter approach enables germplasm genotypic stability to be fully maintained, and its counterpart would be sexual seed collection for long-term storage.

Basic Procedure of *In* Vitro Culture

Withers (1995) has outlined the following steps based on reviews (Sharp *et al.*, 1983, 1984, Beversdorf, 1990, Lindsey, 1991).

- Selecting an appropriate inoculum tissue
- Cutting it to a suitable size
- Removing superficial size
- Sterilizing the surface
- Washing to remove the surface sterilant
- Trimming away non-essential tissue and tissue damaged by sterilization
- Inoculating into the culture vessel containing nutrient medium
- Transferring to incubation conditions

Factors Influencing *In Vitro* Conservation

Roca *et al.* (1989) have described the following factors:

Tissue Culture Capability

Regeneration of whole plants from cell culture systems is often the limiting step in applying in vitro culture techniques to plant species that cannot be propagated by means of pre- formed meristems. Despite the ability to initiate callus from a variety of tissues and organs in many crop species, reproducible regeneration of whole plants remains problematic. Adventitious regeneration via somatic embryogenesis is highly desirable, as the process affords high multiplication rates and results in propagules that possess both root and, shoot axes (Stamp and Henshaw, 1987). Somatic embryos may develop from single cells, and plants with more stable genotypes may be recovered (Evans *et al.*, 1981).

Tissue Culture Viability

The evaluation of viability of in vitro cultures should be carried out systematically. Under slow-growth conditions, in which the cycle term is extended for several months or years, evaluation of cultures should be more frequent than with potentially indefinite conservation under liquid nitrogen, for example. The most important features to be evaluated under slow-growth storage of shoot tip derived cultures are:

- Contamination
- Leaf senescence (the ratio of green leaves to dead leaves)
- Number of green shoots (for further micro- propagation)
- Number of viable (green) nodes relative to stem (green) elongation

- Presence or absence of roots
- Occurrence of callus, etc.

Genotypic Stability

Genetic stability in cultures has long been a matter of concern in the potential application of in vitro techniques for germplasm conservation. The material retrieved from in vitro conservation should genetically represent the material accessed. No in vitro culture system will be acceptable if it introduces a high risk of genetic instability and/or selection among genotypes (Withers, 1988).

In vegetatively propagated crops, morphological criteria have been applied to characterize genotypes, but such differences are difficult to detect in cultures propagated in vitro. It has been pointed out that genetic variation due to chromosomal rearrangement can occur in tissue culture (D'Amato, 1964). A correlation exists between the time during which the plant material is grown as callus and the probability of chromosome changes. This could cause a shift to a variant type in *in vitro* propagation (Schilde-Rentschler and Roca, 1987). The use of axillary buds as a base for multiplication reduces the probability of genetic variation. Therefore, the formation of callus and adventitious buds should be avoided

Monitoring the genetic stability of in vitro plant cultures of crop species is becoming of great interest. Morphological, biochemical, and molecular criteria for the detection of genetic changes have been proposed (Withers and Williams, 1985). Electrophoretic techniques to assess isozyme variability in small samples of tissue obtained from in vitro cultures should be developed as a complementary method of stability evaluation. Furthermore, monitoring genetic stability of cultures by molecular techniques, e.g., variation in fragment length polymorphisms, should be tested.

Data Management

An in vitro gene bank should maintain and continuously update a computerized data base of information related to all aspects of in vitro conservation and associated research areas. The data base should include a large amount of information related to each of the accessions, such as passport data, genotypic characterization, disease indexing, micropropagation, international exchange, equipment and supplies, and other logistics of an *in vitro* conservation operation.

In Vitro Method for Conservation of Germplasm in Cassava-A Case Study

The major advantage of *in vitro* germplasm collecting is its adaptability and flexibility. The need to maintain selected clones and promising hybrids in a disease-free condition for distribution to national programmes stimulated some of the early studies on i*n vitro* germplasm conservation of potato at International Potato

Centre (CIP, Lima, Peru) and cassava at the International Centre for Tropical Agriculture (CIAT, Cali, Colombia).

In Vitro Conservation of Cassava

Cassava (*Manihot esculenta* Crantz) is one of the most important staple food crops of the lowland tropics. It constitutes a major source of calories for nearly 500 million people in over 60 countries. Present world cassava output reaches 130 million tons, most being produced in southeast Asia, Africa and tropical Latin America. Cassava is mainly utilized for human consumption and is available in the fresh market or processed into flour.

The potential danger of genetic erosion of both cultivated and wild *Manihot* germplasm resources and the requirement of genetic variability for improvement of the crop justify cassava germplasm conservation efforts.

Conventional maintenance of cassava germplasm collections is carried out by continuous vegetative field cultivation. New germplasm plantings often use freshly cut stakes from old fields. In addition to high costs, field maintenance often exposes the valuable germplasm to insect attack, disease, and soil or climatic problems. Freshly cut stakes can only be stored for a short time because of premature sprouting and insect or microbial attack.

Vegetative propagation often exposes the cassava crop to a wide range of pests and diseases which can be transmitted via the stakes through successive generations. For instance, cassava bacterial blight, African mosaic disease, superelongation disease, and frog-skin disease can produce great losses.

Maintenance of cassava germplasm by meristem culture can be done through a combination of cryogenic techniques and minimum-growth storage conditions (Roca *et al.*, 1989).

Slow-Growth Storage

Recent research on cassava conservation at CIAT has provided means to store in vitro cultures under minimal conditions of growth. It has been shown (Roca, 1984, 1985) that storage at 22°C decreases the rate of shoot elongation to about one-fifth of that of cultures kept at 28—30°C. Storage temperatures lower than 18°C were detrimental to a number of cassava varieties if the illumination was kept high. However, culture viability could be maintained by lowering the illumination to less than 500 lx. Increasing benzyl adenine (BA) from 0.04 to 0.11 mM and sucrose from 0.050 to 0.12 M also decreased shoot elongation but maintained 90% viability. However, if low-temperature storage was combined with high BA and sucrose levels, the growth of cultures was arrested to such a degree that most of them deteriorated after 3 months. Findings indicated that culture growth was decreased when the total nitrogen content of the medium was lowered to 20 mM at 27—28°C and 40 mM at 20—22°C. Mannitol at 5 - 25 mM was found to be effective not only in arresting growth but in decreasing tissue viability at the lowest storage temperature. However, if Mannitol was added to the medium along with 0.88 and 0.18 M sucrose, culture viability significantly increased at both low and high storage temperatures.

The isozyme variability present in the 100 clones was analyzed and 29 clones that displayed differences in banding pattern were selected. These were the most polymorphic clones, based on esterase (EST) isozyme patterns. A computerized data base has been developed for the management of all the information gathered in the P-IVAG. Passport data, field morpho-agronomic data, *in vitro* morphological characterization, disease indexing, electrophoretic data, micropropagation and subculturing, viability, and other data, such as physical location of cultures, equipment, supplies, and labor needs, are fed into the data base.

International Exchange of Cassava Germplasm

Quarantine regulations for cassava vary from country to country; however, the in vitro culture method has been accepted by various countries as one of the safest means of receiving vegetative cassava material. Between 1979 and 1988 more than 800 pathogen-free cassava strains (varieties and hybrids) were distributed by CIAT to 34 countries (mainly in Latin America and Asia) as in vitro cultures. The in vitro system has also been utilized to introduce to CIAT new cassava germplasm from countries in Latin America. Between 1980 and 1988 nearly 2000 accessions of native cassava cultivars were introduced as meristem cultures.

In Vitro Repository and Cryobank at NBPGR New Delhi

National Bureau of Plant Genetic Resources, New Delhi has taken lead in *in vitro* conservation and cryo-preservation techniques as complementary approach for vegetatively propagated species. *In vitro* slow growth techniques that allow significant increase in the intervals between sub-culturing have been standardized for a large number of species. Considering the significance of *in vitro* and cryobanks for conservation of plant genetic resources, a plant tissue culture lab and cryopreservation facility later renamed as "Tissue Culture and Cryopreservation Unit" (TCCU) has been developed at NBPGR New Delhi.

This unit aims at preservation on a short, medium and long term basis, *in vitro* cultures, pollen and seeds of plants of economic importance to India for which conventional methods of storage are inadequate. High priority has been assigned to plant species of Indian origin and those for which India is a secondary centre of diversity. Work on cryopreservation of difficult to store species using seeds embryos, embryonic axes and pollen and in vegetatively propagated plant species using meristems, shoot tips and somatic embryos has been in progress. The status of *in vitro* conserved germplasm at NBPGR New Delhi is given in Table 15.1 (Chaudhary *et al.*, 2001).

Table 15.1. Status of *in vitro* conserved germplasm (as on 31st March, 2001) at NBPGR, New Delhi

Crops	Storage temp. (°C)	Optimum (months) subculture interval	Accessions (no.)
Tuber and bulb crops			
Allium sativum	25, 4	16-20	97
Cllium spp.	25, 10, 4	12-22	14
Alocasia spp.	25	10	1
Colocasia esculenta	25	8-10	49
Dioscorea spp.	25	8-12	44
Ipomoea batatas	25	8-12	260
Xanthosoma sagittifolium	25	10	3
Spices and industrial Plants			
Curcuma spp.	25	6-10	53
Elettaria spp.	25	15	5
Fagara schinifolia	25	6	1
Humulus lupulus	25	-	8
Piper spp.	25	10-22	7
Vanilla planifolia	25	6	5
Simmondsia chinensis	25	6	12
Zingiber spp.	25	8-24	160
Fruits			
Actinidia chinensis	25	8	3
Musa spp.	25	8-12	350
Fragaria spp.	25	8-10	20
Prunus spp.	25	4	2
Rubus spp.	25	8-10	5
Malus sp.	25	3-4	3
Medicinal and aromatic plants			
Bacopa monneieri	25	6	1
Coleus forskohlii	25	12	7
Digitalis spp.	4	11	5
Eremostachys superba	25	6	2
Gentiana kurroo	4	12	1
Mentha spp.	25	12	22
Picrohiza kuroa	4	12	4
Pogostemon patchouli	25	12	1
Pycnanthemum spp.	25	6	4
Rauvolfia serpentine	25	22	6

Contd...

Table 15.1. Contd...

Rauvolfia canescens	25	22	1
Rheum moorcroftianum	4	12	1
Saussurea lappa	4	11	2
Swertia chirayta	4	6	2
Tylophora indica	25	12	2
Valeriana wallichii	4	12	4
	Total		1138

References

Chaudhary, R., Pandey, R., Mandal, B.B., Sharma, N., Tyagi, R.K., Agarwal, A., Gupta, S., Malik, S.K. and Hussain, Z. 2001. Germplasm conservation : *In vitro* repository and cryobank. *In* B.S. Dhillon, K.S. Varaprasad, Kalyani, Arinivasan, M. Singh, Sunil Archak, Umesh Srivastava and G.D. Sharma (eds.). National Bureau of Plant Genetic Resources: A Compendium of Achievements. NBPGR, New Delhi. pp 209 - 223.

D' Amato, F. 1964. Endopolyploidy as a factor in plant tissue development. Caryologia 17 : 41 - 51.

Evans, D. A., Sharp, W.R. and Flick, C.E. 1981. Growth and behaviour of cell cultures : embryogenesis and organogenesis. *In* T. A. Thorpe (ed.). Plant Tissue Culture: Methods and Applications in Agriculture. Academic Press, New York. pp 45 - 113.

International Board for Plant Genetic Resources. 1983. IBPGR Advisory Committee On *In Vitro Storage*. Report of The First Meeting. International Board for Plant Genetic Resources, Rome.

Ramirez, H., Hussain, A., Roca, W.M. and Bushuk, W. 1987. Isozyme electrophoregrams of sixteen enzymes in five tissues of cassava (*Manihot esculenta* Crantz) varieties. Euphytica 36 : 39 - 48.

Roca, W.M. 1984. Cassava. *In* W.R. Sharp, D.A. Evans, R.V. Amirato and Y. Yamada (eds.). Handbook of Plant Cell Culture. Vol. 2. Macmillan Publishing. Co., New York. pp 269 - 301.

Roca, W.M., Chavez, R., Martin, M.L., Arias, D.I., Mafla, G. and Reyes, R. 1989. *In vitro* methods of germplasm conservation. Genome 31 : 813 - 817.

Withers, L.A. 1980. Tissue Culture Storage for Genetic Conservation. IBPGR, Rome.

Withers, L.A. 1982. Storage of plant tissue. *In* Withers, L.A. and J.T. Williams (eds.). Crop Genetic Resources - The Conservation of Difficult Material. pp 49 - 82. IUBS/IBPGR/IGF, Paris.

Withers, L.A. and Williams, J.T. 1985. Research on long term storage and exchange of *in vitro* plant germplasm. *In* Biotechnology in International Agricultural Research. pp 11 - 24. IRRI, Los Banos.

16

Plant Germplasm Conservation

The methods of plant germplasm preservation or conservation depend upon type of germplasm i.e. seed or non-seed clonal materials.

Storage of Seeds

According to the storage tolerance of seeds to low temperature and dry environments seeds are further divided into storable (orthodox) and non-storable (recalcitrant) types. The orthodox seeds are those seeds which are fully desiccation tolerant. These basically include small seeded grain crops and vegetables. They can be stored for a long time under low temperature and low humidity environment and are separately stored in long, medium and short term storage conditions in accordance with the purpose of storage. On the other hand recalcitrant seeds are intolerant to desiccation. A few typical examples of tropical and subtropical recalcitrant and intermediate seed species are tea (*Camellia sinensis*), jack fruit (*Artocarpus hetrophyllus*), almond (*Prunu amygdalys*), neem (*Azadiracnta indica*), black pepper (*Piper nigrum*), and cardamom (*Glettaria curdumomum*).

Storage of Orthodox Seeds

The orthodox seeds can be stored under long term storage, medium term storage and short term storage conditions.

Long Term Storage

The storage conditions of -18 ± 2°C and 30 ± 3 % RH under long term storage is provided mainly for comprehensive base collection. Careful identification and selection precedes before storage. The base collections of germplasm are stored at low temperature for long term secure conservation. Exceptionally seeds are removed only under following circumstances:

- For regeneration, when their viability has declined below an acceptable standard.

- To provide material for an active collection for regeneration, if the stocks held by the active collection are more than two or three regeneration cycles removed from the original material.
- When stocks of accession are no longer available from an active collection.

It is thus, very clear that such materials will not be used for exchange. They are only preserved for regenerating fresh seed and periodically examined for seed viability level. The purpose of this storage is for the perpetuation of the seed stocks. The life expectancy of seed preserved in long term storage modules will last from 32 to 50 years.

Medium Term Storage

With storage conditions of around 4°C and 40 ± 3 % RH, this storage is provided solely for the seeds of actively used material. This collection also known as active collection is used for regeneration, multiplication, characterization, evaluation and distribution. The seeds will also supplement the base collection in long term storage. Regular germination test and seed regeneration are required for materials in this storage to maintain their viability.

Short Term Storage

The material under short term storage (10 ± 2°C, 40 ± 3 %RH) is known as working collection. They are also referred to as breeder's collection or research collection. This collection may include special genetic stocks such as breeder's lines, mutants, parts and progenies of the breeding materials for research purpose.

Storage of Recalcitrant Type Seed

Such seeds may be stored in low temperature (0 – 10°C) for short period of 1 – 5 years.

Conservation of Vegetatively Propagated Materials

Conservation of clonal material is obtained by:

- Tissue culture
- Cold storage
- Cryopreservation in liquid nitrogen
- Field maintenance

Tissue culture preservation for many clonal materials in artificial media are, more spaces saving and intensively managed than field plantings. Tissue culture also ensures germplsm exchange of virus free material. Bulbs, tubers and rhizomes of certain crops can be stored at 0 – 15°C under high humidity for several months or upto 1 year. It is also a type of short term storage. In cryopreservation (super cooled storage) embryos, tissue and pollen grains can be stored for long periods in liquid nitrogen (-196°C). This practice needs to be standardized as crops require different cooling and thawing treatments and have different viability levels. For field maintenance the germplasm materials such as fruit trees, potatoes and grasses, etc. are grown in nurseries/fields for conservation and such collections are also known as field gene bank.

The Guidelines for Sending the Material to a Gene Bank for Storage

The guidelines applicable to deposition of seeds or propagating materials for base collections in the National Gene Bank of NBPGR, New Delhi are as follows (NBPGR, 2001):

Procedure for Deposition of Seeds or Propagating Materials

(i) All the seed materials of orthodox-seed species (seeds that can be dried to low moisture content without loss of seed viability) be addressed to:

The Head

Germplasm Conservation Division

National Bureau of Plant Genetic Resources Pusa Campus, New Delhi-110 012

(ii) The seed materials of recalcitrant/intermediate seed species to:

The Scientist In-Charge

Tissue Culture and Cryopreservation Unit

National Bureau of Plant Genetic Resources

Pusa Campus, New Delhi-110 012

(iii) Propagules of vegetatively propagating species to the respective Officer- in-Charge of crop-based NAGS with a copy to the Scientist In-Charge, Tissue Culture and Cryopreservation Unit.

The NAGS after proper identification, establishment and regeneration should send a part of the genetic materials listed in to:

The Scientist In-Charge

Tissue Culture and Cryopreservation Unit

National Bureau of Plant Genetic Resources, Pusa Campus, New Delhi - 110 012

(i) Propagules/genetic materials should be supplied in phased manner, keeping in view the appropriate season, conditions suited to the crop.

(ii) It should be supplied with advance information and consent of Tissue Culture and Cryopreservation Unit to facilitate preparation for their processing and regeneration in vitro.

The Seeds and Propagules must Accompany

(i) A declaration to the effect that working stock for supply would be available with the institution associated with the collection and development of the material, until the transfer of germplasm for establishment at NAGS and in base collection.

(ii) Relevant passport data as per the format.

Guidelins for Submitting the Genetic Materials

Orthodox Seeds

i) A minimum number of 4000 seeds in cross-pollinated crop species, 2000 in self- pollinated, and 500 – 1000 in difficult species, such as vegetables, medicinal and aromatic plants, wild relatives, etc. should be submitted.

ii) The seeds should be supplied from freshly harvested crop and should not be more than 90 day old.

iii) The seeds supplied should be sound, physiologically mature and collected from healthy plants.

iv) For providing good quality seeds, and to avoid infections, it is advised to dry the seeds in shade soon after the harvest.

v) The potential viability of seeds should be more than 85 per cent in most crop species, except some species, such as cotton, vegetables, etc.

vi) Seeds should be free from any chemical treatment.

vii) Sees should be packed in good quality paper, muslin cloth or plastic bag(s) with proper identity. If required the bags should be packed in card-board boxes to minimise damage and moisture absorption.

Recalcitrant/Intermediate Seeds

These are generally characterized by large size, more weight and high moisture content (20-80%) at the time of shedding.

i) Preferably more than 1000 seeds should be supplied. However, recognizing the importance and/or rarity of the material, even small quantity is acceptable.

ii) If it should be sent as complete fruit, avoid any injury to the fruit surface. Send in aerated polythene bags/cardboard boxes.

iii) If the fruits are bulky and difficult to transport, the seeds may be extracted without any injury, and transported within 48 hrs, packed in saw dust/ charcoal, peatmoss, etc.

iv) Avoid transporting under high ambient temperature (above 30°C). Store and transport preferably in moist conditions between 15-20°C.

v) Extracted seeds may be treated with suitable fungicide (0.1% Captan/ Thiram powder).

vi) Avoid air-drying and water washing of seeds.

Vegetative Propagules

In case of vegetatively propagated crop species the germplasm materials/ propagules (tubers, bulbs, rhizomes, cuttings, etc.) have to be supplied to the concerned crop-based designated NAGS for initial establishment and conservation. A certificate to this effect has to be obtained from concerned NAGS, which in turn be supplied to the Scientist In-Charge, Tissue Culture and Cryopreservation Unit for information and assigning of national identity number. Additionally, following guidelines need be remembered for safe supply and conservation of germplasm:

i) At least 10-25 propagules (depending on crops) should be supplied to the concerned NAGS for maintenance in their field repository or in vitro repository (if available) with a request for an acknowledgement.

ii) The concerned NAGS should be informed in advance about the supply of material to facilitate processing and establishment of germplasm.

iii) The genetic materials, stocks, propagules of vegetatively propagated crop species are generally being maintained in the form of grafts, propagules, seedlings and plants. While supplying these genetic materials following steps and precautions be remembered (depending on the crop):

a. The grafts, propagules or plants supplied to the NAGS should be free from insects, weeds and disease. The materials should be packed properly in aerated polythene bags. During dry summer, the grafts should be wrapped in moist moss grass to retain the moisture.

b. In case of crops like coconut, the material is sent either as embryos or seedlings. If the embryos need to be transferred from the fields, the embryos embedded in the endosperm should be packed in the sterile plastic bags with sterile moist cotton. These should be kept overnight in refrigerator and transferred in the same box with proper labels on it.

In case of seedlings the embryos should be grown using the sand in plastic bags/boxes. Once the seedlings are established these should be transferred to bigger pots. The healthy, vigorous seedlings should be supplied.

i) The materials should be packed in small wooden/card-board boxes with proper aeration. Also these boxes should be well marked with labels at three or four places "to be handled carefully: seedlings", in order to avoid any damage during transit.

ii) The materials should be sent to the NAGS soon after harvest, to avoid any postal delay in transaction, by speed post.

Note

1. The sample size of seeds and propagules may be revised in exceptional cases in consultation with the Head, Germplasm Conservation Division, NBPGR or the Scientist In-Charge, Tissue Culture and Cryopreservation Unit.
2. On receipt of either the seed materials or acknowledgement receipt from respective NAGS regarding the receipt of genetic materials, the Germplasm Conservation Division, NBPGR or the Tissue Culture and Cryopreservation Unit, as the case may be shall obtain a unique National Identity Number (IC or EC) from Agricultural Research Information System (ARIS) Unit, NBPGR and supply to donor for use.
3. In subsequent correspondence regarding the supplied genetic materials, always use the unique national identity number.

National Gene Bank in India

National Bureau of Plant Genetic Resources (NBPGR) Head Quarters houses the national gene bank (NGB) which is primarily responsible for conservation of germplasm on long term basis. This is one of the well equipped and organized facilities for *ex-situ* PGR conservation in the world. The NGB is the largest repository (infrastructurewise) in the world having a capacity to hold more than 1 million accessions. This gene bank is supported by the laboratory facilities for conducting basic research in germplasm conservation, seed physiology, biochemistry, etc. to support effective and safe conservation of germplasm. It has 12 modules with a capacity to hold about 1 million accessions at -20°C. The present base collection holding in the national gene bank of India are 256781 belonging to 426 species as given in Table 16.1 (Singh *et al.*, 2004).

Table 16.1. Base collection in seed gene bank (September, 2003)

Crop groups	No. of accessions
Cereals	1,09,565
Pseudo cereals	3,412
Millets and forages	38,109
Grain legumes	38,328
Oil seeds	3 1,305
Fibre crops	7,959
Vegetables	15,032
Fruits	172
Spices and condiments	364
Medicinal, aromatic plants and narcotics	2,300
Duplicate safety samples	10,235
Total	**2,56,781**

References

NBPGR, 2001. Guidelines to submit germplasm for conservation as base collections with National Gene Bank. National Bureau of Plant Genetic Resources, Pusa Campus, New Delhi, India.

Singh, A.K., Tyagi, R.K. and Rao, Kameshwar N. 2004. *Ex-situ* approaches for conservation of plant genetic resources : Achievements and challenges. *In* B. S. Dhillon, R.K. Tyagi, Arjun Lal and S. Saxena (eds.). Plant Genetic Resource Management. pp 179 – 193.

17

Genetic Resource Documentation

International Board for Plant Genetic Resources (IBPGR) now known as International Plant Genetic Resources Institute (IPGRI) uses the following items in genetic resources documentation.

i) Passport (accession identifiers and information recorded by collectors)

ii) Characterization (consists of recording those characters which are highly heritable, can be easily seen by the eye and are expressed in all environments)

iii) Preliminary evaluation (consists of recording a limited number of additional traits thought to be desirable by a consensus) of users of the particular crop

Characterization and preliminary evaluation will be the responsibility of the curators while further characterization and evaluation should be carried out by the plant breeders. The data from further evaluation should be fed to the curators who will maintain a data file.

Passport Data

The various items under passport data as per IPGRI proforma are divided into following heads :

1. Accession Data
2. Collection Data
3. Site Data
4. Plant Data

Further Characterization and Evaluation

5. Site Data
6. Plant Data

The details are given below :

Contd...

1. Accession Data

1.1. Accession Number

This number serves as a unique identifier for accessions and is assigned by the curator when an accession is entered into his collection. Once assigned, this number should never be reassigned to another accession in the collection. Even if an accession is lost, its assigned number is still not available for re-use. Letters should occur before the number to identify the genebank or national system (e.g. MG indicates that an accession comes from the gene bank at Bari, Italy; PI indicates an accession within the USA system).

1.2. Donor Name

Name of institution or individual responsible for donating the germplasm.

1.3. Donor Identification Number

Number assigned to accession by the donor

1.4. Other Numbers Associated with the Accession

Any other identification number known to exist in other collection for this accession, e.g. USDA Plant Inventory number (not collector's number, see 2.1). Other numbers can be added as 1.4.3. etc.

1.4.1. Other number 1

1.4.2. Other number 2

1.5. Scientific Name

1.5.1. Genus

1.5.2. Species

1.5.3. Subspecies

1.5.4. Cultivar group

1.6. Pedigree/Cultivar Name

Nomenclature and designations assigned to breeder's material

1.6.1. Pedigree number

1.6.2. Cultivar name

1.7. Acquisition Date

The date at which the accession entered the collection

1.8. Date of Last Regeneration or Multiplication

1.9. Accession Size

Approximate number of seeds of accession in collection

1.10. Number of Times Accession in Collection

Number of regenerations or multiplications since original collection

2. Collection Data

2.1. Collector's Number

Original number assigned by collector of the sample normally composed of the name or initials of the collector(s) followed by a number. This item is essential for identifying duplicates held in different collections and should always accompany sub-samples wherever they are sent.

2.2. Collecting Institute

Institute or person collecting/sponsoring the original sample

2.3.Date of Collection of Original Sample

2.4. Country of Collection or Country Where Cultivar/Variety was Bred

Use the 3-letter abbreviations supported by the Statistical Office of the United Nations. Copies of these abbreviations are available from IBPGR Headquarters and have been published in the FAO/IBPGR Plant Resources Newsletter Number 49.

2.5. Province State

Name of the administrative subdivision of the country in which the sample was collected.

2.6. Location of Collection Site

Name of kilometers and direction from nearest town, village or map grid reference (e.g. TIMBUKTU 7S means 7 km south of Timbuktu).

2.7. Latitude of Collection Site

Degrees and minutes followed by N (north) or S (south), e.g. 103°S

2.8. Longitude of Collection Site

Degrees and minutes followed by E (east) or W (west), e.g. 76.2° SW

2.9. Altitude of Collection Site (m)

Elevation above sea level

2.10. Collection Source

1. Wild
2. Farm land

3. Farm store
4. Backyard
5. Village market
6. Commercial market
7. Institute
8. Other (specify in the Notes descriptor, 11)

2.11. Status of Sample

1. Wild
2. Weedy
3. Breeder's line
4. Breeder's population
5. Primitive cultivar/landrace
6. Advanced cultivar (bred)
7. Other (specify in the Notes descriptor, 11)

2.12. Type of Sample

1. Vegetative
2. Seed
3. Both

2.13. Local/Vernacular Name

Name given by farmer to cultivar/landrace/weed

2.14. Number of Plants Sampled

Approximate number of plants collected in the field to produce this accession

2.15. Photograph

Was a photograph taken of the accession or environment at collection? If so, provide the identification number/system in the Notes descriptor, 11

0 No
+ Yes

2.16 Herbarium Specimen

Was a herbarium specimen collected

0 No
+ Yes

2.16.1. Herbarium Specimen Number

2.17 Distance from Possible Cross-Pollinated Crops (m)

2.18. Normal Crop-Sowing Season

If sown in more than 2 seasons, give the appropriate choices

1. Spring
2. Summer
3. Autumn
4. Winter
5. All year round

2.19. Normal Harvesting Season

If sown in more than 2 seasons, given the approximate choices

1. Spring
2. Summer
3. Autumn
4. Winter
5. All year round

2.20. Organ Used as Primary Product

1. Seed
2. Young plant
3. Mature plant
   3.1. Leaf
   3.2. Stem
   3.3. Head
   3.4. Axillary bud
   3.5. Inflorescence
   3.6. Root

2.21. Organ Used as a Secondary Product

1. Seed
2. Young plant
3. Mature plant
   3.1. Leaf
   3.2. Stem
   3.3. Head
   3.4. Axillary bud
   3.5. Inflorescence
   3.6. Root

2.22. Major Crop Usage

1. Vegetable
   1.1. Unprocessed
   1.2. Processed (specify local name for the processed product)
2. Oil
   2.1. Edible oil
   2.2. Industrial oil
3. Condiment
4. Forage
5. Green manure
6. Other (specify in the Notes descriptor, 11)

2.23. Minor Crop Usage

1. Vegetable
   1.1. Unprocessed
   1.2. Processed (specify local name for the processed product)
2. Oil
   2.1. Edible oil
   2.2. Industrial oil
3. Condiment
4. Forage
5. Green manure
6. Other (specify in the Notes descriptor, 11)

2.24. Other Notes From Collectors

Collectors will record ecological information, for cultivated crops, cultivation practices such as irrigation, season of sowing, etc.

Characterization and Preliminary Evaluation

3. Site Data
4. Plant Data

3. Site Data

3.1. Country of Characterization and Preliminary Evaluation
3.2. Site (Research Institute)
3.3. Name of Person(s) in Care of Characterization
3.4. Sowing Date
3.5. Transplanting Date

3.6. First Harvest Date

3.7. Last Harvest Date

4. Plant Data (Applicable to Cabbage)

4.1. Seedling

Observe at the 5-leaf stage

4.1.1. Seedling Leaf Colour

1. White green
2. Yellow green
3. Light green
4. Green
5. Dark green
6. Purple green
7. Purple
8. Other (specify in the Notes descriptor, 11)

4.1.2. Seedling Leaf Margin Serrations

0 No serration

1 Crenate

2 Dentate

3 Doubly dentate

4 Other (specify in the Notes descriptor, 11)

4.1.3. Intensity of Seedling Leaf Margin Serrations

3 Slight

5 Intermediate

7 High

4.1.4. Seedling Pubescence

0 Glabrous

1 Very sparse (leaf margin only)

3 Sparse

5 Intermediate

7 Abundant

4.2. Mature Plant

Observe immediately before harvest in non-flowering plant

4.2.1. Growth Habit

Angle of outermost fully expanded leaf petioles with soil surface

1 Erect	(> 85°)
3 Open	(-65°)
5 Semi-prostrate	(-45°)
7 Prostrate	(< 30°)

4.2.2. Mature Leaf Shape

1. Orbicular
2. Obovate
3. Elliptic
4. Ovate
5. Pandurate
6. Lanceolate
7. Other (specify in the Notes descriptor, 11)

4.2.3. Mature Leaf Dissection

1. Entire
2. Auricular
3. Lyrate
4. Parted
5. Other (specify in the Notes descriptor, 11)

4.2.4. Mature Leaf Colour

Outer non-wrapper leaves

1. White
2. Green
3. Purple green
4. Purple
5. Other (specify in the Notes descriptor, 11)

4.2.5. Intensity of Mature Leaf Colour

3 Light
5 Intermediate
7 Dark

4.2.6. Petiole Colour

Midrib colour in subspp. *Pekinensis* and *rapa*

1. White
2. Greenish white
3. Pale green
4. Green
5. Purple green
6. Purple
7. Red
8. Other (specify in the Notes descriptor, 11)

4.2.7. Leaf Pubescence

On the underside

0 Glaborous
3 Sparse
5 Intermediate
7 Abundant

4.2.8. Mature Plant Height

From soil level to the tallest point of a standing intact plant

1. < 20 cm
2. 20 - 30 cm
3. 30 - 40 cm
4. 40 - 50 cm
5. 50 - 60 cm
6. 60 - 70 cm
7. > 70 cm

4.3. Flower and Silique

4.3.1. Petal Colour

1. Light yellow
2. Yellow
3. Deep yellow
4. Other (specify in the Notes descriptor, 11)

4.3.2. Petal Shape

1. Round
2. Elliptical
3. Thin elliptical
4. Other (specify in the Notes descriptor, 11)

4.3.3. Silique Colour

Mature, but not dried

1. Whitish green
2. Yellowish green
3. Light green
4. Green
5. Dark green
6. Purple green
7. Other (specify in the Notes descriptor, 11)

4.3.4. Beak Shape

1. Stumpy
2. Intermediate
3. Slender
4. Other (specify in the Notes descriptor, 11)

4.4. Head

Observe at harvest, head-forming types only

4.4.1. Heading Habit

0 Non-heading

5 Semi-heading

7 Heading

4.4.2. Head Shape

1. Round
2. Elliptical
3. Ovate
4. Obovate
5. Oblong
6. Cylindrical
7. Other (specify in the Notes descriptor, 11)

4.4.3. Head-forming Leaf Overlap at Terminal Region

1. Terminal leaves curled outwards
2. Erect
3. Curled inwards
4. Almost touching
5. Slight overlap

6. Intermediate
7. Considerable overlap

4.5. Root

4.5.1. Enlarged root

0 Absent
1 Enlarged
2 Marketable root

4.5.2. Skin Colour Pattern

1. Uniform
2. More than one colour

4.5.3. Base Colour of Skin

1. White
2. Cream
3. Green
4. Pink
5. Red
6. Purple
7. Other (specify in the Notes descriptor, 11)

4.5.4. Intensity of Base Colour of Skin

3 Light
5 Intermediate
7 Dark

4.5.5. Secondary Colour of Skin

1. White
2. Yellow
3. Green
4. Pink
5. Red
6. Purple
7. Grey
8. Other (specify in the Notes descriptor, 11)

4.5.6. Intensity of Secondary Colour of Skin

3 Light
5 Intermediate
7 Dark

4.5.7. Base Colour of Flesh

1. White
2. Greenish white
3. Pink
4. Purple
5. Other (specify in the Notes descriptor, 11)

4.5.8. Secondary Colour of Flesh

1. White
2. Greenish white
3. Pink
4. Purple
5. Other (specify in the Notes descriptor, 11)

4.5.9. Root Shape

1. Oblate
2. Globular
3. Conical
4. Tapered
5. Other (specify in the Notes descriptor, 11)

4.5.10. Shape of Crown

1 Straight
3 Convex
5 Plane
7 Concave

Further Characterization and Evaluation

5. Site Data
6. Plant Data
7. Stress Susceptibility
8. Pest and Disease Susceptibility
9. Alloenzyme Composition
10. Cytological Characters and Identified Genes
11. Notes
12. Descriptors of Nutritional Importance Other than Under 6.3.
13. Future Descriptors

5. Site Data

5.1. Country of further characterization and evaluation
5.2. Site (Research Institute)
5.3. Name of person in charge of evaluation
5.4. Sowing date
5.5. Transplanting date
5.6. First harvest date
5.7. Last harvest date

6. Plant Data

6.1. Seedling

Observe at 5-leaf stage

6.1. 1. Cotyledon length (mm)
6.1. 2. Cotyledon width (mm)
6.1. 3. Cotyledon notch depth

3 Shallow
5 Intermediate
7 Deep

6.1.4. Hypocotyl Colour

1. White
2. Greenish white
3. Light green
4. Green
5. Light purple green
6. Purple green
7. Purple
8. Other (specify in the Notes descriptor, 11)

6.1.5. Anthocyanin Pigment at Seedling Stage

0 Absent
1 Hypocotyl
2 Hypocotyl and petiole
3 Hypocotyl and petiole and blade
4 Other (specify in the Notes descriptor, 11)

6.2. Mature Plant

Observe immediately before harvest in non-flowering plants

6.2.1. Morphotype Uniformity

After discarding obvious off-types

1. Uniform
2. Continuous variation
3. Two or more distinct types

6.2.2. Basal Side Shoot Development

0 Absent

3 Sparse

5 Intermediate

7 Abundant

6.2.3. Weight of Whole Plant Without Root (g)

6.2.4. Number of Non-Wrapper Leaves

6.2.5. Leaf Length (cm)

Measure largest leaf

6.2.6. Leaf Width (cm)

Measure widest point of largest leaf

6.2.7. Petiole Length (cm)

Measure portion where leaf blade intercepts the petiole or is supposed to intercept, but not due to the presence of wing to petiole base of largest leaf

6.2.8. Petiole Width (cm)

Measure widest point of petiole of largest leaf; midrib width in subsp. *Pekinensis* and *rapa*

6.2.9. Petiole Thickness (mm)

Measure thickest point of petiole of largest leaf; midrib thickness in subsp. *Pekinensis* and *rapa*

6.2.10. Cross-Sectional View of Petiole

Take the widest part of the midrib in subspp. *pekinensis* and *rapa*

3 Round

5 Semi-round

7 Flat

6.2.11. Leaf Blade Thickness

3 Thin
5 Intermediate
7 Thick

6.2.12. Leaf Blade Blistering

0 No blistering
3 Slight
5 Intermediate
7 High

6.2.13. Leaf Margin Undulation

0 Absent
3 Low
5 Intermediate
7 High

6.3. Inflorescence Silique and Seed

6.3.1. Bolting behaviour under natural conditions (%) without artificial vernalization

6.3.2. Petal length (mm)

6.3.3. Petal width (mm)

6.3.4. Silique length (cm)

Seed bearing portion only

6.3.5. Silique width (cm)

Seed bearing portion only

6.3.6. Length of beak (cm)

6.3.7. Number of mature seeds per silique

6.3.8. Dominant seed coat colour

After drying

1. Yellow
2. Red brown
3. Brown
4. Dull grey
5. Almost black
6. Other (specify in the Notes descriptor, 11)

6.3.9. 1000 Seed Weight (g)

According to International Seed Testing Association (ISTA) rules; 5 - 6 % moisture content 'wet weight'

6.3.10. Seed Oil Content

Analytical technique to be specified in Notes descriptor, 11

6.4. Head

Observe at harvest, head-forming type only

6.4.1. Head Weight (g)

Non-wrapper leaves removed

6.4.2. Length of Head (cm)

Non-wrapper leaves removed

6.4.3. Head Diameter (cm)

Non-wrapper leaves removed

6.4.4. Axillary Bud Growth

Before trimming

3 inconspicuous
5 Intermediate
7 Clear

6.4.5. Heading Ability Under High Temperature

At mean temperature of 25°C or above

3 Poor
5 Intermediate
7 Excellent

6.5. Root

Observe at harvest on enlarged root-forming types only

6.5.1. Root length (cm)

Storage portion

6.5.2. Root diameter (cm)

Largest point

6.5.3. Root weight (g)

7. Stresses Susceptibility

Scored under artificial and/or natural conditions on a 1-9 scale, where

3 Low susceptibility
5 Medium susceptibility
7 High susceptibility

The conditions should be clearly specified

7.1. Low Temperature Injury
7.2. Growth Rate Under Low Temperature
7.3. High-Temperature Injury
7.4. Drought
7.5. Flooding/Water Logging/Excess Soil Moisture
7.6. Soil Salinity
7.7. Soil Acidity
7.8. Microelement Deficiency
7.9. Microelement Toxicity
7.10. Tip burn
7.11. Internal rot

8. Pest and Disease Susceptibility

Scored on 0-9 scale, where

0 Immunity
3 Low susceptibility
5 Medium susceptibility
7 High susceptibility

8.1. Pests

8.1.1. *Bervicoryne brassicae*	Cabbage aphid
8.1.2. *Hellula undalis*	Cabbage webworm
8.1.3. *Myzus persicae*	Green peach aphid
8.1.4. *Pieris rapae*	Imported cabbage worm
8.1.5. *Plutella xylostella*	Diamondback moth
8.1.6. *Pyllotretra striolula*	Cabbage flea beetle
8.1.7. *Spodoptera litura*	Tobacco cutworm
8.1.8. *Trichoplusia ni*	Cabbage cutworm

8.1.9. Other (specify in Notes descriptor, 11)

8.2. Fungi

8.2.1. *Albugo candida*	White rust
8.2.2. *Alternaria spp.*	Black spots
8.2.3. *Colletotrichum higginsianum*	Anthracnose
8.2.4. *Erysiphe polygoni*	Powdery mildew
8.2.5. *Leptosphaeria maculans*	Black leg
8.2.6. *Peronospora parasitica*	Downy mildew
8.2.7. *Plasmodiophora brassicae*	Club root
8.2.8. *Sclerotinia sclerotiorum*	Sclerotinia rot
8.2.9. Other (specify in the Notes descriptor, 11)	

8.3. Bacteria

8.3.1. *Erwinia carotovora*	Soft rot
8.3.2. *Pseudomonas marginalis*	Soft rot
8.3.3. *Pseudomonas syringar* pv. *Maculicola*	Leaf spot
8.3.4. *Xanthomonas campestris*	Black rot
8.3.5. *Xanthomonas vesticatoria*	Leaf spot
8.3.6. Other specify in the Notes descriptor, 11)	

8.4. Nematodes

8.4.1. *Meloidogyne spp.*	Root knot

9. Alloenzyme Composition

This may prove to be a useful tool for identifying duplicate accessions

10. Cytological Characters and Identified Genes

11. Notes

Give additional information where descriptor state is noted as "Other" as, for example, in descriptors 4.1.2, 4.2.2., 4.2.3, etc. Also include here any other relevant information.

12. Descriptors of Nutritional Importance Other Than Under 6.3

Certain characters which have or may have significant nutritional implication for food, animal feed and feed technology industries and whose assays are technically more demanding should be accumulated by gene banks or laboratories, e.g. erucic acid, glucosinolate, etc.

13. Future Descriptors

The list will remain open for inclusion of future descriptors, such as those pertaining to physiological traits, which can be rapidly screened. Most of the characters in the present list are of a visual morphological nature and an appropriate extension would be to use microscopic evaluation of tissue distribution at critical sites. Fluorescent probes now exist for many compounds, e.g. H_2O, protein probes, Distribution of many of these compounds is potentially important for stress resistance factors (lignin with lodging for example), quality factors (seed protein distribution) and yield (e.g. stem starch and vascular tissue development). These tissue distributions are genetically controlled and not subject to great environmental interaction.

Passport data are information about a germplasm sample and the collecting site recorded at the time of collecting. The passport data may include collector's name and institutions collecting data, collecting numbers, botanical name of a species and any vernacular name's location of collecting site (name of locality and longitude and latitude) sample type (seed, vegetative material, etc.), sample status (wild, cultivated, etc), and site environmental characteristics such as altitude, topography and various soil features. The passport data sheet as recommended by National Bureau of Plant Genetic Resources, New Delhi, India is as follows:

National Bureau of Plant Genetic Resources, New Delhi - 110 012

Passport Data Sheet

Date.......................... Collectors No................................ Accession No.............................

Species Name.. Common Name.....................................

Cultivar/Vernacular Name..Regions Explored.......................

Village/Block................................. District................................State...................................

Latitude.................................... N/S Longitude.................... E/W Altitude................m

Temp...................................... Rainfall..

Source	1. Natural wild 2. Distributed wild 3. Farmer's field 4. Threshing yard 5. Fallow 6. Farm store 7. Market 8. Garden 9. Institute 10.........
Status	1. Wild 2. Weed 3. Landrace 4. Primitive cultivar 5. Breeder's line
Frequency	1. Abundant 2. Frequent 3. Occasional 4. Rare
Material	1. Seeds 2. Fruits 3. Inflorescence 4. Roots 5. Tubers
Sample type	1. Population 2. Pure line 3. Individual plant
Sample method	1. Bulk 2. Random 3. Selective (non-random)
Habitat	1. Cultivated 2. Distributed 3. Partly distributed 4....
Disease symptoms	1. Susceptible 2. Mildly susceptible 3. Tolerant 4. Resistant 5. Immune
Insect/pest/ nematode infection	1. Mild 2. Moderate 3. High

Cultural practices	1. Irrigated 2. Rainfed 3. Arid 4. Wet 5.........
Season	1. Kharif 2. Rabi 3. Spring-summer Approx. sowing date .... Approx harvesting date..............
Associated crop	1. Sole 2. Mixed with ....
Soil colour	1. Black 2. Yellow 3. Red 4. Brown 5..............
Soil texture	1. Sandy 2. Sandy loam 3. Loam 4. Silt loam 5. Clay 6. Silt
Stoniness	1. Stony 2. Pulverized 3...........
Land aspect	1. Level 2. Crest 3. Escarpment 4. Rounded summit 5. Upper summit 6........................
Slope	1. Terrace 2. Lower slope 3. Open depression 4. Closed depression
Topography	1. Swamp 2. Plain 3. Level 4. Undulating 5. Hilly dissected 6. Steeply dissected 7. Mountainous 8. Valley
Agronomic scope	1. Very poor 2. Poor 3. Average 4. Good 5. Very good
Ethnobotanical uses	
Part(s)	1. Stem 2. Leaf 3. Root 4. Fruit 5. Flower 6. Whole plant 7. Seed 8. Others
Kind	1. Food 2. Medicine 3. Fibre 4. Timber 5. Fodder 6. Fuel 7. Insecticide/ Pesticide 8. Others
Informant(s)	Loca; Vaidya 2. House-wife 3. Old folk 4. Grazier/Shepherd 5. Others

Farmer's/Donor's Name....................................Ethnic Group..........

Collector No	Collector No.	Collector No.	Collector No.

Plant Characteristics/Use Addl. Notes

....

....

....

....

....

....

....

....

....

....

Characterization /Plant Descriptors

Plant descriptors are characterization data or observations collected on traits that are highly heritable, can be seen with eye and are fairly consistently expressed in all environments. Examples of such characters which for some crops are included in descriptor list published by International Plant Genetic Resources Institute (IPGRI) are flower colour, pod size, leaf shape, time to flowering, etc. (Perry and Bettencourt, 1995). For an illustration the characterization and preliminary evaluation data sheet/plant descriptor data sheet as prepared by IPGRI 1989 and applicable to mango is given below :

Plant Descriptors for Mango (*Mangifera indica* L.)

4. Plant Data

(Sl. No. 1 is applicable to accession data under passport, Sl. No. 2 to collection data under passport, Sl. No. 3 to site data under characterization and preliminary evaluation and Sl. No. 4 to Plant data)

4.1 Vegetative Characters

4.1.1. Propagation method

1. Grafting (including budding)
2. Inarching
3. Seed
4. Other (specify in Notes, Descriptor 11)

4.1.2. Name of rootstock on which the accession is grafted, if appropriate

4.1.3. Tree habit

1. Erect
2. Intermediate
3. Spreading

4.1.4. Height of mature tree (m)

4.1.5. Leaf shape

1. Oblong -lanceolate
2. Lanceolate
3. Elliptic - oblong

4.1.6. Leaf length (cm)

4.1.7. Leaf width (cm)

4.1.8. Colour of young leaf

1. Light green
2. Light green with brownish tinge

3. Light brick red
4. Reddish brown
5. Deep coppery tan

4.2. Inflorescence

4.2.1. Position of inflorescence
1. Terminal
2. Axillary
3. Both terminal and axillary

4.2.2. Shape of inflorescence
1. Conical
2. Pyramidal
3. Broadly pyramidal

4.2.3. Density of flowers in inflorescence
3. Laxly flowered
7. Densely flowered

4.2.4. Inflorescence length (cm)

4.2.5. Inflorescence colour
1. Light green
2. Green with red patches
3. Light red
4. Dark red
5. Crimson

4.2.6. Hairiness of inflorescence
0 Absent
3 Puberulous
7 Pubescent

4.2.7. Leafy bracts
0 Absent
+ Present

4.2.8. Percentage of hermaphrodite flowers on an average inflorescence

4.3. Flower

4.3.1. Flower diameter (mm)

4.3.2. Type of flower
1. Pentamerous (parts in 5s)

2. Tetramerous (parts in 4s)
3. Both pentamerous and tetramerous

4.3.3. Nature of disc
1. Disc swollen, lobed, broader than ovary
2. Disc narrow, often reduced or absent

4.3.4. Number of stamens
1. 10 – 12, 5-6 fertile (i.e. with anthers)
2. 5, all fertile
3. 5, 3 fertile
4. 4, 1-2 fertile

4.4.1 Fruit length (cm)

4.4.2 Fruit breadth (cm)

4.4.3 Fruit thickness (cm)

4.4.4 Fruit weight (g)

4.4.5 Fruit shape
1. Oblong
2. Elliptic
3. Roundish

4.4.6 Colour of skin of mature fruit

	Reference cultivars
1. Red	Sensation
2. Yellow	Mallika
3. Green-yellow	Himsagar
4. Other	Bombay

4.4.7 Thickness of fruit skin
1. Thin
2. Medium thick
3. Thick
4. Very thick

4.4.8 Skin texture
1. Smooth
2. Rough

4.4.9 Flesh content
Ratio of pulp to skin + stone

4.4.10 Flesh texture

Can be measured with a firmometer or penetrometer

3 Firm

5 Soft

7 Juicy

4.4.11 Adherence of skin to pulp

0 Absent (free)

+ Present (adhering)

4.4.12 Fibre in pulp

0 Absent

+ Present

4.4.13 Quantity of fibre

3 Scarce

7 Abundant

4.4.14 Length of fibres

3 Short

5 Medium

7 Long

4.4.15 Stalk insertion

1. Vertical
2. Oblique

4.5. Stone

4.5.1 Length (cm)

4.5.2 Weight (g)

4.5.3 Veins

1. Level with surface
2. Depressed
3. Elevated

4.5.4 Pattern of venation

1. Parallel
2. Forked

4.5.5 Fibre

0 Absent

+ Present

4.5.6 Length of fibres (mm)

5. Site Data

5.1. Country of further characterization and evaluation
5.2. Site (research institute)
5.3. Name of persons in charge of further characterization and evaluation
5.4. Sowing or planting date (specify)

6. Plant Data

6.1. Vegetative

6.1.1 Leaf texture
1. Coriaceous
2. Thinly coriaceous
3. Thickly coriaceous
4. Membraneous
5. Chartaceous

6.1.2 Leaf tip
1. Obtuse
2. Acute
3. Acuminate

6.1.3 Leaf margin
1. Flat
2. wavy
3. Folded
4. Crinkled

6.1.4 Petiole length (cm)

6.2. Inflorescence

6.2.1 Season of flowering

	Reference cultivars
3 Early	Golapkhas
5 Intermediate	Himsagar
7 Late	Aswina

6.2.2 Duration of flowering (days)
Average over at least 4 years

6.2.3 Regularity of flowering

	Reference cultivars
1. Extremely irregular	Langra

2. Irregular	Bombai
3. Intermediate	Dashehari
4. Regular	Totapari

6.2.4 Secondary flowering

	Reference cultivars
3 Rare	
5 Intermediate	
7 Frequent	Baromasia

6.3. Fruit

Mostly applies to cultivars

6.3.1 Age of tree at first harvest (years)

6.3.2 Maturity period

	Reference cultivars
3 Early	Safdar Pasand
	Gulab Khas
5 Intermediate	Himsagar, Langra
7 Late	Fazil, Chousa

6.3.3 Productivity

Average yield per tree relative to other cultivars

	Reference cultivars
3 Low	Alampur, Baneshan
5 Intermediate	Himsagar
7 High	Totapari

6.3.4 Eating quality

Combined assessment of flavour, acidity, sweetness, aroma and astringency when ripe. Quantitative measurement of TSS, acidity and sugars should be made.

	Reference cultivars
3 Poor	Fazil
5 Intermediate	Bombai
7 Good	Dashehari
9 Excellent	Alphonso

6.3.5 Fruit attractiveness

	Reference cultivars
3 Poor	Bhuto Bombai
5 Intermediate	Himsagar

7 Good Mallika
9 Excellent Sensation

6.3.6 Storage Life (days)
Maximum number of days storage in air under the best commercial conditions

6.3.7 Basal cavity
0 Absent
+ Present

6.3.8 Beak
0 Absent
+ Present

6.3.9 Beak type
0 Absent
1 Point
2 Prominent
3 Mammiliform

6.3.10 Sinus
0 Absent
+ Present

6.3.11 Sinus type
0 Absent
1 Shallow
2 Deep

6.3.12 Groove
0 Absent
+ Present

6.3.13 Shoulders
1. Level
2. Dorsal higher than vertical

6.3.14 Slope of shoulders
1. Sloping abruptly
2. Ending in a long curve
3. Rising and then rounded

7. Stresses Susceptibility

7.1. Drought
7.2. Wind
7.3. Flooding

8. Pest and Disease Susceptibility

Recorded on a 1 – 9 scale, where

3 Low susceptibility

5 Medium susceptibility

7 High susceptibility

8.1. Pests

8.1.1. *Idiocerus* spp.	Mango hopper
8.1.2. *Drosicha mangiferae* Green	Mealy bug
8.1.3. *Dacus dorsalis*	Fruit fly
8.1.4. *Batocera rufomaculata*	
Chulmetia transversa	Stem/shoot borer
8.1.5. *Apsylla cistellata* Buckton	Mango psyllid
8.1.6. *Sternochetus mangoiferae* Fabrcius	Stone weevil

8.2. Fungi

8.2.1. *Colletotrichum goleosporioides* Penz.	Anthracnose
8.2.2. *Glomerella cingulata*	
Oidium mangiferae Berth.	Powdery mildew
8.2.3. *Botrydiploidia theobromae* Pat.	Stem-end rot
8.2.4. Mango malformation	Caused by species of *Fusarium* and mites (*Aceria* spp.) in association with each other

8.3. Bacteria

8.3.1. *Xanthomonas campestris*	Bacterial canker

9. Alloenzyme Composition

This may prove to be a useful tool for identifying duplicate accessions.

10. Cytological Characters and Identified Genes

11. Notes

Give additional information where descriptor state is noted as 'Other' as, for example, in descriptors 2.10, 2.11, etc. Also include here further relevant information.

Reference

IBPGR 1989. Descriptors for mango. International Board for Plant Genetic Resources, Rome.

Perry, M.C. and Bettencourt, E. 1995. Sources of information on existing germplasm collections. *In* L. Guarino, V. Ramantha Rao and R. Reid (eds.). Collecting Plant Genetic Diversity. CAB International. pp 121 - 129.

18

Plant Quarantine

Plant quarantine is a government endeavor enforced through legislative measures to regulate the introduction of planting materials, plant products, soil, living organisms, etc., in order to prevent inadvertent introduction of insect-pests, pathogens and weeds harmful for the agriculture of a country/state/region, and if introduced, prevent their establishment and further spread. Various methods of pest disease control are: exclusion, eradication, protection, therapy, resistance, and biological control. Exclusion or 'keeping out' is fundamental to the concept of plant quarantine while eradication methods are employed to eliminate a newly established pest/pathogen. Plant quarantine may, therefore, be defined as 'Rules and regulations promulgated by governments to regulate the introduction of plants, planting materials, plant products, soil living organisms, etc. with a view to prevent inadvertent introduction of exotic pests, weeds and pathogens harmful to the agriculture or the environment of a country/region, and if introduced, to prevent their establishment and further spread'. Plant quarantine is designed as a safeguard against harmful pests/pathogens exotic to a country or a region (Ram Nath, 1991, Khetrapal *et al.*, 2004).

Plant Quarantine as a National Service

From time to time, the introduced pests/pathogens have devastated crops and even created famine conditions in different parts of the world. The Ireland famine of 1845 was the result of an almost total failure of the potato crop due to the introduction of the late blight pathogen (*Phytophthora infestans*) from Central America. Introduction of powdery mildew (*Uncinula necator*), Phylloxera and the downy mildew (*Plasmopara viticola*) in quick succession about the middle of 19th century from America, virtually annihilated the grape vine industry of France. The chestnut blight (*Endothia parasitica*) was introduced into the US on the nursery stocks imported from the Orient in about 1906. Within 25 years, the American chestnut was almost exterminated as a forest tree causing an estimated loss of 1 billion US dollars. In Sri Lanka, coffee was replaced by tea as a plantation crop because of the widespread epiphytotics of coffee leaf rust (*Hemileia vastatrix*) in 1868. (Ram Nath, 1991).

In India also, several pests and diseases got introduced from time to time, some of which, like late blight of potato, banana bunchy top, bacterial blight and streak diseases of paddy, have since become widespread. Some others like golden nematode and wart disease of potato and downy mildew of onion are still localized in certain parts of the country.

The above examples only highlight the risks involved in inadvertent introduction of serious pests/diseases along with the planting material imported without adequate safeguards. Plant quarantine can provide such safeguards. These measures are of particular importance and relevance to countries like India whose economy is largely based on agriculture. Quarantine not only helps to ward off the threats of exotic pests, but also aims to eliminate and prevent further spread of pests/pathogens (both indigenous and introduced) with restricted distribution within the country (domestic quarantine) Thus, plant quarantine, in real sense, serves as a national service by preventing the introduction of exotic pests/pathogens/weeds and their further spread. However, such endeavours could succeed only with the active support of all— the administrators, general public, farmers, scientists, communication media, customs and others.

Complementary Role of Plant Quarantine

Plant genetic stocks are a global resource meant for the welfare of humanity. Germplasm from centres of origin and crop diversity may possess valuable genes for resistance against pests/pathogens, high yield, early maturity, cold, drought or salinity tolerance and quality traits like increased oil, protein contents, etc. Plant quarantine services are charged with the responsibility of preventing entry of hazardous pests, pathogens and weeds, but to deny entry to the valuable genetic resources would be against national interest. These activities are meant to help agricultural development and they are complementary to each other. Too much conservatism on the part of plant quarantine officials and too liberal an attitude on the part of plant introduction officials, breeders would be harmful. Plant quarantine officials must strive to provide adequate safeguards to allow smooth flow of germplasm resources in a healthy state. They should also try to ensure that the germplasm, when received in quarantine station, is processed promptly and that the delays in release, if any, are purely due to biological considerations alone. The plant quarantine officials should conduct research on developing sensitive and reliable methods of detection and salvaging treatments, or find alternatives to permit introduction of even high risk genera, if introduction of such materials is in the national interest (Ram Nath, 1991).

Pest/Pathogen Detection Techniques

A wide variety of pests and pathogens (insects, mites, nematodes, fungi, bacteria, viruses, viroids/MLOs, spiroplasma, etc.) and weeds are the objects for quarantine consideration. Similarly, planting material also may be introduced in a variety of forms, i.e., true seed, corms, bulbs, rhizomes, suckers, runners, budwood, scions, cuttings and rooted plants. Therefore, detection techniques would vary depending on the type of material, the host species and the type of pests/pathogens involved. Many a times, more than one technique would have to be used. Detection techniques may broadly be classified into two groups:

- Generalized tests which would reveal a wide range of pests/pathogens
- Specialized or specific tests which are used to detect specific pests/pathogens (Neergaard, 1977)

Detection of Pest/Pathogen During Quarantine

Generalized Tests

A very widely used method is the inspection of dry seed with the naked eye or under the low power of microscope. This method would reveal a wide range of free moving insects, their eggs and larval stages, mites on or with the seed, weeds, soil, infected/infested plant debris, fungal fructifications like sclerotia, smut and bunt balls, nematode galls, discoloured or deformed seeds mixed with seed; oospore or bacterial crusts, acervuli, pycnidia, sclerotia and even free spores of rusts, smuts and many other fungi on the seed surface. Examination of dry seed under UV or NUV light may reveal infections of certain fungi and bacteria through emission of fluorescence of different colours. Examination of seed washings may reveal surface contamination by rusts, smuts, downy mildews and a large number of other fungi.

Most commonly used incubation methods for the detection of fungi are the common moist blotter and agar tests wherein seeds are incubated on these media for a specific length of time (generally about a week) at a suitable temperature under alternating light and dark cycles. These two media reveal a wide range of internally seed-borne fungal and some bacterial pathogens in a wide variety of crops. Seedling symptom test and the grow-out test are quite versatile and reveal the symptoms produced by any category of plant pathogens including fungi, bacteria and viruses. Grow-out test is the simplest of the tests extensively used for the detection of viruses. However, some viruses may be carried symptomlessly in the plant and, therefore, it should be used in combination with other test like indexing on indicator test plants and serology (Ram Nath, 1991).

Specialized Tests

Insects

X-ray radiography has been used very successfully all over the world for the detection of hidden infestation (with no apparent sign of infestation on the seed surface) of insects, particularly seed infesting chalcids and bruchids. Seed transparency test (boiling the seeds in lactophenol to make them transparent) may also be used for the detection of hidden infestation and extraction of the insects for identification. X-ray radiography is also very effective in salvaging infested seed lots (Ram Nath, 1991).

Nematodes

For the detection of seed-borne nematodes, seeds are soaked in water for about 24 hours. This makes the nematodes active, which then come out of the seed into the water, or the seeds may be teased out with the help of forceps and a needle and examined for detection of nematodes under a stereomicroscope. In rooted plants, the accompanying soil and plant debris may similarly be soaked in water and nematodes may be extracted for identification using nematological sieves or tissue paper.

Fungi, Bacteria and Viruses

Serological tests are very effective for the detection and identification of viruses and bacterial pathogens and are being used in various plant quarantine stations with great success. Phage-plague technique is still more sensitive for bacterial pathogens as even strains of bacteria can be identified. Indicator test plants are also very helpful as they may reveal pathogenic races within a species of a fungus, bacterium and specific strains within a virus.

Modifications of the generalized incubation tests (agar and blotter tests) have also been used for the detection of specific plant pathogens. Deep-freezing blotter test and 2,4-D blotter test are very efficient for detection of black-leg pathogen (*Phoma lingam*) in crucifer crops. Potato-dextrose- agar is useful for the detection of *Septoria nodorum* in wheat while PCNB agar is a selective medium for detection of Fusarium species in cereals.

In the case of vegetative propagules, laboratory methods may suffice for the detection of insects and mites, nematodes, majority of fungi and certain bacteria. However, for the detection of systemic fungal pathogens, bacteria, viruses, viroids and MLOs, isolation growing for a season or a year or more in quarantine glass-houses/net-houses is required. Availability of glass-houses/net-houses in large number is an expensive proposition, but the quarantine safeguards afforded by them to any country are worth that expenditure (Ram Nath, 1991).

Salvaging of Infested Material During Quarantine

Following steps are followed (Ram Nath, 1991) :

Fumigation

Fumigation of the material under atmospheric or under reduced pressure has been found acceptable as a quarantine treatment against insects and mites. Fumigants like methyl bromide, HCN, phosphide and EDCT (ethylene dichloride + carbon tetrachioride mixture) are commonly used.

Heat Treatment

Hot water treatment or hot air treatment are also used in quarantine for eradication of insects, mites, nematodes, fungi, bacteria and viruses. The basic principle involved is that treatment temperature should be sufficiently high to kill the associated pest pathogen but not the host. However, in most cases, margin of safety is very narrow and, therefore, the temperature should be very accurately controlled. Some recommended hot water treatments are:

- Against nematodes- Flower bulbs, 44° C for 24 min., chrysanthemum, 45°C for 25 min., potato tubers, 45° C for 5 min.
- Against insects and mites - Narcissus bulbs, 44° C for 180 min, strawberry runners, 46° C for 10 min.
- Against viruses- Grape vine, 45° C for 120-180 min., sugarcane setts, 50°C for 120 min., potato tubers, 50° C for 17 min..
- Against fungi- Celery seed, 50° C for 25 min., wheat seed, 52—54° C for 10 min.

Chemical Treatments

Chemicals may be applied as dust, slurry, spray or as dip. It should be ensured that dosage of chemical should be enough to eradicate the inoculum but should not kill the host and the chemical should not be hazardous to personnel handling the treated seed. Treatment should be given on arrival and only after ascertaining the health status of the material. The choice of the chemical and dosage to be used should be made depending upon the pest/pathogen involved. Seeds treated at origin are not only difficult to examine but are hazardous to inspect also. Heavily, treated seed, which makes inspection difficult, should be denied entry.

Tissue Culture

Tissue culture as a safeguard in quarantine has been advocated by Kahn (1979). This method reduces the pest/pathogen introduction risk in two ways: (i) the size of the consignment is very much reduced since the introductions are represented by meristem tips, excised buds or embryos, and (ii) the aseptic plantlet system has built-in pest/pathogen detection capability. All insects, mites, nematodes and most fungi can be eliminated. Symptoms on young seedlings, and growth of the organisms on the agar medium, if any, may be visible through the transparent culture tubes, and these could be discarded. Tissue culture in combination with thermotherapy and chemotherapy is an excellent safeguard from quarantine angle. However, certain systemically infecting pathogens like rusts, downy mildews, bacteria, viruses, viroids and MLOs, may still get transported. Therefore, as an additional safeguard, the tissue culture material could be passed through post-entry quarantine isolation growing and index tested for the suspected pathogens. Indeed, tissue culture technology provides an exciting prospect for large scale exchange of genetic stocks with very little pest/pathogen introduction risk.

Quarantine Regulations

Plant quarantine regulations are promulgated by the national and the state governments to prevent the introduction and spread of harmful pests and pathogens. Plant quarantine will be justified only when the pest has no natural means of spread and when they are based on biological considerations only, i.e., pest/pathogen introduction risks and the available safeguards.

In general, risks are more with the introduction of vegetative propagules than with true seed. In case of true seed, risks are more with deep-seated infections than with the surface borne contamination of pests/pathogens. Again, risks are far greater with pathogens like viruses, downy mildews, smuts and many bacteria carried inside the seed without any external symptoms. When vegetative propagules are introduced, rooted plants, and other underground plant parts like rhizomes, suckers, runners, etc. carry higher risks than bud-wood, scions and un-rooted cuttings. In any case, bulk introductions are always risky as thorough examination and treatment in such cases is very difficult and planting area is far too large to prevent the establishment and spread of the introduced pest/disease.

Based on these factors, plant quarantine regulates the introductions as follows:

- *Complete embargo/prohibition:* When the pest risk is very high, the safeguards available in the country are not adequate and, therefore, import is prohibited.
- *Post-entry quarantine:* The risk is very high but adequate safe guards in the form of post-entry isolation growing facilities are available.
- *Restricted:* Pest risk is not high and import permit is required stipulating conditions for entry, inspection and treatment.
- *Unrestricted:* Import permit is not required, and material may enter without restriction.

While formulating quarantine regulations, local conditions like crop spectrum and environmental conditions are also to be considered. Since quarantine regulations are designed to break the life cycle of the pest/pathogen involved, the presence of alternate or collateral hosts in the country of import and their introduction should also be taken into account.

Plant quarantine activities in India are carried out under the Destructive Insects and Pests Act (DIP Act) of 1914, and the rules and regulations framed from time to time thereunder by the Govt. of India. Seed was not covered under the DIP Act until 1984, when the Govt. of India brought forward a comprehensive Plants, Fruits and Seeds (Regulation of Import into India) Order, 1984' which came into force in June 1985 (Anonymous, 1985).

With a view to provide the farmers the best planting materials available in the world for maximising productivity per unit area and to encourage the private seed industry in India not only to meet the internal requirements but also to develop export potential for high quality planting materials, the Government of India announced a 'New Policy on Seed Development' in September 1988. The new policy covers the import of seeds, planting materials of wheat, paddy, coarse cereals, oilseeds, pulses, vegetables, flowers, ornamentals and fruit crops; procedures for their import and the related plant quarantine procedures/ requirements.

This policy also states that absolutely no compromise shall be made with the requirements of plant quarantine procedures to prevent entry into the country of exotic pests, diseases and weeds detrimental to Indian agriculture. Therefore, to meet the plant quarantine requirements in respect to the provisions of the new seed policy, the Govt. of India brought forward, through Gazette Notification, the updated Plants, Fruits and Seeds (Regulation of Import into India) Order, 1989 (Anonymous, 1989).

The main features of the existing plant quarantine regulations in India are as follows:

- No consignment of seeds/planting materials shall be imported into India without a valid 'Import Permit', which is to be issued by a competent authority, to be notified by the Central Government from time to time in the Official Gazette.
- No consignment of seeds/planting materials shall be imported into India unless accompanied by a 'Phytosanitary Certificate', issued by the official Plant Quarantine Service of the source country.
- All consignments of plants and seeds for sowing/propagation/planting purposes shall be imported into India through land customs station, seaport, airport at Amritsar, Bombay, Calcutta, Delhi and Madras, and such other entry points as may be specifically notified by the Central Government from time to time, where these shall be inspected and, if necessary, fumigated, disinfested/disinfected by authorized plant quarantine officials, before quarantine clearance.
- Seeds/planting materials for isolation growing under detention, shall be grown in post-entry quarantine facility approved and certified by the Designated Inspection Authority (DIA) to conform to the conditions laid down by the Plant Protection Adviser to the Govt. of India.
- Hay, straw or any other materials of plant origin shall not be used as packing material.
- Import of soil, earth, sand, compost, and plant debris accompanying seeds /planting materials shall not be permitted. However, soil can be imported for research purposes under a special permit issued by the Plant Protection Adviser to the Govt. of India.

The DIP Act empowers the Central Government to make rules for regulating the import of seeds/planting materials into India and also the movement of the materials from one State to another within the country. The State Governments are also empowered to enact rules/regulations to regulate the movement of materials from one region/area to another within a State.

The Directorate of Plant Protection, Quarantine & Storage, headed by the Plant Protection Adviser to the Govt. of India, is primarily responsible for enforcing the quarantine rules and regulations framed under the DIP Act in the country. For this purpose, plant quarantine/and fumigation stations have been established at various international airports, seaports and land custom stations where the incoming consignments are inspected, fumigated or otherwise disinfested/ disinfected before release to indentors. Consignments of plants/seeds for sowing/ planting/propagation purposes, however, can only be imported through Amritsar, Bombay, Calcutta, Delhi and Madras quarantine stations, where facilities in respect to well equipped laboratories, quarantine green-houses and trained scientific and technical manpower are being strengthened to adequately meet the quarantine needs. Quarantine and fumigation stations under the Directorate of Plant Protection, Quarantine & Storage handle bulk imports for commerce and for planting.

As the National Bureau of Plant Genetic Resources (NBPGR), New Delhi has been designated as the national nodal agency for exchange of germplasm material of agri-horticultural and agri-silvicultural crops for research purposes in the country, it has also been entrusted with the quarantine responsibilities in respect to germplasm of these crops. The Director of NBPGR has been empowered to issue 'Permits' for import of seeds/planting materials for research purpose.

NBPGR has a separate Division of Plant Quarantine to meet the quarantine requirements in respect of the germplasm materials being exchanged through it. The Division has trained scientific and technical staff representing the disciplines of entomology, nematology and plant pathology, well equipped laboratories, greenhouses and post-entry isolation growing field facilities to discharge its quarantine responsibilities efficiently. In case of certain crops, after laboratory examination at NBPGR, the exotic material is passed on to the specific crop-based institutes for post-entry isolation growing, before it is released to the indentors. These institutes have established adequate post-entry isolation growing facilities and required expertise is also available with them. These are Central Potato Research Institute, Shimla; Central Tuber Crops Research Institutes, Trivandrum; Central Tobacco Research Institute, Rajahmundri; Sugarcane Breeding Institute, Coimbatore; and Central Plantation Crops Research Institute, Kasaragod. NBPGR has established a regional Plant Quarantine Station at Hyderabad to fulfill the quarantine requirements of the International Crops Research Institute for Semi-Arid Tropics (ICRISAT), Directorate of Rice Research and other research organisations in the region. It is also proposed to establish quarantine facilities for temperate fruit crops at NBPGR's Regional Station, Bhowali and an off-shore Quarantine Station at Port Blair in the Andaman Group of Islands for vegetatively propagated tropical crops. During the last 12 years or so, a large number of exotic insects and mites, plant parasitic nematodes, plant pathogens and weeds have been intercepted from the imported germplasm materials, many of which are of major quarantine significance and are not yet known to occur in the country. While processing the germplasm for quarantine clearance, all out efforts are made to salvage the infested/infected materials so that valuable exotic germplasm could be made available in a healthy state for exploitation in crop improvement programmes in the country.

The Directorate of Plant Protection, Quarantine and Storage (DPPQ&S) Ministry of Agriculture, Government of India, is the nodal organization for implementation of plant quarantine regulations. It directly deals in quarantine of bulk imports meant for planting or consumption. It operates through its network of 29 stations comprising the National Plant Quarantine Station at New Delhi, Regional Plant Quarantine Stations a Chennai, Mumbai, Kolkata and Amritsar and 24 other Plant Quarantine Stations. Of the 29 stations, 10 are located at airports, 10 at seaports and 9 at land frontiers.

The National Bureau of Plant Genetic Resources (NBPGR) at New Delhi, being the nodal institution for genetic resources management, has been authorized by the government of India to issue import permit and to undertake quarantine processing of germplasm including transgenic planting material meant for research by both public and private sectors. In case of import of transgenic material the indentor has to obtain and import clearance from the Review Committee on Genetic Manipulation of Department of Biotechnology prior to applying for import permit with NBPGR. For importing the germplasm of certain crops, namely, coconut seeds and plants (all Cocos species), coffee seeds and plants (all species of Coffea), cotton seeds, forest seeds (all species of Pinus, Ulmus and Castanea), potato, sugarcane (all species of Saccharum) and tobacco (all species of Nicotiana) request has to be routed through the Directors of the respective ICAR crop-based institutes to the Director, NBPGR, New Delhi, to ensure the effective implementation of the regulations laid by the Government.

Post entry quarantine isolation facilities are available at NBPGR, New Delhi and Hyderabad, the latter being used for International Crops Research Institute for Semi-Arid Tropics (ICRISAT) mandate crops by NBPGR Regional Station, Hyderabad. Post-entry quarantine examination is also conducted at times in the premises of importing organization including State Agriculture Universities (SAUs) and private companies directly under the supervision of quarantine staff or by the Designed Inspection Authorities approved by DPPQ&S. However, in such cases post-entry quarantine specifications and isolation standards are mentioned by NBPGR (Khetrapal *et al.*, 2004).

Exchange Procedures of Transgenic Plants

The transgenics are greatly under research focus by biotechnology seed companies in recent years. The commercial acreage under transgenic is estimated to be about 80 million hectares globally during 2004. The major transgenic crops under commercial planting are - corn, cotton and soybean. In 2004, there were 14 biotech megacountries (countries growing 50,000 ha or more of biotech crops), 9 developing countries and 5 industrial countries. Those countries in order of acreage were U.S., Argentina, Canada, Brazil, China, Paraguay, India, South Africa, Uruguay, Australia, Romania, Mexico, Spain and the Philippines. During the nine year period from 1996 - 2004, the global area of biotech crops increased more than 47 fold, from 1.68 million ha in 1996 to 80 million ha in 2004. In India, the transgenic has been allowed in cotton where the commercial area during 2005 has been estimated to be 1.2 million ha. and the cotton transgenic area in India during 2006 has gone upto 3.2 mha.

According the Department of Biotechnology (DBT), Government of India in collaboration with National Bureau of Plant Genetic Resources has set procedures to regulate the import of transgenic lines for research purpose. Any transgenic material to be imported requires a permission which is issued by Government of India after screening by Review Committee on Genetic Manipulation (RCGM). RCGM examines the desirability of import of transgenic lines from the bioasfety point of view under the Environment Protection Act- 1986 in accordance with the fact that all transgenic plants are regulated items.

Under the PFS Order (1989) which governs quarantine regulations, NBPGR has been designated as the competent authority to issue import permits for import of seeds by public and private sector agencies for research purpose. The provisions of the New Policy on Seed Development and PFS Order are applicable to import of transgenic seeds as well. The Department of Biotechnology and the Ministry of Environment and Forest have separate set of prescribed procedures for providing permission for import of transgenic materials with the stipulation that all imports would be routed through NBPGR for national accessioning, quarantine clearance, distribution to the users because the issuance of import permit of transgenic material is extremely important from the point of view of their potential impact on environment and on agriculture in the country.

Issuance of Import Permit for Transgenics

With the enactment of New Policy on Seed Development, the Government of India has made it obligatory for all plant breeders and researchers intending to import seed/planting material to fulfill two mandatory requirements, i.e. (i) Import Permit before importing any material, and (ii) Phytosanitary certificate from country of origin. These two documents must accompany with every seed/plant consignment imported from abroad. The Director, NBPGR has been authorised to issue Import Permit and receive imported materials from custom authorities for its quarantine inspection and clearance. The scientists/institutes desirous of importing the materials have to apply on a prescribed application form. Before obtaining import permit the indentor should furnish the information and provide the undertaking as well as certificate from the supplier as per para 4 and 5 of the permission accorded by the DBT, from safety point of view. The details of para 4 and 5 are given below (Singh *et al.*, 2001).

Para No. 4

- No transgenic material is permitted for experimentation in open environment without prior authorization from the Government of India.
- Full account of transgenic plants raised from the imported seeds is to be kept in a bound book, which should be available for inspection by the authority in case such a need arises.
- All transgenic materials prescribed by the indentors may be available for inspection, whenever required.
- All the unwanted transgenic materials may be destroyed by burning after the experiments are conducted.
- All precautions would be taken to prevent the escape of the genetic material into the open environment and shall follow the Recombinant DNA Safety Guidelines of the Government of India.

Para No. 5

- The supplier of the transgenic material shall certify that the transgenic has the genes as has been described in the permission.
- The supplier shall also certify that these transgenic materials do not contain any embryogenesis deactivator gene sequence (terminator gene/gene restriction use technology).

After fulfilling the above requirements, the Director, NBPGR issues the import permit in duplicate for research purpose only. After obtaining import permit the indentor sends it to the concerned organization which has agreed to supply the required transgenic material for research purpose only.

The second mandatory requirement is that of Phytosanitary Certificate which is to be issued by the National Plant Protection Agency of the donor country. It should also be ensured that the consignment must be addressed to the Director, NBPGR, Pusa Campus, New Delhi.

Registration, National Accessioning and Import

All indents for import of transgenics are registered for assigning the case number and then forwarded to the Plant Quarantine (PQ) Division without opening along with Import Quarantine (IQ) form for detailed quarantine inspection and clearance. After clearance from PQ Division, the samples are first arranged taxonomically indicating their genus, species, common name and cultivar name, etc. for national accessioning in the national record. Each introduction/accession is assigned an EC (Exotic Collection) number which remains unchanged with information like name and address of donors, characteristics of the germplasm, relevant references, date of arrival, condition of the material and distribution of the materials. All assembled healthy plant material is regularly transmitted to the various researchers to make use of these valuable genetic resources.

References

Anonymous, 1985. The Plants, Fruits and Seeds (Regulation of Import Into India) Order, 1984. Ministry of Agriculture and Cooperation, Department of Agriculture, Government of India.

Anonymous, 1989. The Plants, Fruits and Seeds (Regulation of Import Into India Order, 1989. Ministry of Agriculture, Department of Agriculture and Cooperation, Government of India.

Kahn, R.P. 1979. Tissue culture application in plant quarantine. Pp 185 - 201. *In* R. Maramorsch and H. Hirumi (eds.).Practical Tissue Culture Application. Academic Press, New York, USA.

Khetrapal, R.K., Nath, Ram and Raghunathan, V. 2004. Safe movement of plant germplasm. *In* : B. S. Dhillon, R.K. Tyagi, Arjun Lal and S. Saxena (eds.). Plant Genetic Resource Management. pp. 172 - 178.

Neegaard, P. 1977. Methods for detection of seed borne fungi and bacteria, pp 33 – 38. *In* W. B. Hewitt and L. Chaiarappa (eds.).Plant Health and Quarantine in International Transfer of Genetic Resources. CRC Press, Cleveland, USA.

Ram Nath 1991. Plant quarantine. *In* R.S. Paroda and R.K. Arora (eds.). Plant Genetic Resources - Conservation and Management - Concepts and Approaches. pp 121-139. International Board for Plant Genetic Resources, Regional Office for South and Southeast Asia, New Delhi.

Singh, R.V., Chand, Deep and Tyagi, Vandana. 2001. Exchange procedures of transgenic plants. *In* G.J. Randhawa, R.K. Khetarpal, R.K. Tyagi, B.S. Dhillon (eds.). Transgenic Crops and Biosafety Concerns. NBPGR, New Delhi. pp 96 - 99.

19

International Agreements and National Legislations on PGR

No country is self-sufficient in plant genetic resources, and until recently this was thought to be sufficient to ensure the free flow of germplasm among nations. However, the concepts of ownership, sovereignty and intellectual property rights (IPR) have increasingly been invoked of late in discussions of how best to conserve and use plant genetic resources. This has resulted from the growing realization that germplasm has real economic value. At the same time, awareness has grown that the repercussions of overexploitation of natural resources, of the extinction of species and the disappearance of crop landraces, of environmental damage and of habitat destruction can be global, transcending national boundaries and short-term financial considerations (Anon. 1995). Therefore important international agreements/legislations on plant genetic resources management have been enacted from time to time. The important ones are as follows :

Convention on Biological Diversity (CBD)

The Convention on Biological Diversity was a result of a negotiating process, under the auspices of the United Nations Environmental Programme (UNEP) that formally began in February 1991 and concluded in May 1992. Following its adoption in May 1992, the treaty was opened to signature at the United Nations Conference on Environment and Development (UNCED), held in Rio de Janeiro, Brazil in June 1992. It attracted the signatures of 150 States at the Conference - indicating worldwide acceptance of the Treaty.

The idea of the global convention on biodiversity had its origin from the World Conservation Union (UCN), which played a key role, on the recommendation of the 1984 and 1987 UN General Assemblies, in preparing draft for the treaty on biodiversity. In 1987, the United Nations Environmental Programme established an Ad Hoc Working Group of Experts on Biological Diversity which investigated the desirability of an umbrella Convention which would rationalise activities with respect to biodiversity and also address others which could fall within such a Convention. The Ad Hoc Working Group concluded that it was technically and legally not feasible to consolidate and absorb existing Conventions, and instead it recommended a new global treaty. In 1991 an Ad Hoc Working Group of Technical and Legal experts was established to prepare an international legal instrument for conservation and sustainable use of biological diversity. It had to also take into account, "the need to share costs and benefits between developed and developing countries" as well as "ways and means to support innovation of local people". This Ad Hoc Working Group in 1991 took form of the Intergovernmental Negotiating Committee, which later finalised the negotiations on the Convention on Biological Diversity.

The institutional arrangements of the Convention were established as per Article 23, 24 and 25 of the text of the Convention of Parties (COP) and specific terms of reference are set out in Article 23.4. Article 24 establishes the secretariat and specifies its functions. Article 25 establishes the Subsidiary Body on Scientific Technical and Technological Advice (SBSTTA) and sets out its terms of reference.

In addition, Article 21 enables for the establishment of a financial mechanism for provision of financial resources to developing countries, whilst Article 39 puts in place interim financial arrangements. On condition that it was restructured in accordance with Article 21, Global Environmental Facility (GEF) was designated to be the mechanism for interim financial arrangements until such time that the COP decided on an institutional structure. The first and second COP maintained the status quo as per Article 39. The third COP adopted a Memorandum of Understanding with GEF, which formalized the relationship between GEF and the Convention. The details of the CBD are as follows (http://www.biodiv.org/default.aspx)

Preamble

The Contracting Parties,

Conscious of the intrinsic value of biological diversity and of the ecological, genetic, social, economic, scientific, educational, cultural, recreational and aesthetic values of biological diversity and its components,

Conscious also of the importance of biological diversity for evolution and for maintaining life sustaining systems of the biosphere,

Affirming that the conservation of biological diversity is a common concern of humankind,

Reaffirming that States have sovereign rights over their own biological resources,

Reaffirming also that States are responsible for conserving their biological diversity and for using their biological resources in a sustainable manner,

Concerned that biological diversity is being significantly reduced by certain human activities,

Aware of the general lack of information and knowledge regarding biological diversity and of the urgent need to develop scientific, technical and institutional capacities to provide the basic understanding upon which to plan and implement appropriate measures,

Noting that it is vital to anticipate, prevent and attack the causes of significant reduction or loss of biological diversity at source,

Noting also that where there is a threat of significant reduction or loss of biological diversity, lack of full scientific certainty should not be used as a reason for postponing measures to avoid or minimize such a threat,

Noting further that the fundamental requirement for the conservation of biological diversity is the in-situ conservation of ecosystems and natural habitats and the maintenance and recovery of viable populations of species in their natural surroundings,

Noting further that ex-situ measures, preferably in the country of origin, also have an important role to play,

Recognizing the close and traditional dependence of many indigenous and local communities embodying traditional lifestyles on biological resources, and the desirability of sharing equitably benefits arising from the use of traditional knowledge, innovations and practices relevant to the conservation of biological diversity and the sustainable use of its components,

Recognizing also the vital role that women play in the conservation and sustainable use of biological diversity and affirming the need for the full participation of women at all levels of policy-making and implementation for biological diversity conservation,

Stressing the importance of, and the need to promote, international, regional and global cooperation among States and intergovernmental organizations and the non-governmental sector for the conservation of biological diversity and the sustainable use of its components,

Acknowledging that the provision of new and additional financial resources and appropriate access to relevant technologies can be expected to make a substantial difference in the world's ability to address the loss of biological diversity,

Acknowledging further that special provision is required to meet the needs of developing countries, including the provision of new and additional financial resources and appropriate access to relevant technologies,

Noting in this regard the special conditions of the least developed countries and small island States,

Acknowledging that substantial investments are required to conserve biological diversity and that there is the expectation of a broad range of environmental, economic and social benefits from those investments,

Recognizing that economic and social development and poverty eradication are the first and overriding priorities of developing countries,

Aware that conservation and sustainable use of biological diversity is of critical importance for meeting the food, health and other needs of the growing world population, for which purpose access to and sharing of both genetic resources and technologies are essential,

Noting that, ultimately, the conservation and sustainable use of biological diversity will strengthen friendly relations among States and contribute to peace for humankind,

Desiring to enhance and complement existing international arrangements for the conservation of biological diversity and sustainable use of its components, and

Determined to conserve and sustainably use biological diversity for the benefit of present and future generations,

Have agreed as follows:

Article 1. Objectives

The objectives of this Convention, to be pursued in accordance with its relevant provisions, are the conservation of biological diversity, the sustainable use of its components and the fair and equitable sharing of the benefits arising out of the utilization of genetic resources, including by appropriate access to genetic resources and by appropriate transfer of relevant technologies, taking into account all rights over those resources and to technologies, and by appropriate funding.

Article 2. Use of Terms

For the purposes of this Convention:

"Biological diversity" means the variability among living organisms from all sources including, inter alia, terrestrial, marine and other aquatic ecosystems and the ecological complexes of which they are part; this includes diversity within species, between species and of ecosystems.

"Biological resources" includes genetic resources, organisms or parts thereof, populations, or any other biotic component of ecosystems with actual or potential use or value for humanity.

"Biotechnology" means any technological application that uses biological systems, living organisms, or derivatives thereof, to make or modify products or processes for specific use.

"Country of origin of genetic resources" means the country which possesses those genetic resources in in-situ conditions.

"Country providing genetic resources" means the country supplying genetic resources collected from in-situ sources, including populations of both wild and domesticated species, or taken from ex-situ sources, which may or may not have originated in that country.

"Domesticated or cultivated species" means species in which the evolutionary process has been influenced by humans to meet their needs.

"Ecosystem" means a dynamic complex of plant, animal and micro-organism communities and their non-living environment interacting as a functional unit.

"Ex-situ conservation" means the conservation of components of biological diversity outside their natural habitats.

"Genetic material" means any material of plant, animal, microbial or other origin containing functional units of heredity.

"Genetic resources" means genetic material of actual or potential value.

"Habitat" means the place or type of site where an organism or population naturally occurs.

"In-situ conditions" means conditions where genetic resources exist within ecosystems and natural habitats, and, in the case of domesticated or cultivated species, in the surroundings where they have developed their distinctive properties.

"In-situ conservation" means the conservation of ecosystems and natural habitats and the maintenance and recovery of viable populations of species in their natural surroundings and, in the case of domesticated or cultivated species, in the surroundings where they have developed their distinctive properties.

"Protected area" means a geographically defined area which is designated or regulated and managed to achieve specific conservation objectives.

"Regional economic integration organization" means an organization constituted by sovereign States of a given region, to which its member States have transferred competence in respect of matters governed by this Convention and which has been duly authorized, in accordance with its internal procedures, to sign, ratify, accept, approve or accede to it.

"Sustainable use" means the use of components of biological diversity in a way and at a rate that does not lead to the long-term decline of biological diversity, thereby maintaining its potential to meet the needs and aspirations of present and future generations.

"Technology" includes biotechnology.

Article 3. Principle

States have, in accordance with the Charter of the United Nations and the principles of international law, the sovereign right to exploit their own resources pursuant to their own environmental policies, and the responsibility to ensure that activities within their jurisdiction or control do not cause damage to the environment of other States or of areas beyond the limits of national jurisdiction.

Article 4. Juridictional Scope

Subject to the rights of other States, and except as otherwise expressly provided in this Convention, the provisions of this Convention apply, in relation to each Contracting Party:

(a) In the case of components of biological diversity, in areas within the limits of its national jurisdiction; and

(b) In the case of processes and activities, regardless of where their effects occur, carried out under its jurisdiction or control, within the area of its national jurisdiction or beyond the limits of national jurisdiction.

Article 5. Cooperation

Each Contracting Party shall, as far as possible and as appropriate, cooperate with other Contracting Parties, directly or, where appropriate, through competent international organizations, in respect of areas beyond national jurisdiction and on other matters of mutual interest, for the conservation and sustainable use of biological diversity.

Article 6. General Measures for Conservation and Susrainable Use

Each Contracting Party shall, in accordance with its particular conditions and capabilities:

(a) Develop national strategies, plans or programmes for the conservation and sustainable use of biological diversity or adapt for this purpose existing strategies, plans or programmes which shall reflect, inter alia, the measures set out in this Convention relevant to the Contracting Party concerned; and

(b) Integrate, as far as possible and as appropriate, the conservation and sustainable use of biological diversity into relevant sectoral or cross-sectoral plans, programmes and policies.

Article 7. Identification and Monitoring

Each Contracting Party shall, as far as possible and as appropriate, in particular for the purposes of Articles 8 to 10:

(a) Identify components of biological diversity important for its conservation and sustainable use having regard to the indicative list of categories set down in Annex I;

(b) Monitor, through sampling and other techniques, the components of biological diversity identified pursuant to subparagraph (a) above, paying particular attention to those requiring urgent conservation measures and those which offer the greatest potential for sustainable use;

(c) Identify processes and categories of activities which have or are likely to have significant adverse impacts on the conservation and sustainable use of biological diversity, and monitor their effects through sampling and other techniques; and

(d) Maintain and organize, by any mechanism data, derived from identification and monitoring activities pursuant to subparagraphs (a), (b) and (c) above.

Article 8. In-situ Conservation

Each Contracting Party shall, as far as possible and as appropriate:

(a) Establish a system of protected areas or areas where special measures need to be taken to conserve biological diversity;

(b) Develop, where necessary, guidelines for the selection, establishment and management of protected areas or areas where special measures need to be taken to conserve biological diversity;

(c) Regulate or manage biological resources important for the conservation of biological diversity whether within or outside protected areas, with a view to ensuring their conservation and sustainable use;

(d) Promote the protection of ecosystems, natural habitats and the maintenance of viable populations of species in natural surroundings;

(e) Promote environmentally sound and sustainable development in areas adjacent to protected areas with a view to furthering protection of these areas;

(f) Rehabilitate and restore degraded ecosystems and promote the recovery of threatened species, inter alia, through the development and implementation of plans or other management strategies;

(g) Establish or maintain means to regulate, manage or control the risks associated with the use and release of living modified organisms resulting from biotechnology which are likely to have adverse environmental impacts that could affect the conservation and sustainable use of biological diversity, taking also into account the risks to human health;

(h) Prevent the introduction of, control or eradicate those alien species which threaten ecosystems, habitats or species;

(i) Endeavour to provide the conditions needed for compatibility between present uses and the conservation of biological diversity and the sustainable use of its components;

(j) Subject to its national legislation, respect, preserve and maintain knowledge, innovations and practices of indigenous and local communities embodying traditional lifestyles relevant for the conservation and sustainable use of biological diversity and promote their wider application with the approval and involvement of the holders of such knowledge, innovations and practices and encourage the equitable sharing of the benefits arising from the utilization of such knowledge, innovations and practices;

(k) Develop or maintain necessary legislation and/or other regulatory provisions for the protection of threatened species and populations;

(l) Where a significant adverse effect on biological diversity has been determined pursuant to Article 7, regulate or manage the relevant processes and categories of activities; and

(m) Cooperate in providing financial and other support for in-situ conservation outlined in subparagraphs (a) to (l) above, particularly to developing countries.

Article 9. Ex-situ Conservation

Each Contracting Party shall, as far as possible and as appropriate, and predominantly for the purpose of complementing in-situ measures:

(a) Adopt measures for the ex-situ conservation of components of biological diversity, preferably in the country of origin of such components;

(b) Establish and maintain facilities for ex-situ conservation of and research on plants, animals and micro- organisms, preferably in the country of origin of genetic resources;

(c) Adopt measures for the recovery and rehabilitation of threatened species and for their reintroduction into their natural habitats under appropriate conditions;

(d) Regulate and manage collection of biological resources from natural habitats for ex-situ conservation purposes so as not to threaten ecosystems and in-situ populations of species, except where special temporary ex-situ measures are required under subparagraph (c) above; and

(e) Cooperate in providing financial and other support for ex-situ conservation outlined in subparagraphs (a) to (d) above and in the establishment and maintenance of ex- situ conservation facilities in developing countries.

Article 10. Sustainable Use of Components of Biological Diversity

Each Contracting Party shall, as far as possible and as appropriate:

(a) Integrate consideration of the conservation and sustainable use of biological resources into national decision-making;

(b) Adopt measures relating to the use of biological resources to avoid or minimize adverse impacts on biological diversity;

(c) Protect and encourage customary use of biological resources in accordance with traditional cultural practices that are compatible with conservation or sustainable use requirements;

(d) Support local populations to develop and implement remedial action in degraded areas where biological diversity has been reduced; and

(e) Encourage cooperation between its governmental authorities and its private sector in developing methods for sustainable use of biological resources.

Article 11. Incentive Measures

Each Contracting Party shall, as far as possible and as appropriate, adopt economically and socially sound measures that act as incentives for the conservation and sustainable use of components of biological diversity.

Article 12. Research and Training

The Contracting Parties, taking into account the special needs of developing countries, shall:

(a) Establish and maintain programmes for scientific and technical education and training in measures for the identification, conservation and sustainable use of biological diversity and its components and provide support for such education and training for the specific needs of developing countries;

(b) Promote and encourage research which contributes to the conservation and sustainable use of biological diversity, particularly in developing countries, inter alia, in accordance with decisions of the Conference of the Parties taken in consequence of recommendations of the Subsidiary Body on Scientific, Technical and Technological Advice; and

(c) In keeping with the provisions of Articles 16, 18 and 20, promote and cooperate in the use of scientific advances in biological diversity research in developing methods for conservation and sustainable use of biological resources.

Article 13. Public Education and Awareness

The Contracting Parties shall:

(a) Promote and encourage understanding of the importance of, and the measures required for, the conservation of biological diversity, as well as its propagation through media, and the inclusion of these topics in educational programmes; and

(b) Cooperate, as appropriate, with other States and international organizations in developing educational and public awareness programmes, with respect to conservation and sustainable use of biological diversity.

Article 14. Impact Assessment and Minimizing Adverse Impacts

1. Each Contracting Party, as far as possible and as appropriate, shall:

   (a) Introduce appropriate procedures requiring environmental impact assessment of its proposed projects that are likely to have significant adverse effects on biological diversity with a view to avoiding or minimizing such effects and, where appropriate, allow for public participation in such procedures;

   (b) Introduce appropriate arrangements to ensure that the environmental consequences of its programmes and policies that are likely to have significant adverse impacts on biological diversity are duly taken into account;

   (c) Promote, on the basis of reciprocity, notification, exchange of information and consultation on activities under their jurisdiction or control which are likely to significantly affect adversely the biological diversity of other States or areas beyond the limits of national jurisdiction, by encouraging the conclusion of bilateral, regional or multilateral arrangements, as appropriate;

   (d) In the case of imminent or grave danger or damage, originating under its jurisdiction or control, to biological diversity within the area under jurisdiction of other States or in areas beyond the limits of national jurisdiction, notify immediately the potentially affected States of such danger or damage, as well as initiate action to prevent or minimize such danger or damage; and

   (e) Promote national arrangements for emergency responses to activities or events, whether caused naturally or otherwise, which present a grave and imminent danger to biological diversity and encourage international cooperation to supplement such national efforts and, where appropriate and agreed by the States or regional economic integration organizations concerned, to establish joint contingency plans.

2. The Conference of the Parties shall examine, on the basis of studies to be carried out, the issue of liability and redress, including restoration and compensation, for damage to biological diversity, except where such liability is a purely internal matter.

Article 15. Access to Genetic Resources

1. Recognizing the sovereign rights of States over their natural resources, the authority to determine access to genetic resources rests with the national governments and is subject to national legislation.

2. Each Contracting Party shall endeavour to create conditions to facilitate access to genetic resources for environmentally sound uses by other Contracting Parties and not to impose restrictions that run counter to the objectives of this Convention.

3. For the purpose of this Convention, the genetic resources being provided by a Contracting Party, as referred to in this Article and

   Articles 16 and 19, are only those that are provided by Contracting Parties that are countries of origin of such resources or by the Parties that have acquired the genetic resources in accordance with this Convention.

4. Access, where granted, shall be on mutually agreed terms and subject to the provisions of this Article.

5. Access to genetic resources shall be subject to prior informed consent of the Contracting Party providing such resources, unless otherwise determined by that Party.

6. Each Contracting Party shall endeavour to develop and carry out scientific research based on genetic resources provided by other Contracting Parties with the full participation of, and where possible in, such Contracting Parties.

7. Each Contracting Party shall take legislative, administrative or policy measures, as appropriate, and in accordance with Articles 16 and 19 and, where necessary, through the financial mechanism established by Articles 20 and 21 with the aim of sharing in a fair and equitable way the results of research and development and the benefits arising from the commercial and other utilization of genetic resources with the Contracting Party providing such resources. Such sharing shall be upon mutually agreed terms.

Article 16. Access to and Transfer of Technology

1. Each Contracting Party, recognizing that technology includes biotechnology, and that both access to and transfer of technology among Contracting Parties are essential elements for the attainment of the objectives of this Convention, undertakes subject to the provisions of this Article to provide and/or facilitate access for and transfer to other Contracting Parties of technologies that are relevant to the conservation and sustainable use of biological diversity or make use of genetic resources and do not cause significant damage to the environment.

2. Access to and transfer of technology referred to in paragraph 1 above to developing countries shall be provided and/or facilitated under fair and most favourable terms, including on concessional and preferential terms where mutually agreed, and, where necessary, in accordance with the financial mechanism established by Articles 20 and 21. In the case of technology subject to patents and other intellectual property rights, such access and transfer shall be provided on terms which recognize and are consistent with the adequate and effective protection of intellectual property rights. The application of this paragraph shall be consistent with paragraphs 3, 4 and 5 below.

3. Each Contracting Party shall take legislative, administrative or policy measures, as appropriate, with the aim that Contracting Parties, in particular those that are developing countries, which provide genetic resources are provided access to and transfer of technology which makes use of those resources, on mutually agreed terms, including technology protected by patents and other intellectual property rights, where necessary, through the provisions of Articles 20 and 21 and in accordance with international law and consistent with paragraphs 4 and 5 below.
4. Each Contracting Party shall take legislative, administrative or policy measures, as appropriate, with the aim that the private sector facilitates access to, joint development and transfer of technology referred to in paragraph 1 above for the benefit of both governmental institutions and the private sector of developing countries and in this regard shall abide by the obligations included in paragraphs 1, 2 and 3 above.
5. The Contracting Parties, recognizing that patents and other intellectual property rights may have an influence on the implementation of this Convention, shall cooperate in this regard subject to national legislation and international law in order to ensure that such rights are supportive of and do not run counter to its objectives.

Article 17. Exchange of Information

1. The Contracting Parties shall facilitate the exchange of information, from all publicly available sources, relevant to the conservation and sustainable use of biological diversity, taking into account the special needs of developing countries.
2. Such exchange of information shall include exchange of results of technical, scientific and socio-economic research, as well as information on training and surveying programmes, specialized knowledge, indigenous and traditional knowledge as such and in combination with the technologies referred to in Article 16, paragraph 1. It shall also, where feasible, include repatriation of information.

Article 18. Technical and Scientific Cooperation

1. The Contracting Parties shall promote international technical and scientific cooperation in the field of conservation and sustainable use of biological diversity, where necessary, through the appropriate international and national institutions.
2. Each Contracting Party shall promote technical and scientific cooperation with other Contracting Parties, in particular developing countries, in implementing this Convention, inter alia, through the development and implementation of national policies. In promoting such cooperation, special attention should be given to the development and strengthening of national

capabilities, by means of human resources development and institution building.

3. The Conference of the Parties, at its first meeting, shall determine how to establish a clearing-house mechanism to promote and facilitate technical and scientific cooperation.
4. The Contracting Parties shall, in accordance with national legislation and policies, encourage and develop methods of cooperation for the development and use of technologies, including indigenous and traditional technologies, in pursuance of the objectives of this Convention. For this purpose, the Contracting Parties shall also promote cooperation in the training of personnel and exchange of experts.
5. The Contracting Parties shall, subject to mutual agreement, promote the establishment of joint research programmes and joint ventures for the development of technologies relevant to the objectives of this Convention.

Article 19. Handling of Biotechnology and Distribution of its Benefits

1. Each Contracting Party shall take legislative, administrative or policy measures, as appropriate, to provide for the effective participation in biotechnological research activities by those Contracting Parties, especially developing countries, which provide the genetic resources for such research, and where feasible in such Contracting Parties.
2. Each Contracting Party shall take all practicable measures to promote and advance priority access on a fair and equitable basis by Contracting Parties, especially developing countries, to the results and benefits arising from biotechnologies based upon genetic resources provided by those Contracting Parties. Such access shall be on mutually agreed terms.
3. The Parties shall consider the need for and modalities of a protocol setting out appropriate procedures, including, in particular, advance informed agreement, in the field of the safe transfer, handling and use of any living modified organism resulting from biotechnology that may have adverse effect on the conservation and sustainable use of biological diversity.
4. Each Contracting Party shall, directly or by requiring any natural or legal person under its jurisdiction providing the organisms referred to in paragraph 3 above, provide any available information about the use and safety regulations required by that Contracting Party in handling such organisms, as well as any available information on the potential adverse impact of the specific organisms concerned to the Contracting Party into which those organisms are to be introduced.

Article 20. Financial Resources

1. Each Contracting Party undertakes to provide, in accordance with its capabilities, financial support and incentives in respect of those national

activities which are intended to achieve the objectives of this Convention, in accordance with its national plans, priorities and programmes.

2. The developed country Parties shall provide new and additional financial resources to enable developing country Parties to meet the agreed full incremental costs to them of implementing measures which fulfil the obligations of this Convention and to benefit from its provisions and which costs are agreed between a developing country Party and the institutional structure referred to in Article 21, in accordance with policy, strategy, programme priorities and eligibility criteria and an indicative list of incremental costs established by the Conference of the Parties. Other Parties, including countries undergoing the process of transition to a market economy, may voluntarily assume the obligations of the developed country Parties. For the purpose of this Article, the Conference of the Parties, shall at its first meeting establish a list of developed country Parties and other Parties which voluntarily assume the obligations of the developed country Parties. The Conference of the Parties shall periodically review and if necessary amend the list. Contributions from other countries and sources on a voluntary basis would also be encouraged. The implementation of these commitments shall take into account the need for adequacy, predictability and timely flow of funds and the importance of burden-sharing among the contributing Parties included in the list.

3. The developed country Parties may also provide, and developing country Parties avail themselves of, financial resources related to the implementation of this Convention through bilateral, regional and other multilateral channels.

4. The extent to which developing country Parties will effectively implement their commitments under this Convention will depend on the effective implementation by developed country Parties of their commitments under this Convention related to financial resources and transfer of technology and will take fully into account the fact that economic and social development and eradication of poverty are the first and overriding priorities of the developing country Parties.

5. The Parties shall take full account of the specific needs and special situation of least developed countries in their actions with regard to funding and transfer of technology.

6. The Contracting Parties shall also take into consideration the special conditions resulting from the dependence on, distribution and location of, biological diversity within developing country Parties, in particular small island States.

7. Consideration shall also be given to the special situation of developing countries, including those that are most environmentally vulnerable, such as those with arid and semi- arid zones, coastal and mountainous areas.

Article 21. Financial Mechanism

1. There shall be a mechanism for the provision of financial resources to developing country Parties for purposes of this Convention on a grant or concessional basis the essential elements of which are described in this Article. The mechanism shall function under the authority and guidance of, and be accountable to, the Conference of the Parties for purposes of this Convention. The operations of the mechanism shall be carried out by such institutional structure as may be decided upon by the Conference of the Parties at its first meeting. For purposes of this Convention, the Conference of the Parties shall determine the policy, strategy, programme priorities and eligibility criteria relating to the access to and utilization of such resources. The contributions shall be such as to take into account the need for predictability, adequacy and timely flow of funds referred to in Article 20 in accordance with the amount of resources needed to be decided periodically by the Conference of the Parties and the importance of burden-sharing among the contributing Parties included in the list referred to in Article 20, paragraph 2. Voluntary contributions may also be made by the developed country Parties and by other countries and sources. The mechanism shall operate within a democratic and transparent system of governance.
2. Pursuant to the objectives of this Convention, the Conference of the Parties shall at its first meeting determine the policy, strategy and programme priorities, as well as detailed criteria and guidelines for eligibility for access to and utilization of the financial resources including monitoring and evaluation on a regular basis of such utilization. The Conference of the Parties shall decide on the arrangements to give effect to paragraph 1 above after consultation with the institutional structure entrusted with the operation of the financial mechanism.
3. The Conference of the Parties shall review the effectiveness of the mechanism established under this Article, including the criteria and guidelines referred to in paragraph 2 above, not less than two years after the entry into force of this Convention and thereafter on a regular basis. Based on such review, it shall take appropriate action to improve the effectiveness of the mechanism if necessary.
4. The Contracting Parties shall consider strengthening existing financial institutions to provide financial resources for the conservation and sustainable use of biological diversity.

Article 22. Relationship with Other International Conventions

1. The provisions of this Convention shall not affect the rights and obligations of any Contracting Party deriving from any existing international agreement, except where the exercise of those rights and obligations would cause a serious damage or threat to biological diversity.

2. Contracting Parties shall implement this Convention with respect to the marine environment consistently with the rights and obligations of States under the law of the sea.

Article 23. Conference of the Parties

1. A Conference of the Parties is hereby established. The first meeting of the Conference of the Parties shall be convened by the Executive Director of the United Nations Environment Programme not later than one year after the entry into force of this Convention. Thereafter, ordinary meetings of the Conference of the Parties shall be held at regular intervals to be determined by the Conference at its first meeting.
2. Extraordinary meetings of the Conference of the Parties shall be held at such other times as may be deemed necessary by the Conference, or at the written request of any Party, provided that, within six months of the request being communicated to them by the Secretariat, it is supported by at least one third of the Parties.
3. The Conference of the Parties shall by consensus agree upon and adopt rules of procedure for itself and for any subsidiary body it may establish, as well as financial rules governing the funding of the Secretariat. At each ordinary meeting, it shall adopt a budget for the financial period until the next ordinary meeting.
4. The Conference of the Parties shall keep under review the implementation of this Convention, and, for this purpose, shall:

   (a) Establish the form and the intervals for transmitting the information to be submitted in accordance with Article 26 and consider such information as well as reports submitted by any subsidiary body;

   (b) Review scientific, technical and technological advice on biological diversity provided in accordance with Article 25;

   (c) Consider and adopt, as required, protocols in accordance with Article 28;

   (d) Consider and adopt, as required, in accordance with Articles 29 and 30, amendments to this Convention and its annexes;

   (e) Consider amendments to any protocol, as well as to any annexes thereto, and, if so decided, recommend their adoption to the parties to the protocol concerned;

   (f) Consider and adopt, as required, in accordance with Article 30, additional annexes to this Convention;

   (g) Establish such subsidiary bodies, particularly to provide scientific and technical advice, as are deemed necessary for the implementation of this Convention;

(h) Contact, through the Secretariat, the executive bodies of conventions dealing with matters covered by this Convention with a view to establishing appropriate forms of cooperation with them; and

(i) Consider and undertake any additional action that may be required for the achievement of the purposes of this Convention in the light of experience gained in its operation.

5. The United Nations, its specialized agencies and the International Atomic Energy Agency, as well as any State not Party to this Convention, may be represented as observers at meetings of the Conference of the Parties. Any other body or agency, whether governmental or non-governmental, qualified in fields relating to conservation and sustainable use of biological diversity, which has informed the Secretariat of its wish to be represented as an observer at a meeting of the Conference of the Parties, may be admitted unless at least one third of the Parties present object. The admission and participation of observers shall be subject to the rules of procedure adopted by the Conference of the Parties.

Article 24. Secretariat

1. A secretariat is hereby established. Its functions shall be:

(a) To arrange for and service meetings of the Conference of the Parties provided for in Article 23;

(b) To perform the functions assigned to it by any protocol;

(c) To prepare reports on the execution of its functions under this Convention and present them to the Conference of the Parties;

(d) To coordinate with other relevant international bodies and, in particular to enter into such administrative and contractual arrangements as may be required for the effective discharge of its functions; and

(e) To perform such other functions as may be determined by the Conference of the Parties.

2. At its first ordinary meeting, the Conference of the Parties shall designate the secretariat from amongst those existing competent international organizations which have signified their willingness to carry out the secretariat functions under this Convention.

Article 25. Subsidiary Body on Scientific, Technical and Technological Advice

1. A subsidiary body for the provision of scientific, technical and technological advice is hereby established to provide the Conference of the Parties and, as appropriate, its other subsidiary bodies with timely advice relating to the implementation of this Convention. This body shall be open to participation

by all Parties and shall be multidisciplinary. It shall comprise government representatives competent in the relevant field of expertise. It shall report regularly to the Conference of the Parties on all aspects of its work.

2. Under the authority of and in accordance with guidelines laid down by the Conference of the Parties, and upon its request, this body shall:

   (a) Provide scientific and technical assessments of the status of biological diversity;

   (b) Prepare scientific and technical assessments of the effects of types of measures taken in accordance with the provisions of this Convention;

   (c) Identify innovative, efficient and state-of-the-art technologies and know-how relating to the conservation and sustainable use of biological diversity and advise on the ways and means of promoting development and/or transferring such technologies;

   (d) Provide advice on scientific programmes and international cooperation in research and development related to conservation and sustainable use of biological diversity; and

   (e) Respond to scientific, technical, technological and methodological questions that the Conference of the Parties and its subsidiary bodies may put to the body.

3. The functions, terms of reference, organization and operation of this body may be further elaborated by the Conference of the Parties.

Article 26. Reports

Each Contracting Party shall, at intervals to be determined by the Conference of the Parties, present to the Conference of the Parties, reports on measures which it has taken for the implementation of the provisions of this Convention and their effectiveness in meeting the objectives of this Convention.

Article 27. Settlement of Disputes

1. In the event of a dispute between Contracting Parties concerning the interpretation or application of this Convention, the parties concerned shall seek solution by negotiation.

2. If the parties concerned cannot reach agreement by negotiation, they may jointly seek the good offices of, or request mediation by, a third party.

3. When ratifying, accepting, approving or acceding to this Convention, or at any time thereafter, a State or regional economic integration organization may declare in writing to the Depositary that for a dispute not resolved in accordance with paragraph 1 or paragraph 2 above, it accepts one or both of the following means of dispute settlement as compulsory:

(a) Arbitration in accordance with the procedure laid down in Part 1 of Annex II;

(b) Submission of the dispute to the International Court of Justice.

4. If the parties to the dispute have not, in accordance with paragaph 3 above, accepted the same or any procedure, the dispute shall be submitted to conciliation in accordance with Part 2 of Annex II unless the parties otherwise agree.

5. The provisions of this Article shall apply with respect to any protocol except as otherwise provided in the protocol concerned.

Article 28. Adoption of Protocols

1. The Contracting Parties shall cooperate in the formulation and adoption of protocols to this Convention.

2. Protocols shall be adopted at a meeting of the Conference of the Parties.

3. The text of any proposed protocol shall be communicated to the Contracting Parties by the Secretariat at least six months before such a meeting

Article 29. Amendment of the Convention or Protocols

1. Amendments to this Convention may be proposed by any Contracting Party. Amendments to any protocol may be proposed by any Party to that protocol.

2. Amendments to this Convention shall be adopted at a meeting of the Conference of the Parties. Amendments to any protocol shall be adopted at a meeting of the Parties to the Protocol in question. The text of any proposed amendment to this Convention or to any protocol, except as may otherwise be provided in such protocol, shall be communicated to the Parties to the instrument in question by the secretariat at least six months before the meeting at which it is proposed for adoption. The secretariat shall also communicate proposed amendments to the signatories to this Convention for information.

3. The Parties shall make every effort to reach agreement on any proposed amendment to this Convention or to any protocol by consensus. If all efforts at consensus have been exhausted, and no agreement reached, the amendment shall as a last resort be adopted by a two-third majority vote of the Parties to the instrument in question present and voting at the meeting, and shall be submitted by the Depositary to all Parties for ratification, acceptance or approval.

4. Ratification, acceptance or approval of amendments shall be notified to the Depositary in writing. Amendments adopted in accordance with paragraph 3 above shall enter into force among Parties having accepted them on the

ninetieth day after the deposit of instruments of ratification, acceptance or approval by at least two thirds of the Contracting Parties to this Convention or of the Parties to the protocol concerned, except as may otherwise be provided in such protocol. Thereafter the amendments shall enter into force for any other Party on the ninetieth day after that Party deposits its instrument of ratification, acceptance or approval of the amendments.

5. For the purposes of this Article, "Parties present and voting" means Parties present and casting an affirmative or negative vote.

Article 30. Adoption and Amendment of Annexes

1. The annexes to this Convention or to any protocol shall form an integral part of the Convention or of such protocol, as the case may be, and, unless expressly provided otherwise, a reference to this Convention or its protocols constitutes at the same time a reference to any annexes thereto. Such annexes shall be restricted to procedural, scientific, technical and administrative matters.

2. Except as may be otherwise provided in any protocol with respect to its annexes, the following procedure shall apply to the proposal, adoption and entry into force of additional annexes to this Convention or of annexes to any protocol:

   (a) Annexes to this Convention or to any protocol shall be proposed and adopted according to the procedure laid down in Article 29;

   (b) Any Party that is unable to approve an additional annex to this Convention or an annex to any protocol to which it is Party shall so notify the Depositary, in writing, within one year from the date of the communication of the adoption by the Depositary. The Depositary shall without delay notify all Parties of any such notification received. A Party may at any time withdraw a previous declaration of objection and the annexes shall thereupon enter into force for that Party subject to subparagraph (c) below;

   (c) On the expiry of one year from the date of the communication of the adoption by the Depositary, the annex shall enter into force for all Parties to this Convention or to any protocol concerned which have not submitted a notification in accordance with the provisions of subparagraph (b) above.

3. The proposal, adoption and entry into force of amendments to annexes to this Convention or to any protocol shall be subject to the same procedure as for the proposal, adoption and entry into force of annexes to the Convention or annexes to any protocol.

4. If an additional annex or an amendment to an annex is related to an amendment to this Convention or to any protocol, the additional annex or

amendment shall not enter into force until such time as the amendment to the Convention or to the protocol concerned enters into force.

Article 31. Right to Vote

1. Except as provided for in paragraph 2 below, each Contracting Party to this Convention or to any protocol shall have one vote.

2. Regional economic integration organizations, in matters within their competence, shall exercise their right to vote with a number of votes equal to the number of their member States which are Contracting Parties to this Convention or the relevant protocol. Such organizations shall not exercise their right to vote if their member States exercise theirs, and vice versa.

Article 32. Relationship Between this Convention and Its Protocols

1. A State or a regional economic integration organization may not become a Party to a protocol unless it is, or becomes at the same time, a Contracting Party to this Convention.

2. Decisions under any protocol shall be taken only by the Parties to the protocol concerned. Any Contracting Party that has not ratified, accepted or approved a protocol may participate as an observer in any meeting of the parties to that protocol.

Article 33. Signature

This Convention shall be open for signature at Rio de Janeiro by all States and any regional economic integration organization from 5 June 1992 until 14 June 1992, and at the United Nations Headquarters in New York from 15 June 1992 to 4 June 1993.

Article 34. Ratification, Acceptance or Approval

1. This Convention and any protocol shall be subject to ratification, acceptance or approval by States and by regional economic integration organizations. Instruments of ratification, acceptance or approval shall be deposited with the Depositary.

2. Any organization referred to in paragraph 1 above which becomes a Contracting Party to this Convention or any protocol without any of its member States being a Contracting Party shall be bound by all the obligations under the Convention or the protocol, as the case may be. In the case of such organizations, one or more of whose member States is a Contracting Party to this Convention or relevant protocol, the organization and its member States shall decide on their respective responsibilities for the performance of their obligations under the Convention or protocol, as the case may be. In such cases, the organization and the member States shall

not be entitled to exercise rights under the Convention or relevant protocol concurrently.

3. In their instruments of ratification, acceptance or approval, the organizations referred to in paragraph 1 above shall declare the extent of their competence with respect to the matters governed by the Convention or the relevant protocol. These organizations shall also inform the Depositary of any relevant modification in the extent of their competence.

Article 35. Accession

1. This Convention and any protocol shall be open for accession by States and by regional economic integration organizations from the date on which the Convention or the protocol concerned is closed for signature. The instruments of accession shall be deposited with the Depositary.
2. In their instruments of accession, the organizations referred to in paragraph 1 above shall declare the extent of their competence with respect to the matters governed by the Convention or the relevant protocol. These organizations shall also inform the Depositary of any relevant modification in the extent of their competence.
3. The provisions of Article 34, paragraph 2, shall apply to regional economic integration organizations which accede to this Convention or any protocol.

Article 36. Entry Into Force

1. This Convention shall enter into force on the ninetieth day after the date of deposit of the thirtieth instrument of ratification, acceptance, approval or accession.

2. Any protocol shall enter into force on the ninetieth day after the date of deposit of the number of instruments of ratification, acceptance, approval or accession, specified in that protocol, has been deposited.

3. For each Contracting Party which ratifies, accepts or approves this Convention or accedes thereto after the deposit of the thirtieth instrument of ratification, acceptance, approval or accession, it shall enter into force on the ninetieth day after the date of deposit by such Contracting Party of its instrument of ratification, acceptance, approval or accession.

4. Any protocol, except as otherwise provided in such protocol, shall enter into force for a Contracting Party that ratifies, accepts or approves that protocol or accedes thereto after its entry into force pursuant to paragraph 2 above, on the ninetieth day after the date on which that Contracting Party deposits its instrument of ratification, acceptance, approval or accession, or on the date on which this Convention enters into force for that Contracting Party, whichever shall be the later.

5. For the purposes of paragraphs 1 and 2 above, any instrument deposited by a regional economic integration organization shall not be counted as additional to those deposited by member States of such organization.

Article 37. Reservations

No reservations may be made to this Convention.

Article 38. Withdrawals

1. At any time after two years from the date on which this Convention has entered into force for a Contracting Party,that Contracting Party may withdraw from the Convention by giving written notification to the Depositary.
2. Any such withdrawal shall take place upon expiry of one year after the date of its receipt by the Depositary, or on such later date as may be specified in the notification of the withdrawal.
3. Any Contracting Party which withdraws from this Convention shall be considered as also having withdrawn from any protocol to which it is party.

Article 39. Financial Interim Arrangements

Provided that it has been fully restructured in accordance with the requirements of Article 21, the Global Environment Facility of the United Nations Development Programme, the United Nations Environment Programme and the International Bank for Reconstruction and Development shall be the institutional structure referred to in Article 21 on an interim basis, for the period between the entry into force of this Convention and the first meeting of the Conference of the Parties or until the Conference of the Parties decides which institutional structure will be designated in accordance with Article 21.

Article 40. Secretariat Interim Arrangements

The secretariat to be provided by the Executive Director of the United Nations Environment Programme shall be the secretariat referred to in Article 24, paragraph 2, on an interim basis for the period between the entry into force of this Convention and the first meeting of the Conference of the Parties.

Article 41. Depositary

The Secretary-General of the United Nations shall assume the functions of Depositary of this Convention and any protocols.

Article 42. Authentic Texts

The original of this Convention, of which the Arabic, Chinese, English, French, Russian and Spanish texts are equally authentic, shall be deposited with the Secretary- General of the United Nations.

IN WITNESS WHEREOF the undersigned, being duly authorized to that effect, have signed this Convention.

Done at Rio de Janeiro on this fifth day of June, one thousand nine hundred and ninety- two.

Annex I. Identification and Monitoring

1. Ecosystems and habitats: containing high diversity, large numbers of endemic or threatened species, or wilderness; required by migratory species; of social, economic, cultural or scientific importance; or, which are representative, unique or associated with key evolutionary or other biological processes;
2. Species and communities which are: threatened; wild relatives of domesticated or cultivated species; of medicinal, agricultural or other economic value; or social, scientific or cultural importance; or importance for research into the conservation and sustainable use of biological diversity, such as indicator species; and
3. Described genomes and genes of social, scientific or economic importance.

Annex II - Part 1. Arbitration

Article 1

The claimant party shall notify the secretariat that the parties are referring a dispute to arbitration pursuant to Article 27. The notification shall state the subject-matter of arbitration and include, in particular, the articles of the Convention or the protocol, the interpretation or application of which are at issue. If the parties do not agree on the subject matter of the dispute before the President of the tribunal is designated, the arbitral tribunal shall determine the subject matter. The secretariat shall forward the information thus received to all Contracting Parties to this Convention or to the protocol concerned.

Annex II - Part 2. Conciliation

Article 1

A conciliation commission shall be created upon the request of one of the parties to the dispute. The commission shall, unless the parties otherwise agree, be composed of five members, two appointed by each Party concerned and a President chosen jointly by those members.

Article 2

In disputes between more than two parties, parties in the same interest shall appoint their members of the commission jointly by agreement. Where two or more parties have separate interests or there is a disagreement as to whether they are of the same interest, they shall appoint their members separately.

Article 3

If any appointments by the parties are not made within two months of the date of the request to create a conciliation commission, the Secretary-General of the United Nations shall, if asked to do so by the party that made the request, make those appointments within a further two-month period.

Article 4

If a President of the conciliation commission has not been chosen within two months of the last of the members of the commission being appointed, the Secretary-General of the United Nations shall, if asked to do so by a party, designate a President within a further two-month period.

Article 5

The conciliation commission shall take its decisions by majority vote of its members. It shall, unless the parties to the dispute otherwise agree, determine its own procedure. It shall render a proposal for resolution of the dispute, which the parties shall consider in good faith.

Article 6

A disagreement as to whether the conciliation commission has competence shall be decided by the commission.

International Treaty on Plant Genetic Resources for Food and Agriculture

The FAO International Treaty on Plant Genetic Resources for Food and Agriculture enters into force on 29 June 2004. This legally binding instrument is crucial for sustainable agriculture. It provides a framework for national, regional and international efforts to conserve and sustainably use plant genetic resources for food and agriculture - and for sharing the benefits equitably, in line with the Convention on Biological Diversity. The Treaty contains two important and unique elements. First, it recognizes the enormous contribution made by farmers in all regions of the world towards the conservation and development of plant genetic resources and identifies ways of protecting and promoting Farmers' Rights. Second, it establishes a multilateral system of access and benefit sharing. This will ensure that countries have access to some of the most important plant genetic resources needed for food security. It identifies a range of benefits to be shared on a multilateral basis - benefits like information exchange, technology transfer and access, building capacity at local levels, and monetary and other benefits of commercialization. Benefits are targeted mainly to developing countries and countries with economies in transition, to help ensure that they will have the capacity to conserve and sustainably use their own genetic resources as well as any they may obtain under the multilateral system.

The Treaty is vital in ensuring the continued availability of the plant genetic resources that countries will need to feed their people. We must conserve for future generations the genetic diversity that is essential for food and agriculture. The relevant questions dealt with under this treaty are as follows :

What Are "Plant Genetic Resources for Food and Agriculture"?

The Treaty defines them as "any genetic material of plant origin of actual or potential value for food and agriculture".

What Are the Treaty's Objectives?

Its objectives are the conservation and sustainable use of plant genetic resources for food and agriculture and the fair and equitable sharing of benefits derived from their use, in harmony with the Convention on Biological Diversity, for sustainable agriculture and food security.

What Is the Multilateral System for Access and Benefit-Sharing?

Through the Treaty, countries agree to establish an efficient, effective and transparent Multilateral System to facilitate access to plant genetic resources for food and agriculture, and to share the benefits in a fair and equitable way. The Multilateral System applies to over 64 major crops and forages. The Governing Body of the Treaty, which will be composed of the countries that have ratified it, will set out the conditions for access and benefit-sharing in a "Material Transfer Agreement".

What Are The Conditions for Access in The Multilateral System?

Resources may be obtained from the Multilateral System for utilization and conservation in research, breeding and training. When a commercial product is developed using these resources, the Treaty provides for payment of an equitable share of the resulting monetary benefits, if this product may not be used without restriction by others for further research and breeding. If others may use it, payment is voluntary.

How Will Benefits Be Shared?

The Treaty provides for sharing the benefits of using plant genetic resources for food and agriculture through information-exchange, access to and the transfer of technology, and capacity-building. It also foresees a funding strategy to mobilize funds for activities, plans and programmes the help, above all, small farmers in developing countries. This funding strategy also includes the share of the monetary benefits paid under the Multilateral System.

How Does the Treaty Protect Farmers' Rights?

The Treaty recognizes the enormous contribution that farmers and their communities have made and continue to make to the conservation and development of plant genetic resources. This is the basis for Farmers' Rights, which include the protection of traditional knowledge, and the right to participate equitably in benefit-sharing and in national decision-making about plant genetic resources. It gives governments the responsibility for implementing these rights.

Who Benefits From the Treaty and How?

All benefit, in many ways:

- Farmers and their communities, through Farmers' Rights;
- Consumers, because of a greater variety of foods, and of agriculture products, as well as increased food security;
- The scientific community, through access to the plant genetic resources crucial for research and plant breeding;
- International Agricultural Research Centres, whose collections the Treaty puts on a safe and long-term legal footing;
- Both the public and private sectors, which are assured access to a wide range of genetic diversity for agricultural development; and
- The environment, and future generations, because the Treaty will help conserve the genetic diversity necessary to face unpredictable environmental changes, and future human needs.

When Will the Treaty Come Into Force?

The Treaty will come into force ninety days after forty governments have ratified it. Governments that have ratified it will make up its Governing Body. At its first meeting, this Governing Body will address important questions, such as the level, form and manner of monetary payments on commercialization, a standard Material Transfer Agreement for plant genetic resources, mechanisms to promote compliance with the Treaty, and the funding strategy. Countries may therefore consider it important to be among the first to ratify, so as to ensure that their national interests can be taken into account at the Governing Body's first meeting.

What's Next?

Each country that ratifies will then develop the legislation and regulations it needs to implement the Treaty.

Biological Diversity Act-2002 Enacted by Government of India

Title	The Biological Diversity Act, 2002
Level	National
Country	India
Area of activity	General
Type of measure	Legislation

Scope	Biodiversity
Coverage	-Access to genetic resources -Equitable sharing of benefits arising out of the utilization of genetic resources - Intellectual property rights related to genetic resources and/or protection of traditional knowledge, innovations and practices related to genetic resources
Status of measures	Entered into force
Date of entry into force	5/2/2003
	Additional Information
Text of measure	http://envfor.nic.in/divisions/biodiv/act/bio_div_act_2002.pdf

The details of this act are as follows :

The Biological Diversity Act 2002

WHEREAS India is rich in **biological diversity** and associated traditional and contemporary knowledge system relating thereto;

AND WHEREAS India is a party to the United Nations Convention on **Biological**

Diversity signed at Rio de Janeiro on the 5th day of June, 1992;

AND WHEREAS the said Convention came into force on the 29th December, 1993;

AND WHEREAS the said Convention reaffirms the sovereign rights of the States over their **biological** resources;

AND WHEREAS the said Convention has the main objective of conservation of **biological diversity**, sustainable use of its components and fair and equitable sharing of the benefits arising out of utilisation of genetic resources;

AND WHEREAS it is considered necessary to provide for conservation, sustainable utilisation and equitable sharing of benefits arising out of utilisation of genetic resources and also to give effect to the said Convention;

BE it enacted by Parliament in the Fifty-third Year of the Republic of India as follows:-

Chapter I

Preliminary

1. Short Title, extent and commencement

(1) This Act may be called the **Biological Diversity** Act, 2002.

(2) It extends to the whole of India.

(3) It comes into force on 2 May 2003.

Provided that different dates may be appointed for different provisions of this Act and any reference in any such provision to the commencement of this Act shall be construed as a reference to the coming into force of that provision.

2. Definitions

In this Act, unless the context otherwise requires,-

(a) "benefit claimers" means the conservers of **biological** resources, their by-products, creators and holders of knowledge and information relating to the use of such **biological** resources, innovations and practices associated with such use and application;

(b) "**biological diversity**" means the variability among living organisms from all sources and the ecological complexes of which they are part and includes **diversity** within species or between species and of ecosystems;

(c) "**biological** resources" means plants, animals and micro-organisms or parts thereof, their genetic material and by-products (excluding value added products) with actual or potential use or value but does not include human genetic material;

(d) "bio-survey and bio-utilisation" means survey or collection of species, subspecies, genes, components and extracts of **biological** resource for any purpose and includes characterisation, inventorisation and bioassay;

(e) "Chairperson" means the Chairperson of the National Biodiversity Authority or, as the case may be, of the State Biodiversity Board;

(f) "commercial utilisation" means end uses of **biological** resources for commercial utilisation such as drugs, industrial enzymes, food flavours, fragrance, cosmetics, emulsifiers, oleoresins, colours, extracts and genes used for improving crops and livestock through genetic intervention, but does not include conventional breeding or traditional practices in use in any agriculture, horticulture, poultry, dairy farming, animal husbandry or bee keeping;

(g) "fair and equitable benefit sharing" means sharing of benefits as determined by the National Biodiversity Authority under section 21;

(h) "local bodies" means Panchayats and Municipalities, by whatever name called, within the meaning of clause (1) of article 243B and clause (1) of article 243Q of the Constitution and in the absence of any Panchayats or Municipalities, institutions of self-government constituted under any Central Act or State Act;

(i) "member" means member of the National Biodiversity Authority or a State Biodiversity Board and includes the Chairperson;

(j) "National Biodiversity Authority" means the National Biodiversity Authority established under section 8;

(k) "prescribed" means prescribed by rules made under this Act;

(l) "regulations" means regulations made under this Act;

(m) "research" means study or systematic investigation of any **biological** resource or technological application, that uses **biological** systems, living, organisms or derivatives thereof to make or modify products or processes for any use.

(n) "State Biodiversity Board" means the State Biodiversity Board established under section 22;

(o) "sustainable use" means the use of components of **biological diversity** in such manner and at such rate that does not lead to the long-term decline of the **biological diversity** thereby maintaining its potential to meet the needs and aspirations of present and future generations.

(p) "value added products" means products which may contain portions or extracts of plants and animals in unrecognisable and physically inseparable form.

Chapter II

Regulation of Biological Diversity

3. Certain persons not to undertake Biodiversity related activities without prior approval of National Biodiversity Authority.

(1) No person referred to in sub-section (2) shall without previous approval of the National Biodiversity Authority obtain any **biological** resource occurring in India or knowledge associated thereto for research or for commercial utilisation or for bio-survey and bio-utilisation.

(2) The persons who shall be required to take the approval of the National Biodiversity Authority under sub-section (1) are the following, namely:-

(a) a person who is not a citizen of India;

(b) a citizen of India, who is a non-resident as defined in clause (30) of section 2 of the Income-tax Act, 1961;

(c) a body corporate, association or organisation-

(i) not incorporated or registered in India; or

(ii) incorporated or registered in India under any law for the time being in force which has any non-Indian participation in its share capital or management.

4. Results of research not to be transferred to certain persons without approval of National Biodiversity Authority.

No person shall without the previous approval of the National Biodiversity Authority, transfer the results of any research relating to any **biological** resources occurring or obtained from India for monetary consideration or otherwise to any person who is not a citizen of India or a body corporate or organisation which is not registered or incorporated in India or which has any non-Indian participation in its share capital or management.

Explanation—For the purposes of this section, "transfer" does not include publication ofresearch papers or dissemination of knowledge in any seminar or workshop, if such publication is as per the guidelines issued by the Central Government.

5. Sections 3 and 4 not to apply to certain collaborative research projects

(1) The provisions of sections 3 and 4 shall not apply to collaborative research. projects involving transfer or exchange of **biological** resources or information relating thereto between institutions, including Government sponsored institutions of India, and such institutions in other countries, if such collaborative research projects satisfy the conditions specified in sub-section (3).

(2) All collaborative research projects, other than those referred to in sub-section (1) which are based on agreements concluded before the commencement of this Act and in force shall, to the extent the provisions of agreement are inconsistent with the provisions of this Act or any guidelines issued under clause (a) of sub-section (3), be void.

(3) For the purposes of sub-section (1) collaborative research projects shall,-

(a) conform to the policy guidelines issued by the Central Government in this behalf;

(b) be approved by the Central Government.

6. Application for intellectual property rights not to be made without approval of National Biodiversity Authority.

(1) No person shall apply for any intellectual property right by whatever name called in or outside India for any invention based on any research or information on a **biological** resource obtained from India without obtaining the previous approval of the National Biodiversity Authority before making such application:

Provided that if a person applies for a patent, permission of the National Biodiversity Authority may be obtained after the acceptance of the patent but before the sealing of the patent by the patent authority concerned.

Provided further that the National Biodiversity Authority shall dispose of the application for permission made to it within a period of ninety days from the date of receipt thereof.

(2) The National Biodiversity Authority may, while granting the approval under this section, impose benefit sharing fee or royalty or both or impose conditions including the sharing of financial benefits arising out of the commercial utilisation of such rights.

(3) The provisions of this section shall not apply to any person making an application for any rights under any law relating to protection of plant varieties enacted by Parliament.

(4) Where any right is granted under law referred to in sub-section (3), the concerned authority granting such right shall endorse a copy of such document granting the right to the National Biodiversity Authority.

*7. Prior Intimation to State Biodiversity Board before obtaining **biological** resource for certain purposes*

No person who is a citizen of India or a body corporate, association or organisation which is registered in India shall obtain any **biological** resource for commercial utilisation or bio-survey and bio-utilisation except after giving prior intimation to the State Biodiversity Board concerned:

Provided that the provisions of this section shall not apply to the local people and communities of the area, including growers and cultivators of biodiversity, and *vaids and hakims,* who have been practising indigenous medicine.

Chapter III

National Biodiversity Authority

8. Establishment of National Biodiversity Authority.

(1) With effect from such date as the Central Government may, by notification in the Official Gazette, appoint, there shall be established for the purposes of this Act, a body to be called the National Biodiversity Authority.

(2) The National Biodiversity Authority shall be a body corporate by the name aforesaid, having perpetual succession and a common seal, with power to acquire, hold and dispose of property, both movable and immovable, and to contract, and shall by the said name sue and be sued.

(3) The head office of the National Biodiversity Authority shall be at Chennai and the National Biodiversity Authority may, with the previous approval of the Central Government,establish offices at other places in India.

(4) The National Biodiversity Authority shall consist of the following Members, namely:

(a) a Chairperson, who shall be an eminent person having adequate knowledge and experience in the conservation and sustainable use of **biological diversity** and in matters relating to equitable sharing of benefits, to be appointed by the Central Government;

(b) three members *ex officio* to be appointed by the Central Government, one representing the Ministry of Tribal Affairs and two representing the Ministry of Environment and Forests of whom one shall be the Additional Inspector General of Forests or the Inspector General of Forests;

(c) seven members *ex officio* to be appointed by the Central Government to represent respectively the Ministries of the Central Government dealing with-

(i) Agricultural Research and Education;
(ii) Biotechnology;
(iii) Ocean Development.,
(iv) Agriculture and Cooperation;
(v) Indian Systems of Medicine and Homoeopathy;
(vi) Science and Technology;
(vii) Scientific and Industrial Research;

(d) five non-official members to be appointed from amongst specialists and scientists having special knowledge of, or experience in, matters relating to conservation of **biological diversity**, sustainable use of **biological** resources and equitable sharing of benefits arising out of the use of **biological** resources, representatives of industry, conservers, creators and knowledge holders of **biological** resources.

9. Conditions of service of Chairperson and Members.

The term of office and conditions of service of the Chairperson and the other members other than *ex officio* members shall be such as may be prescribed by the Central Government.

10. Chairperson to be Chief Executive of National Biodiversity Authority.

The Chairperson shall be the Chief Executive of the National Biodiversity Authority and shall exercise such powers and perform such duties, as may be prescribed.

11. Removal of Members

The Central Government may remove from the National Biodiversity Authority any member who, in its opinion has,-

(a) been adjudged as an insolvent; or

(b) been convicted of an offence which involves moral turpitude; or

(c) become physically or mentally incapable of acting as a member; or

(d) so abused his position as to render his continuance in office detrimental to the public interest; or

(e) acquired such financial or other interest as is likely to affect prejudicially his functions as a member.

12. Meetings of National Biodiversity Authority

(1) The National Biodiversity Authority shall meet at such time and place and shall observe such rules of procedure in regard to the transaction of business at its meetings (including the quorum at its meetings) as may be prescribed.

(2) The Chairperson of the National Biodiversity Authority shall preside at the meetings of the National Biodiversity Authority.

(3) If for any reason the Chairperson is unable to attend any meeting of the National Biodiversity Authority, any member of the National Biodiversity Authority chosen by the members present at the meeting shall preside at the meeting.

(4) All questions which come before any meeting of the National Biodiversity Authority shall be decided by a majority of the votes of the members present and voting and in the event of equality of votes, the Chairperson or, in his absence, the person presiding, shall have and exercise a second or casting vote.

(5) Every member who is in any way, whether directly, indirectly or personally, concerned or interested in a matter to be decided at the meeting shall disclose the nature of his concern or interest and after such disclosure, the member concerned or interested shall not attend that meeting.

(6) No act or proceeding of the National Biodiversity Authority shall be invalidated merely by reason of-

(a) any vacancy in, or any defect in the constitution of the National Biodiversity Authority; or

(b) any defect in the appointment of a person acting as a member; or

(c) any irregularity in the procedure of the National Biodiversity Authority not affecting the merits of the case.

13. Committees of National Biodiversity Authority

(1) National Biodiversity Authority may constitute a committee to deal with agro-biodiversity.

Explanation: For the purposes of this sub-section "agro-biodiversity" means **biological diversity** of agriculture related species and their wild relatives.

(2) Without prejudice to the provisions of sub-section (1), the National Biodiversity Authority may constitute such number of committees as it

deems fit for the efficient discharge of its duties and performance of its functions under this Act.

(3) A committee constituted under this section, shall co-opt such number of persons, who are not members of the National Biodiversity Authority, as it may think fit and the persons so co-opted shall have the right to attend the meetings of the committee and take part in its proceedings but shall not have the right to vote.

(4) The persons appointed as members of the committee under sub-section (2) shall be entitled to receive such allowances or fees for attending the meetings of the committee as may be fixed by the Central Government.

14. Officers and Employees of National Biodiversity Authority.

(1) The National Biodiversity Authority may appoint such officers and other employees as it considers necessary for the efficient discharge of its functions under this Act.

(2) The terms and conditions of service of such officers and other employees of the National Biodiversity Authority shall be such as may be specified by regulations.

15. Authentication of Orders and Decisions of National Biodiversity Authority.

All orders and decisions of the National Biodiversity Authority shall be authenticated by the signature of the Chairperson or any other member authorised by the National Biodiversity Authority in this behalf and all other instruments executed by the National

Biodiversity Authority shall be authenticated by the signature of an officer of the National Biodiversity Authority authorised by it in this behalf.

16. Delegation of Powers

The National Biodiversity Authority may, by general or special order in writing, delegate to any member, officer of the National Biodiversity Authority or any other person subject to such conditions, if any, as may he specified in the order, such of the powers and functions under the Act (except the power to settle disputes under section 50 and the power to make regulations under section (62) as it may deem necessary.

17. Expenses of National Biodiversity Authority to be Defrayed Out of the Consolidated Fund of India.

The salaries and allowances payable to the members and the administrative expenses of the National Biodiversity Authority including salaries, allowances and pension payable to,or in respect of the officers and other employees of the National Biodiversity Authority shall be defrayed out of the Consolidated Fund of India.

Chapter IV

Functions And Powers of The Nanonal Biodiversity Authority

18. Functions of National Biodiversity Authority.

(1) It shall be the duty of the National Biodiversity Authority to regulate activities referred to in sections 3,4 and 6 and by regulations, issue guidelines for access to and equitable benefit sharing.

(2) The National Biodiversity Authority may grant approval for undertaking any activity referred to in sections 3, 4 and 6.

(3) The National Biodiversity Authority may-

(a) advise the Central Government on matters relating to the conservation of biodiversity, sustainable use of its components and equitable sharing of benefits arising out of the utilisation of **biological** resources;

(b) advise the State Governments in the selection of areas of biodiversity importance to be notified under sub-section (l) of Section 37 as heritage sites and measures for the management of such heritage sites;

(c) perform such other functions as may be necessary to carry out the provisions of this Act.

(4) The National Biodiversity Authority may, on behalf of the Central Government take any measures necessary to oppose the grant of intellectual property rights in any country outside India on any **biological** resource obtained from India or knowledge associated with such **biological** resource which is derived from India.

Chapter V

Approvals By The National Biodiversity Authority

19. Approval by National Biodiversity Authority for Undertaking Certain Activities.

(1) Any person referred to in sub-section *(2)* of section 3 who intends to obtain any **biological** resource occurring in India or knowledge associated thereto for research or for commercial utilisation or for bio-survey and bio-utilisation or transfer the results of any research relating to **biological** resources occurring in, or obtained from, India, shall make application in such form and payment of such fees as may be prescribed, to the National Biodiversity Authority.

(2) Any person who intends to apply for a patent or any other form of intellectual property protection whether in India or outside India may make an application in such form and in-such manner as may be prescribed to the National Biodiversity Authority.

(3) On receipt of an application under sub-section (1) or sub-section (2), the National Biodiversity Authority may, after making such enquiries as it may

deem fit and if necessary after consulting an expert committee constituted for this purpose, by order, grant approval subject to any regulations made in this behalf and subject to such terms and conditions as it may deem fit, including the imposition of charges by way of royalty or for reasons to be recorded in writing reject the application:

Provided that no such order for rejection shall be made without giving an opportunity of being heard to the person affected.

(4) The National Biodiversity Authority shall give public notice of every approval granted by it under this section.

20. Transfer of Biological Resource or Knowledge

(1) No person who has been granted approval under section 19 shall transfer any **biological** resource or knowledge associated thereto which is the subject matter of the said approval except with the permission of the National Biodiversity Authority.

(2) Any person who intends to transfer any **biological** resource or knowledge associated thereto referred to in sub-section (1) shall make an application in such form and in such manner as may be prescribed to the National Biodiversity Authority.

(3) On receipt of an application under sub-section (2), the National Biodiversity Authority may, after making such enquires as deemed fit and if necessary after consulting an expert committee constituted for this purpose, by order, grant approval subject to such terms and conditions as it may deem fit, including the imposition of charges by way of royalty or for reasons to be recorded in writing reject the application:

Provided that no such order for rejection shall be made without giving an opportunity of being heard to the person affected.

(4) The National Biodiversity Authority shall give public notice of every approval granted by it under this section.

21. Determination of Equitable Benefit Sharing by National Biodiversity Authority

(1) The National Biodiversity Authority shall while granting approvals under section 19 or section 20 ensure that the terms and conditions subject to which approval is granted secures equitable sharing of benefits arising out of the use of accessed **biological** resources, their by-products, innovations and practices associated with their use and applications and knowledge relating thereto in accordance with mutually agreed terms and conditions between the person applying for such approval, local bodies concerned and the benefit claimers.

(2) The National Biodiversity Authority shall subject to any regulations made in this behalf determine the benefit sharing which shall be given effect in all or any of the following manner, namely: –

(a) grant of joint ownership of intellectual property rights to the National Biodiversity Authority, or where benefit claimers are identified, to such benefit claimers,

(b) transfer of technology;

(c) location of production, research and development units in such areas which will facilitate better living standards to the benefit claimers;

(d) association of Indian scientists, benefit claimers and the local people with research and development in **biological** resources and bio-survey and bio-utilisation;

(e) setting up of venture capital fund for aiding the cause of benefit claimers;

(f) payment of monetary compensation and other non-monetary benefits to the benefit claimers as the National Biodiversity Authority may deem fit.

(3) Where any amount of money is ordered by way of benefit sharing, the National Biodiversity Authority may direct the amount to be deposited in the National Biodiversity Fund:

Provided that where **biological** resource or knowledge was a result of access from specific individual or group of individuals or organisations, the National Biodiversity Authority may direct that the amount shall be paid directly to such individuals or group of individuals or organizations in accordance with the terms of any agreement and in such manner as it deems fit.

(4) For the purposes of this section, the National Biodiversity Authority shall, in consultation with the Central Government, by regulations, frame guidelines.

Chapter VI

State Biodiversity Board

22. Establishment of State Biodiversity Board

(1) With effect from such date as the State Government may, by notification in the Official Gazette, appoint in this behalf, there shall be established for the purposes of this Act, a Board for the State to be known as the __________ (name of the State) Biodiversity Board.

(2) Notwithstanding anything contained in this section, no State Biodiversity Board shall be constituted for a Union territory and in relation to a Union territory, the National Biodiversity Authority shall exercise the powers and perform the functions of a State Biodiversity Board for that Union territory:

Provided that in relation to any Union Territory, the National Biodiversity Authority may delegate all or any of its powers or functions under this sub-section to such person or group of persons as the Central Government may specify.

(3) The Board shall be a body corporate by the name aforesaid, having perpetual succession and a common seal, with power to acquire, hold and dispose of property, both movable and immovable, and to contract, and shall by the said name sue and be sued.

(4) The Board shall consist of the following members, namely.-

(a) a Chairperson who shall be an eminent person having adequate knowledge and experience in the conservation and sustainable use of **biological diversity** and in matters relating to equitable sharing of benefits, to be appointed by the State Government;

(b) not more than five *ex officio* members to be appointed by the State Government to represent the Departments of the State Government;

(c) not more than five members to be appointed from amongst experts in matters relating to conservation of **biological diversity**, sustainable use of **biological** resources and equitable sharing of benefits arising out of the use of **biological** resources.

(5) The head office of the State Biodiversity Board shall be at such place as the State Government may, by notification in the Official Gazette, specify.

23. Functions of State Biodiversity Board

The functions of the State Biodiversity Board shall be to,—

(a) advise the State Government, subject to any guidelines issued by the Central Government, on matters relating to the conservation of biodiversity, sustainable use of its components and equitable sharing of the benefits arising out of the utilisation of **biological** resources;

(b) regulate by granting of approvals or otherwise requests for commercial utilization or bio-survey and bio-utilisation of any **biological** resource by Indians;

(c) perform such other functions as may be necessary to carry out the provisions of this Act or as may be prescribed by the State Government.

24. Power of State Biodiversity Board to Restrict Certain Activities Violating The Objectives of Conservation, etc.

(1) Any citizen of India or a body corporate, organisation or association registered in India intending to undertake any activity referred to in section 7 shall give prior intimation in such form as may be prescribed by the State Government to the State Biodiversity Board.

(2) On receipt of an intimation under sub-section (1), the State Biodiversity Board may, in consultation with the local bodies concerned and after making such enquires as it may deem fit by order, prohibit or restrict any such activity if it is of opinion that such activity is detrimental or contrary to the objectives of conservation and sustainable use of biodiversity or equitable sharing of benefits arising out of such activity:

Provided that no such order shall be made without giving an opportunity of being heard to the person affected.

25. Provisions of Sections 9 to 17 to Apply With Modifications to State Biodiversity Board

The provisions of sections 9 to 17 shall apply to a State Biodiversity Board and shall have effect subject to the following modifications, namely.-

(a) references to the Central Government shall be construed as references to the State Government;

(b) references to the National Biodiversity Authority shall be construed as references to the State Biodiversity Board;

(c) reference to the Consolidated Fund of India shall be construed as reference to the Consolidated Fund of the State.

Chapter VII

Finance, Accounts And Audit Of National Biodiversity Authority

26. National Biodiversity Fund

The Central Government may, after due appropriation made by Parliament by law in this behalf, pay to the National Biodiversity Authority by way of grants or loans such sums of money as the Central Government may think fit for being utilised for the purposes of this Act,

27. Application of National Biodiversity Fund

(1) There shall be constituted a Fund to be called the National Biodiversity Fund and there shall be credited thereto—

(a) any grants and loans made to the National Biodiversity Authority under section 26,

(b) all charges and royalties received by the National Biodiversity Authority under this Act; and

(c) all sums received by the National Biodiversity Authority from such other sources as may be decided upon by the Central Government.

(2) The Fund shall be applied for-

(a) channelling benefits to the benefit claimers;

(b) conservation of **biological** resources and development of areas from where such **biological** resources or knowledge associated thereto has been accessed;

(c) socio-economic development of areas referred to in clause (b) in consultation with the local bodies concerned.

28. Annual Report of National Biodiversity Authority

The National Biodiversity Authority shall prepare, in such form and at such time each financial year as may be prescribed, its annual report, giving a full account of its activities during the previous financial year and furnish, to the Central Government, before such date as may be prescribed, its audited copy of accounts together with auditors' report thereon.

29. Budget, Accounts and Audit

(1) The National Biodiversity Authority shall prepare a budget, maintain proper accounts and other relevant records (including the accounts and other relevant records of the National Biodiversity Fund) and prepare an annual statement of account in such form as may be prescribed by the Central Government in consultation with the Comptroller and Auditor-General of India.

(2) The accounts of the National Biodiversity Authority shall be audited by the Comptroller and Auditor-General of India at such intervals as may be specified by him and any expenditure incurred in connection with such audit shall be payable by the National Biodiversity Authority to the Comptroller and Auditor-General of India.

(3) The Comptroller and Auditor-General of India and any other person appointed by him in connection with the audit of the accounts of the National Biodiversity Authority shall have the same rights and privileges and authority in connection with such audit as the Comptroller and Auditor-General generally has in connection with the audit of the

Government accounts and, in particular, shall have the right to demand the production of books, accounts, connected vouchers and other documents and papers and to inspect any of the offices of the National Biodiversity Authority.

(4) The accounts of the National Biodiversity Authority as certified by the Comptroller and Auditor-General of India or any other person appointed by him in this behalf together with the audit report thereon shall be forwarded annually to the Central Government.

30. Annual Report to Be Laid Before Parliament

The Central Government shall cause the annual report and auditors report to be laid, as soon as may be after they are received, before each House of Parliament.

Chapter VIII

Finance, Accounts And Audit Of State Biodiversity Board

31. Grant of Money By State Government to State Biodiversity Board

The State Government may, after due appropriation made by the State Legislature by law in this behalf, pay to the State Biodiversity Board by way of grants or loans such sums of money as the State Government may think fit for being utilized for the purposes of this Act.

32. State Biodiversity Fund

(1) There shall be constituted a Fund to be called the State Biodiversity Fund and there shall be credited thereto

(a) any grants and loans made to the State Biodiversity Board under section 31;

(b) any grants or loans made by the National Biodiversity Authority;

(c) all sums received by the State Biodiversity Board from such other sources as may be decided upon by the State Government.

(2) The State Biodiversity Fund shall be applied for –

(a) the management and conservation of heritage sites;

(b) compensating or rehabilitating any section of the people economically affected by restriction imposed under section 37;

(c) conservation of **biological** resources;

(d) socio-economic development of areas from where such **biological** resources or knowledge associated thereto has been accessed subject to any approval granted under section 24, in consultation with the local bodies concerned.,

(e) meeting the expenses incurred for purposes authorized by this Act.

33. Annual Report of State Biodiversity Board

The State Biodiversity Board shall prepare, in such form and at such time each financial year as may be prescribed, its annual report, giving a full account of its activities during the previous financial year, and submit a copy thereof to the State Government.

34. Audit of Accounts of State Biodiversity Board

The accounts of the State Biodiversity Board shall be maintained and audited in such manner as may, in consultation with the Accountant-General of the State, be prescribed and the National Biodiversity Authority shall furnish, to the State Government, before such date as may be prescribed, its audited copy of accounts together with auditors' report thereon.

35. Annual Report of State Biodiversity Board to Be Laid Before State Legislature

The State Government shall cause the annual report and auditor's report to be laid, as soon as may be after they are received, before the House of State Legislature.

Chapter IX

Duties of The Central and The State Governments

36. Central Government to develop National Strategies, Plans, etc. for Conservation, etc., of Biological Diversity.

(1) The Central Government shall develop national strategies, plans, programmes for the conservation and sustainable use of **biological diversity** including measures for identification and monitoring areas rich in **biological** resources, promotion of *in situ* conservation and *ex situ* conservation of **biological** resources, incentives for research, training, public education to increase awareness with respect to biodiversity.

(2) Where the Central Government has reason to believe that any area rich in **biological diversity, biological** resources and their habitats is being threatened by overuse, abuse or neglect, it shall issue directives to the concerned State Government any technical and other assistance that is possible to be provided or needed

(3) The Central Government shall, as far as practicable wherever it deems appropriate, integrate the conservation and sustainable use of **biological diversity** into relevant sectoral or cross-sectoral plans, programmes and policies.

(4) The Central Government shall undertake measures,—

(i) wherever necessary, for assessment of environmental impact of that project which is likely to have adverse effect on **biological diversity**, with a view to avoid or minimise such effects and where appropriate provide for public participation in such assessment,

(ii) to regulate, manage or control the risks associated with the use and release of living modified organisms resulting from biotechnology likely to have adverse impact on the conservation and sustainable use of **biological diversity** and human health.

(5) The Central Government shall endeavour to respect and protect the knowledge of local people relating to **biological diversity**, as recommended by the National Biodiversity Authority through such measures, which may include registration of such knowledge at the local, State or national levels, and other measures for protection, including *sui generis* system.

Explanation: For the purposes of this section,

(a) "*ex situ* conservation" means the conservation of components of **biological diversity** outside their natural habitats;

(b) "*in situ* conservation" means the conservation of ecosystems and natural habitats and the maintenance and recovery of viable populations of species in their natural surroundings and, in the case of domesticated or cultivated species, in the surroundings where they have developed their distinctive properties.

37. Biodiversity Heritage Sites

(1) Without prejudice to any other law for the time being in force, the State Government may from time to time in consultation with the local bodies, notify in the Official Gazette, areas of biodiversity importance as biodiversity heritage sites under this Act.

(2) The State Government, in consultation with the Central Government, may frame rules for the management and conservation of all the heritage sites.

(3) The State Government shall frame schemes for compensating or rehabilitating any person or section of people economically affected by such notification.

38. Power of Central Government to Notify Threatened Species

Without prejudice to the provisions of any other law for the time being in force, the Central Government may from time to time notify, any species which is on the verge ofextinction or likely to become extinct in the near future as a threatened species and prohibit or regulate collection thereof for any purpose and take appropriate steps to rehabilitate and preserve those species.

39. Power of Central Government to Designate Repositories

(1) The Central Government may in consultation with the National Biodiversity Authority, designate institutions as repositories under this Act for different categories of **biological** resources.

(2) The repositories shall keep in safe custody the **biological** material including voucher specimens deposited with them.

(3) Any new taxon discovered shall be notified to the repositories or any institution designated for this purpose and deposit the voucher specimens with such repository or institution.

*40. Power of Central Government to Exempt Certain **Biological** Resources.*

Notwithstanding anything contained in this Act, the Central Government may in consultation with the National Biodiversity Authority, by notification in the Official Gazette, declare that the provisions of this Act shall not apply to any items, including **biological** resources normally traded as commodities.

Chapter X

Biodiversity Management Committees

41. Constitution of Biodiversity Management Committees

(1) Every local body shall constitute a Biodiversity Management Committee within its area for the purpose of promoting conservation, sustainable use and documentation of **biological diversity** including preservation of habitats, conservation of land races, folk varieties and cultivars, domesticated stocks and breeds of animals and micro organisms and chronicling of knowledge relating to **biological diversity**.

Explanation : For the purposes of this sub-section,-

(a) "cultivar" means a variety of plant that has originated and persisted under cultivation or was specifically bred for the purpose of cultivation;

(b) "folk variety" means a cultivated variety that was developed, grown and exchanged informally among farmers;

(c) "landrace" means primitive cultivar that was grown by ancient farmers and their successors.

(2) The National Biodiversity Authority and the State Biodiversity Boards shall consult the Biodiversity Management Committees while taking any decision relating to the use of **biological** resources and knowledge associated with such resources occurring within the territorial jurisdiction of the Biodiversity Management Committee.

(3) The Biodiversity Management Committees may levy charges by way of collection fee from any person for accessing or collecting any **biological** resource from areas falling within its territorial jurisdiction.

Chapter XI

Local Biodiversity Fund

42. Grants to Local Biodiversity Fund

The State Government may, after due appropriation made by State Legislature by law in this behalf, pay to the Local Biodiversity Funds by way of grants or loans such sums of money as the State Government may think fit for being utilised for the purposes of this Act.

43. Constitution of Local Biodiversity Fund

(1) There shall be constituted a Fund to be called the Local Biodiversity Fund at every area notified by the State Government where any institution of self-government is functioning and there shall be credited thereto-

(a) any grants and loans made under section 42;

(b) any grants or loans made by the National Biodiversity Authority;

(c) any grants or loans made by the State Biodiversity Boards;

(d) fee referred to in sub-section (3) of section 41 received by the Biodiversity Management Committee;

(e) all sums received by the Local Biodiversity Fund from such other sources as may be decided upon by the State Government.

44. Application of Local Biodiversity Fund

(1) Subject to the provisions of sub-section (2), the management and the custody of the Local Biodiversity Fund and the purposes for which such Fund shall be applied as may be prescribed by the State Government.

(2) The Fund shall be used for conservation of biodiversity in the areas falling within the jurisdiction of the concerned local body and for the benefit of the community in so far such use is consistent with conservation of biodiversity.

45. Annual Report of Biodiversity Management Committees

The person holding the custody of the Local Biodiversity Fund shall prepare, in such form and during each financial year at such time as may be prescribed, its annual report, giving a full account of its activities during the previous financial year, and submit a copy thereof to the State Government.

46. Audit of Accounts of Biodiversity Management Committees

The accounts of the Local Biodiversity Fund shall be maintained and audited in such manner as may, in consultation with the Accountant-General of the State, be prescribed and the person holding the custody of the Local Biodiversity Fund shall furnish, to the State Government, before such date as may be prescribed, its audited copy of accounts together with auditors' report thereon.

47. Annual Report, etc. of Biodiversity Management Committee to be Submitted to District Magistrate

Every local body constituting a Biodiversity Management Committee under sub-section (l) of section 41, shall cause, the annual report and audited copy of accounts together with auditor's report thereon referred to in sections 45 and 46, respectively and relating to such Committee to be submitted to the District Magistrate having jurisdiction over the area of the local body.

Chapter XII

Miscellaneous

48. National Biodiversity Authority to be Bound by the Directions Given by Central Government

(1) Without prejudice to the foregoing provisions of this Act, the National Biodiversity Authority shall, in the discharge of its functions and duties

under this Act, be bound by such directions on questions of policy as the Central Government may give in writing to it from time to time.

Provided that the National Biodiversity Authority shall, as far as practicable, be given opportunity to express its views before any direction is given under this sub-section.

(2) The decision of the Central Government whether a question is one of policy or not shall be final.

49. Power of State Government to give directions

(1) Without prejudice to the foregoing provisions of this act, the State Biodiversity Board shall, in the discharge of its functions and duties under this Act, be bound by such directions on questions of policy as the State Government, may give in writing to it from time to time:

Provided that the State Biodiversity Board shall, as far as practicable, be given an opportunity to express its views before any direction is given under this sub-section.

(2) The decision of the State Government whether a question is one of policy or not shall be final.

50. Settlements of disputes between State Biodiversity Boards.

(1) If a dispute arises between the National Biodiversity Authority and a State Biodiversity Board, the said Authority or the Board, as the case may be, may prefer an appeal to the Central Government within such time as may be prescribed.

(2) Every appeal made under sub-section (1) shall be in such form as may be prescribed by the Central Government.

(3) The procedure for disposing of an appeal shall be such as may be prescribed by the Central Government:

Provided that before disposing of an appeal, the appellant shall he given a reasonable opportunity of being heard.

(4) If a dispute arises between the State Biodiversity Boards, the Central Government shall refer the same to the National Biodiversity Authority.

(5) While adjudicating any dispute under sub-section (4), the National Biodiversity Authority shall be guided by the principles of natural justice and shall follow such procedure as may be prescribed by the Central Government.

(6) The National Biodiversity Authority shall have, for the purposes of discharging its functions under this section, the same powers as are vested

in a civil court under the Code of Civil Procedure, 1908 (5 of 1908) in respect of the following matters, namely:

(a) summoning and enforcing the attendance of any person and examining him on oath;

(b) requiring the discovery and production of documents;

(c) receiving evidence on affidavits;

(d) issuing commissions for the examination of witnesses or documents;

(e) reviewing its decisions;

(f) dismissing an application for default or deciding it *ex parte;*

(g) setting aside any order of dismissal of any application for default or any order passed by it *ex parte;*

(h) any other matter which may be prescribed.

(7) Every proceeding before the National Biodiversity Authority shall be deemed to be a judicial proceeding within the meaning of sections 193 and 228 and for the purpose of section 196 (45 of 1860) of the Indian Penal Code and the National Biodiversity Authority shall be deemed to be a civil court for all the purposes of section 195 and Chapter XXVI of the Code of Criminal Procedure, 1973. (2 of 1974)

51. Members, Officers, etc., of National Biodiversity Authority and State Biodiversity Board deemed to be public servants.

All members, officers and other employees of the National Biodiversity Authority or the State Biodiversity Board shall be deemed, when acting or purporting to act in pursuance of any of the provisions of this Act, to be public servants within the meaning of section 21 of the Indian Penal Code (45 of 1860)

52. Appeal

Any person aggrieved by any determination of benefit sharing or order of the National Biodiversity Authority or a State Biodiversity Board under this Act, may file an appeal to the High Court within thirty days from the date of communication to him, of the determination or order of the National Biodiversity Authority or the State Biodiversity Board as the case maybe;

Provided that the High Court may, if it is satisfied that the appellant was prevented by sufficient cause from filing the appeal within the said period, allow it to be filed within a further period not exceeding sixty days.

53. Decree

Every determination of benefit sharing or order made by the National Biodiversity Authority or a State Biodiversity Board under this Act or the order made by the High Court in any appeal against any determination or order of the by the National Biodiversity Authority or a State Biodiversity Board shall, on a certificate issued by any officer of the National Biodiversity Authority or a State Biodiversity Board or the Registrar of the High Court, as the case may be, be deemed to be decree of the civil court and shall be executed in the same manner as a decree of that court.

Explanation - For the purposes of this section and section 52, the expression "State Biodiversity Board" includes the person or group of persons to whom the powers or functions under sub-section (2) of section 22 have been delegated under the proviso to that sub-section and the certificate relating to such person or group of persons under this section shall be issued by such person or group of persons, as the case may be.

54. Protection of action taken in good faith.

No suit, prosecution or other legal proceedings shall lie against the Central Government or the State Government or any officer of the Central Government or the State Government or any member, officer or employee of the National Biodiversity Authority or the State Biodiversity Board for anything which is in good faith done or intended to be done under this Act or the rules or regulations made thereunder.

55. Penalties

(1) Whoever contravenes or attempts to contravene or abets the contravention of the provisions of section 3, section 4, or section 6 shall be punishable with imprisonment for a term which may extend to five years, or with fine which may extend to ten lakh rupees and where the damage caused exceeds ten lakhs such fine may be commensurate with the damage caused, or with both,

(2) Whoever contravenes or attempts to contravene or abets the contravention of the provisions of section 7 or any order made under sub-section (1) of section 24 shall be punishable with imprisonment for a term which may extend to three years, or with fine which may extend to five lakh rupees, or with both.

56. Penalty for contravention of directions or orders of Central Government, State Government, National Biodiversity Authority and State Biodiversity Boards.

If a person contravenes any direction given or order made by the Central Government, the State Government, the National Biodiversity Authority or the State Biodiversity Board for which no punishment has been separately provided under this Act, he shall be punished with a fine which may extend to one lakh rupees and in case of a second or subsequent offence, with fine which may extend to two lakh rupees and in the case of continuous contravention with additional fine which may extend to two lakh rupees everyday during which the default continues.

57. Offences by companies

(1) Where an offence or contravention under this Act has been committed by a company, every person who at the time the offence or contravention was committed was in charge of, and was responsible to the company for the conduct of the business of the company, as well as the company, shall be deemed to be guilty of the offence or contravention and shall be liable to be proceeded against and punished accordingly:

Provided that nothing contained in this sub-section shall render any such person liable to any punishment provided in this Act, if he proves that the offence or contravention was committed without his knowledge or that he had exercised all due diligence to prevent the commission of such offence or contravention.

(2) Notwithstanding anything contained in sub-section (1), where an offence or contravention under this Act has been committed by a company and it is proved that the offence or contravention has been committed with the consent or connivance of, or is attributable to any neglect on the part of any director, manager, secretary or other officer of the company, such director, manager, secretary or other officer shall also be deemed to be guilty of the offence or contravention and shall be liable to be proceeded against and punished accordingly.

Explanation : For the purposes of this section,

(a) "company" means any body corporate and includes a firm or other association of individuals; and

(b) "director", in relation to a firm, means a partner in the firm.

58. Offences to be cognizable and non-bailable

The offences under this Act shall be cognizable and non-bailable.

59. Act to have effect in addition to other Acts

The provisions of this Act shall be in addition to, and not in derogation of, the provisions in any other law for the time being in force, relating to forests or wildlife.

60. Power of Central Government to give directions to State Government

The Central Government may give directions to any State Government as to the carrying into execution in the State of any of the provisions of this Act or of any rule or regulation or order made thereunder.

61. Cognizance of offences

No court shall take cognizance of any offence under this Act or rules and regulations made thereunder save on a complaint made by the National Biodiversity Authority or State Biodiversity Board, as the case may be.

62. Power of Central Government to make rules

(1) The Central Government may, by notification in the Official Gazette, make rules for carrying out the purposes of this Act.

(2) In particular, and without prejudice to the generality of the foregoing power, such rules may provide for all or any of the following matters, namely:—

(a) terms and conditions of service of the Chairperson and member under section 9;

(b) powers and duties of the Chairperson under section 10;

(c) procedure under sub-section (1) of section 12 in regard to transaction of business at meetings;

(d) form of application and payment of fees for undertaking certain activities under sub-section (1) of section 19;

(e) form of application and the manner for transfer of **biological** resource or knowledge under sub-section (2) of section 20;

(f) form in which, and the time of each financial year at which, the annual report shall be prepared under section 28;

(g) form in which the annual statement of account shall be prepared under section 29;

(h) the time within which and the form in which, an appeal may be preferred, the procedure for disposing of an appeal and the procedure for adjudication. under section 50;

(i) any other matter which is to be, or may be, prescribed, or in respect of which provision is to be made, by rules.

(3) Every rule made under this section and every regulation made under this Act shall be laid, as soon as may be after it is made, before each House of Parliament, while it is in session, for a total period of thirty days which may be comprised in one session or in two or more successive sessions, and if, before the expiry of the session immediately following the session or the successive sessions aforesaid, both Houses agree in making any modification in the rule or regulation or both Houses agree that the rule or regulation should not be made, the rule or regulation shall thereafter have effect only in such modifier form or be of no effect, as the case may be; so, however, that any such modification or annulment shall be without prejudice to the validity of anything previously done under that rule or regulation.

63. Power of State Government to make rules

(1) The State Government may, by notification in the Official Gazette, make rules for carrying out the purposes of this Act.

(2) In particular, and without prejudice to the generality of the foregoing power, such rules may provide for all or any of the following matter, namely:—

(a) the other functions to be performed by the State Biodiversity Board under clause(c) of section 23;

(b) the form in which the prior intimation shall be given under sub-section (1) of section 24;

(c) the form in which, and the time of each financial year at which, the annual report shall be prepared under section 33;

(d) the manner of maintaining and auditing the accounts of the State Biodiversity Board and the date before which its audited copy of the accounts together with auditor's report thereon shall be furnished under section 34;

(e) management and conservation of national heritage sites under section 37;

(f) the manner of management and custody of the Local Biodiversity Fund and the purposes for which such Fund shall be applied under subsection(1) of section 44;

(g) the form of annual report and the time at which such report shall be prepared during each financial year under section 45;

(i) any other matter which is to be, or may be, specified.

(3) Every rule made by the State Government under this section shall be laid, as soon as may be after it is made, before each House of the State Legislature where it consists of two Houses, or where such Legislature consists of one House, before that House.

64. Power to make regulations

The National Biodiversity Authority shall with the previous approval of the Central Government by notification in the Official Gazette, make regulations for carrying out the purposes of this Act.

65. Power to remove difficulties

(1) If any difficulty arises in giving effect to the provisions of this Act, the Central Government may, by order, not inconsistent with the provisions of this Act, remove the difficulty:

Provided that no such order shall be made after the expiry of a period of two years from the commencement of this Act.

(2) Every order made under this section shall be laid, as soon as may be after it is made, before each House of Parliament.

Biological Diversity Rules-2004

Ministry of Environment and Forests Notification

New Delhi, the 15th April 2004

G.S.R. 261(E) - In exercise of the powers conferred by Section 62 of the Biological Diversity Act, 2002, and in super session of the National Biodiversity Authority (salary, allowances and conditions of service of Chairperson and other Members) Rules, 2003, except as respect to things done or omitted to be done before super session, the Central Government hereby makes the following rules namely: -

1. Short title and commencement

(1) These rules maybe called the Biological Diversity Rules, 2004.

(2) Thus shall come into force on 15th April, 2004.

2. Definitions

In these rules, unless the context otherwise requires,

(a) "Act" means the Biological Diversity Act 2002 (18 of 2003);

(b) "Authority" means the National Biodiversity Authority established under sub-section (1) of Section 8;

(c) "Biodiversity Management Committee" means a Biodiversity Management Committee established by a local body under subsection (1) of Section 41;

(d) "Chairperson" means the chairperson of the National Biodiversity Authority or as the case maybe, of the State Biodiversity Board;

(e) "fee" means any fee stipulated in the schedule;

(f) "Form" means from annexed to these rules;

(g) "Member" means a member of the National Biodiversity Authority or a State Biodiversity Board and includes the chairperson as the case may be;

(h) "section" means a section of the Act;

(i) "Secretary" means the full time Secretary of the Authority;

(j) words and expressions used but not defined in these rules and defined in the Act shall have the meaning respectively assigned to them in the Act;

3. Manner of Selection and Appointment of the Chairperson

(1) The Chairperson of the Authority shall be appointed by the Central Government.

(2) Every appointment of Chairperson under sub-section (1) shall be made either on deputation basis or by selection from outside the Central Government. In case the appointment is through deputation, the applicant should not be below the rank of Additional Secretary to the Government of India.

4. Term of Office of the Chairperson

(1) The chairperson of the Authority shall hold the office for a term of three years' and shall be eligible for re-appointment.

(2) Provided that no Chairperson shall hold office as such after he attains the age of sixty-five years or his term of office expires which is earlier.

(3) The Chairperson may resign form his office by giving at least one month notice in writing to the Central Government.

5. Pay and Allowances of Chairperson

(1) A chairperson shall be entitled to a fixed pay of Rs. 26000/- per month. In case of retired person is appointed as Chairperson, his pay shall be fixed in accordance with the orders of the Central Government as applicable to such persons.

(2) A Chairperson shall be entitled to such allowances, leave, pension, provident fund, house and other perquisites, etc. to be decided by the Central Government from time to time.

6. Term of Office and Allowances of Non-official Members

(1) Every non-official member of the Authority shall hold his office for a term not exceeding three years at a time form the date of publication of his appointment in the Official Gazette.

(2) Every non-official member attending the meeting of the Authority shall be entitled to sitting allowance, travelling expenses, daily allowance and such other allowances as are applicable to non-official member of commissions and committees of the Central Government attending the meeting(s) of such Commissions or Committees.

7. Filling up Vacancies of Non-Official Members

(1) A non-official member of the Authority may resign his office at any time by giving in writing under his hand addressed to the Central government and the seat of that member in the Authority shall become vacant.

(2) A casual vacancy of a non-official member in the Authority shall be filled up by a fresh nomination and the person nominated to fill the vacancy shall hold office only for the remainder of the term of the member in whose place he was nominated.

8. Removal of the Members of the Authority

No member of the Authority shall be removed from his office on any ground specified in section 11, without a due and proper enquiry by an officer not below the rank of a Secretary to the Government of India appointed by the Central government and without giving such member a reasonable opportunity of being heard.

9. Secretary of the Authority

(1) The Authority shall appoint a Secretary to it.

(2) The terms and conditions of the appointment of the Secretary shall be determined by the Authority by regulation.

(3) The Secretary shall be responsible for coordinating and convening the meetings of the Authority, maintenance of the records of the proceedings of the Authority and such other matters as may be assigned to him by the Authority.

10. Meetings of the Authority

(1) The Authority shall meet at least four times in a year normally after a period of three months at the Head quarters of the Authority or at such place as may be decided by the Chairperson.

(2) The Chairperson shall, upon a written request from not less than five Members of the Authority or upon a direction of the Central Government, call a special meeting of the Authority.

(3) The members shall be given at least fifteen days' notice for holding an ordinary meeting and at least three days' notice for holding a special meeting specifying the purpose, the time and the place at which such meeting is to be held.

(4) Every meeting shall be presided over by the Chairperson and in his absence, by a presiding officer to be elected by the members present from amongst themselves.

(5) The decision of the Authority at a meeting shall, if necessary, be taken by a simple majority of the Members present and voting and the Chairperson or in his absence, the Member presiding shall have a second or casting vote.

(6) Each member shall have one vote.

(7) The quorum at every meeting of the Authority shall be five.

(8) No Member shall be entitled to bring forward for the consideration of a meeting any matter of which he has not given ten days' notice unless the Chairperson in his discretion permits him to do so.

(9) Notice of the meeting or sending it by registered post to his last known place of residence or business or in such other manner as the Secretary of the Authority may, in the circumstances of the case, think fit.

11. Appointment of Expert Committee by the Authority and Their Entitlements

(1) the Authority may constitute any number of Committees for such purposes as it may deem fit consisting wholly of members or wholly of other persons or partly of members or partly of other persons.

(2) The members of the Committee other than the members of the Authority shall be paid such fees and allowances for attending the meetings as the Authority may deem fit.

12. General Functions of the Authority

The Authority may perform the following functions, namely:

(i) lay down the procedure and guidelines to govern the activities provided under sections 3, 4, and 6;

(ii) advise the Central Government on any matter concerning conservation of biodiversity, sustainable use of its components and fair and equitable sharing of benefits arising out of the use of biological resource and knowledge;

(iii) coordinate the activities of the State Bio-diversity Boards;

(iv) provide technical assistance and guidance to the State Bio-diversity Boards;

(v) commission studies and sponsor investigation and research;

(vi) engage consultants, for a specific period, not exceeding three years, for providing technical assistance to the Authority in the effective discharge of its functions;

Provided that if it is necessary and expedient to engage any consultant beyond the period of three years, the Authority shall seek prior approval of the Central Government for such an engagement;

(vii) collect, compile and publish technical and statistical data, manuals, codes or guides relating to conservation of bio-diversity, sustainable use of its components and fair and equitable sharing of benefits arising out of the use of biological resource and knowledge;

(viii) organize through mass media a comprehensive programme regarding conservation of bio-diversity, sustainable use of its components and fair and equitable sharing of benefits arising out of the use of biological resource and knowledge;

(ix) plan and organize training or personnel engaged or likely to be engaged in programmes for the conservation of bio-diversity and sustainable use of its components;

(x) prepare the annual Budget of the Authority incorporating its own receipts as also the devaluation from the Central Government provided that the allocation by the Central government shall be operated in accordance with the budget provisions approved by the Central Government;

(xi) recommend creation of posts to the Central Government, for effective discharge of the functions by the Authority and to create such posts, provided that no such post whether permanent/temporary or of any nature, would be created without prior approval of the Central Government;

(xii) approve the method of recruitment to the officers and servants of the Authority;

(xiii) take steps to build up database and to create information and documentation system for biological resources and associated traditional knowledge through bio-diversity registers and electronics databases to ensure effective management, promotion and sustainable uses;

(xiv) give directions to State Bio-diversity Boards and the Bio-diversity Management Committees in writing for effective implementation of the Act;

(xv) report to the Central Government about the functioning of the Authority and implementation of the Act;

(xvi) recommend, modify, collection of benefit sharing fee under sub section (1) of Section 6 or Changes of royalties under sub-section (2) of section 19 in respect of biological resources form time to time;

(xvii) sanction grants-in-aid and grants to the State Bio-diversity Board and Biodiversity Management Committees for specific purposes;

(xviii) undertake physical inspection of any area in connection with the implementation of the Act;

(xix) take necessary measures including appointment of legal experts to oppose grant of intellectual property right in any country outside India on any biological resource and associated knowledge obtained from India in an illegal manner;

(xx) do such other functions as may be assigned or directed by the Central Government from time to time.

13. Powers and Duties of Chairperson

(1) The Chairperson shall have the overall control of the day-to-day activities of the Authority.

(2) Subject to the provisions of Section 10, the Chairperson shall have the powers of general superintendence over the officers and staff of the Authority and he may issue necessary directions for the conduct and management of the affairs of the Authority.

(3) The Chairperson shall be in charge of all the confidential papers and records of the Authority and shall be responsible for their safe custody.

(4) All orders and instructions to be issued by the Authority shall be under the signature of the Chairperson or of any other officer authorized by the Chairperson in this behalf.

(5) The Chairperson, either himself or through an officer of the authority authorized for the purpose, may sanction and disburse all payments against the approved budget.

(6) The Chairperson shall have full powers for granting administrative and technical sanction to all estimates.

(7) The Chairperson shall convene and preside over all the meetings of the Authority and shall ensure that all decisions taken by the Authority are implemented in proper manner.

(8) The Chairperson shall exercise such other powers and perform such other functions as may be delegated to him from time to time by the Authority or the Central Government.

14. Procedure for Access to Biological Resources and Associated Traditional Knowledge

(1) Any person seeking approval of the Authority for access to biological resources and associated knowledge for research or for commercial utilization shall make an application in Form 1.

(2) Every application under sub-rule (1) shall be accompanied by a fee of ten thousand rupees in the form of a cheque or demand draft drawn in favour of the Authority.

(3) The Authority shall after consultation with the concerned local bodies and collecting such additional information from the applicant and other sources, as it may deem necessary, dispose of the application, as far as possible, within a period of six months from the date of its receipts.

(4) On being satisfied with the merit of the application, the Authority may grant the approval for access to biological resources and associated knowledge subject to such term and conditions as it may deem fit to impose.

(5) The approval to access shall be in the form of a written agreement duly signed by the authorized officer of the Authority and the applicant.

(6) The form of the agreement referred to in sub-rule (5) shall be laid down by the Authority and shall include the following; namely:

(i) general objectives and purpose of the application for seeking approval;

(ii) description of the biological resources and traditional knowledge including accompanying information;

(iii) intended uses of the biological resources (research, breeding, commercial utilization, etc.);

(iv) conditions under which the applicant may seek intellectual property rights;

(v) quantum of monetary and other incidental benefits. If need be, a commitment to enter into a fresh agreement particularly in case if the biological material is taken for research purposes, and also in case of any other change in use thereof subsequently;

(vi) restriction to transfer the accessed biological resources and the traditional knowledge to any third party without prior approval of Authority;

(vii) to adhere to a limit set by the Authority on the quantity and specification of the quality of the biological resources for which the applicant is seeking access;

(viii) guarantee to deposit a reference sample of the biological material sought to be accessed with the repositories identified in Section 39;

(ix) submitting to the Authority a regular status report of research and other developments;

(x) commitment to abide with the provisions of Act and rules and other related legislations in force in the country;

(xi) commitment to facilitate measures for conservation and sustainable use of biological resources accessed;

(xii) commitment to minimize environmental impacts of collecting activities;

(xiii) legal provisions such as duration of the agreement, notice to terminated the agreement, independent enforceability of individual clauses, provision to the extent that obligations in benefit sharing clauses survive the termination of the agreement, events limiting liability (natural calamities), arbitration, any confidentiality clause.

(7) The conditions for access may specifically provide measures for conservation and protection of biological resources to which the access is being granted.

(8) The Authority may for reasons to be recorded in writing reject as application it considers that the request cannot be acceded to.

(9) No application shall be rejected unless the applicant is given a reasonable opportunity of being heard.

(10) The Authority shall take steps to widely publicize the approvals granted, through print or electronic media and shall periodically monitor compliance of conditions on which the approval was accorded.

15. Revocation of Access or Approval

(1) The Authority may either on the basis of any complaint or *suo moto* withdraw the approval granted for access under rule 15 and revoke the written agreement under the following conditions, namely:

(i) on the basis of reasonable belief that the person to whom the approval was granted has violated any of the provisions of the Act or the condition on which the approval was granted;

(ii) when the person who has been granted approval has failed to comply with the terms of the agreement;

(iii) on failure to comply with any of the conditions of access granted;

(iv) on account of overriding public interest or for protection of environment and conservation of biological diversity.

(2) The Authority shall send a copy of every order of revocation issued by it to the concerned State Biodiversity Board and the Biodiversity Management Committees for prohibiting the access and also to assess the damage, if any caused and take steps to recover the damage.

16. Restriction on Activities Related to Access to Biological Resources

(1) The Authority if it deems necessary and appropriate shall take the steps to restrict or prohibit the request for access to biological resources for the following reasons; namely:

(i) the request for access is for any endangered taxa;

(ii) the request for access is for any endemic and rare species;

(iii) the request for access may likely to result in adverse effect on the livelihoods of the local people;

(iv) the request to access may result in adverse environmental impact which may be difficult to control and mitigate;

(v) the request for access may cause genetic erosion or affecting the ecosystem function;

(vi) use of resources for purposes contrary to national interest and other related international agreements entered into by India.

17. Procedure for Seeking Approval for Transferring Results of Research

(1) Any person desirous of transferring results of research relating to biological resources obtained from India for monetary consideration to foreign nationals, companies and Non Resident Indians (NRIs), shall make an application to the Authority in the Form II.

(2) Every application under sub-rule (i) shall be accompanied by a fee of five thousand rupees in the form of a Bank draft or Cheque drawn in favour of the Authority.

(3) Every application under sub-rule (i) shall be decided upon by the Authority as far as possible within a period of three months from the receipt of the same.

(4) On being satisfied that the applicant has fulfilled all the requirements, the Authority may grant the approval for transferring the results of research subject to such terms and conditions as it may deem fit to impose in each case.

(5) The approval for transfer shall be granted in the form of a written agreement duly signed by an authorized officer of the Authority and the applicant. The form of the agreement shall be such as may be decided by the Authority.

(6) The Authority may for reasons to be recorded in writing reject an application if it considers that the application cannot be allowed; provided that the application shall be rejected unless the applicant has been given a reasonable opportunity of being heard.

18. Procedure for Seeking Prior Approval Before Applying for Intellectual Property Protection

(1) Any person desirous of applying for a patent or any other intellectual property based on research on biological material and knowledge obtained from India shall make an application Form III.

(2) Every application under sub-rule (1) shall be accompanied by paying a fee of five hundred rupees.

(3) The Authority after due appraisal of the application and after collecting any additional information, on the basis of merit shall decide on the application, as far as possible within a period of three months or receipt of the same.

(4) On being satisfied that the applicant has fulfilled all the necessary requirements, the Authority may grant approval for applying for a patent or any other IPR subject to such terms and conditions as it may deem fit to impose in each case.

(5) The approval shall be granted in the form of a written agreement duly signed by an authorized officer of the Authority and the applicant. The form of the agreement may be decided by the Authority.

(6) The Authority may reject the application if it considers that the request cannot be acceded to after recording the reasons. Before passing order of rejection, the applicant shall be given an opportunity of hearing.

19. Procedure for Third Party Transfer Under Sub-Section (2) of Section 20

(1) The persons who have been granted approval for access to biological resources and associated knowledge, intend to transfer the accessed biological resource or knowledge to any other person or organization shall make an application to the Authority in From IV.

(2) Every application under sub-rule (1) shall be accompanied by a fee of ten thousand rupees in the form of Bank draft or cheque drawn in favour of the Authority.

(3) The Authority shall after collecting any additional information, decide upon the application as far as possible within a period of six months of receipt of the same.

(4) On being satisfied that the applicant has fulfilled all the necessary requirements, the Authority may grant approval for third party transfer subject to such terms and conditions it may deem fit to impose in each case.

(5) The approval as may be granted under sub-rule (4) in the form of a written agreement duly signed by an authorized officer of the Authority and the applicant. The form of the agreement shall be such as may be decided by the Authority.

(6) The Authority may for reasons to be recorded in writing reject the application if it considers that the request cannot be acceded to, provided that no application shall be rejected unless the applicant has been given an opportunity of being heard.

20. Criteria for Equitable Benefit Sharing (Section 21)

(1) The Authority shall by notification in the Official Gazette formulate the guidelines and describe the benefit sharing formula.

(2) The guidelines shall provide for monetary and other benefits such as royalty, joint ventures, technology transfer, product development, education and awareness raising activities, institutional capacity building and venture capital fund.

(3) The formula for benefit sharing shall be determined on a case-by-case basis.

(4) The Authority while granting approval to any person for access or for transfer of results of research or applying for patent and IPR or for third party transfer of the accessed biological resource and associated knowledge may impose terms and conditions for ensuring equitable sharing of the benefits arising out of the use of accessed biological material and associated knowledge.

(5) The quantum of benefits shall be mutually agreed upon between the persons applying for such approval and the Authority in consultation with the local bodies and benefit claimers and may be decided in due regard to the defined parameters of access, the extent of use, the sustainability aspect, impact and expected outcome levels, including measures ensuring conservation and sustainable use of biological diversity.

(6) Depending upon each case, the Authority shall stipulate the time frame for assessing benefit sharing on short, medium, and long term benefits.

(7) The Authority shall stipulate that benefits shall ensure conservation and sustainable use of biological diversity.

(8) Where biological resources or knowledge is accessed form a specific individual or a group of individuals or organizations, the Authority may take steps to ensure that the agreed amount is paid directly to them through the district administration. Where such individuals or group of individuals or organizations cannot be identified, the monetary benefits shall be deposited in the National Biodiversity Fund.

(9) Five percent of the assessed benefits shall be earmarked for the Authority or Board as the case may be, towards the administrative and service charges.

(10) The Authority shall monitor the flow of benefits as determined under sub rule (4) in a manner determined by it.

21. Application of National Biodiversity Fund

(1) The National Biodiversity Fund shall be operated by the Chairperson or by such other officer of the Authority as may be authorized in this regard.

(2) The National Biodiversity Fund shall have two separate heads of accounts, one relating to the receipts from the Central Government and the other concerning the fee, license fee, royalty and other receipts of the Authority.

22. Constitution of Biodiversity Management Committees

(1) Every local body shall constitute a Biodiveristy Management Committee (BMC's within it's area of jurisdiction.

(2) The Biodiversity Management Committee as constituted under sub-rule (1) shall consist of a Chairperson and not more than six persons nominated by the local body, of whom not less than one third should be women and not less than 18% should belong to the Scheduled Castes/Scheduled Tribes.

(3) The Chairperson of the Biodiversity Management Committee shall be elected from amongst the members of the committee in a meeting to be chaired by the Chairperson of the local body. The Chairperson of the local body shall have the casting votes in case of a tie.

(4) The Chairperson of the Biodiversity Management Committee shall have a tenure of three years.

(5) The local Member of Legislative Assembly/Member of Legislative Council and Member of Parliament would be special invitees to the meetings of the Committee.

(6) The main function of the BMC is to prepare People's Biodiversity Register in consultation with local people. The Register shall contain comprehensive information on availability and knowledge of local biological resources, their medicinal or any other use or any other traditional knowledge associated with them.

(7) The other functions of the BMC are to advise on any matter referred to it by the State Biodiversity Board or Authority for granting approval, to maintain data about the local vaids and practitioners using the biological resources.

(8) The Authority shall take steps to specify the form of the People's Biodiversity Registers, and the particulars it shall contain and the format for electronic database.

(9) The Authority and the State Biodiversity Boards shall provide guidance and technical support to the Biodiversity Management Committees for preparing People's Biodiversity Registers.

(10) The People's Biodiversity Registers shall be maintained and validated by the Biodiversity Management Committee.

(11) The Committee shall also maintain a Register giving information about the details of the access to biological resources and traditional knowledge granted, details of the collection fee imposed and details of the benefits derived and the mode of their sharing.

23. Appeal for Settlement of Disputes Under Section 50

(1) If a dispute arises between the Authority or a State Biodiversity Board or between one Board and other Board(s) on account of implementation of any order or direction or on any issue of policy decision, either of the aggrieved parties i.e, Authority or that Board, as the case may be, prefer an appeal to the Central Government under section 50, in Form V to the Secretary, Ministry of Environment and Forests, Government of India.

(2) In case the dispute arises between a State Biodiversity Board and another State Biodiversity Board or Boards, the aggrieved Board or Boards, shall prefer the point or points of dispute to the Central Government which shall refer the same to the Authority.

(3) The memorandum of appeal shall state the facts of the case, the grounds relied upon by the appellant, for preferring the appeal and the relief sought for.

(4) The memorandum of appeal shall be accompanied by an authenticated copy of the order, direction or policy decision, as the case may be, by which the appellant is aggrieved and shall be duly signed by the authorized representative of the appellant.

(5) The memorandum of appeal shall be submitted in quadruplicate, either in person or through a registered post with acknowledgement due, within 30 days from the date of the orders, direction or policy decision, impugned provided that if the Central Government is satisfied that there was good and sufficient reason for the delay in preferring the appeal, it may, for reason

to be recorded in writing, allow the appeal to be preferred after the expiry of the aforesaid period of 30 days but before the expiry of 45 days form the date of the orders impugned, direction or policy decision, as the case may be.

(6) The notice for hearing of the appeal shall be given in Form VI by a registered post with an acknowledgement due.

(7) The Central Government shall, after hearing the appellant and the other parties, dispose of the appeal.

(8) In disposing of an appeal, it may vary or modify or cancel impugned order, direction or policy, as the case may be.

(9) In adjudicating a dispute, the Authority shall be guided by the principles or natural justice and as far as practicable, follow the same procedures which the Central Government is required to follow under this rule.

24. Manner of Giving Notice Under Section 61

(1) The manner of giving notice, under clause (b) of section 61, shall be as follows namely:

(i) The notice shall be in writing in Form VII.

(ii) The person giving the notice may send it to:

(a) If the alleged offense has taken place in a Union territory, to the Chairperson of the National Biodiversity Authority; and

(b) If the alleged offence has been taken place in a State, to the Chairperson of the State Biodiversity Board.

(2) The notice referred to in sub-rule (1) shall be sent by registered post acknowledgement due; and

(3) The period of thirty days mentioned in clause (b) of section 61 shall be reckoned from the date, the notice is received by the Authorities mentioned in sub-rule (1)

FORM I

(See rule 14)

Application Form for Access to Biological Resources and Associated Traditional Knowledge

Part A

(i) Full particulars of the applicant

(ii) Name

(iii) Permanent address

(iv) Address of the contact person/agent, if any, in India

(v) Profile of the organization (personal profile in case the applicant is an individual). Please attach relevant documents of authentication):

(vi) Nature of business

(vii) Turnover of the organization in US$

2. Details and specific information about nature of access sought and biological material and associated knowledge to be accessed

   a) Identification (scientific name) of biological resources and its traditional use:

   b) Geographical location of proposed collection:

   c) Description / nature of traditional knowledge:

   d) Any identified individual/community holding the traditional knowledge:

   e) Quantity of biological resources to be collected (give the schedule):

   f) Time span in which the biological resources is proposed to be collected:

   g) Name and number of persons authorized by the company for making the selection:

   h) The purpose for which the access is requested including the type and extent of research, commercial use being derived and expected to be derived from it:

   i) Whether any collection of the resource endangers any component of biological diversity and the risks which may arise from the access:

3. Details of any national institution which will participate in the Research and Development activities.

4. Primary destination of accessed resource and identity of the location where the R&D will be carried out.

5. The economic and other benefits including those arriving out of any IPR, patent obtained out of accessed biological resources and knowledge that are intended, or may accrue to the applicant or to the country that he/she belongs.

6. The biotechnological, scientific, social or any other benefits obtained out of accessed biological resources and knowledge that are intended, or may accrue to the applicant or to the country that he/she belongs.

7. Estimation of benefits that would flow to India/communities arising out of the use of accessed bio-resources and traditional knowledge.

8. Proposed mechanism and arrangements for benefit sharing.

9. Any other information considered relevant.

PART B

Declaration

I/We declare that:

- ❐ Collection of proposed biological resources shall not adversely affect the sustainability of the resources;
- ❐ Collection of proposed biological resources shall not entail any environmental impact;
- ❐ Collection of proposed biological resources shall not pose any risk to ecosystems;
- ❐ Collection of proposed biological resources shall not adversely affect the local communities;

I/We further declare the Information provided in the application form is true and correct and

I/we shall be responsible for any incorrect/wrong information.

Signed

Name

Title

Place

Date

FORM II

(See rule 17)

Application for Seeking Prior Approval of National Biodiversity Authority for Transferring the Results or Research to Foreign Nationals, Companies. NRI's, for Commercial Purposes.

1. Full particulars of the applicant

   i) Name

   ii) Address

   iii) Professional profile

   iv) Organizational affiliation (Please attach relevant documents of authentication

2. Details of the results of research conducted
3. Details of the biological resources and/or associated knowledge used in the research
4. Geographical location from where the biological resources used in the research are collected.

5. Details of any traditional knowledge used in the research and any identified individual/community holding the traditional knowledge.
6. Details of institution where R&D activities carried out.
7. Details of the individual/organization to whom the research results are intend to transfer.
8. Details of economic, biotechnological, scientific or any other benefits that are intended, or may accrue to the individual/ organization due to commercialization of transferred research results.
9. Details of economic, biotechnological, scientific or any other benefits that are intended, or may accrue to the applicant seeking approval for transfer of results or research.
10. Details of any agreement or MOU between by the proposed recipient and applicant seeking approval for transfer of results of research.

Declaration

I/We declare the Information provided in the application form is true and correct and I/we shall be responsible for any incorrect/wrong information.

Signed

Name

Title

Place

Date

FROM III

(See rule 18)

Application for Seeking Prior Approval of National Biodiversity Authority for Applying for Intellectual Property Right

1. Full particulars of the applicant

   i) Name

   ii) Address

   iii) Professional profile

   iv) Organizational affiliation (Please attach relevant documents of authentication)
2. Details of the invention on which IPRs sought
3. Details of the biological resources and/or associated knowledge used in the invention

4. Geographical location from where the biological resources used in the invention are collected
5. Details of any traditional knowledge used in the invention and any identified individual/community holding the traditional knowledge
6. Details of institution where Research and Development activities carried out.
7. Details of economic, biotechnological, scientific or any other benefits that are intended, or may accrue to the applicant due commercialization of the invention.

Declaration

I/We declare the Information provided in the application form is true and correct and I/we shall be responsible for any incorrect/wrong information.

Signed
Name
Title
Place
Date

FORM IV

(See rule 19)

Application form for Seeking Approval of National Biodiversity Authority for Third Party Transfer of the Accessed Biological Resources and Associated Traditional Knowledge.

1. Full particulars of the applicant

   (i) Name

   (ii) Address

   (iii) Professional profile

   (iv) Organizational affiliation (Please attach relevant documents of authentication

2. Details of the biological material and traditional knowledge accessed.
3. Details of the access contract entered (Copy to be enclosed).
4. Details of benefits and mechanism/arrangements for benefit sharing already implemented.
5. Full particulars of the third party to whom the accessed material/knowledge is intended to transfer.

6. The purpose of the intended third party transfer.
7. Details of economic, social, biotechnological, scientific or any other benefits that are intended, or may accrue to the third party due to transfer of accessed biological material and knowledge.
8. Details of any agreement to be entered between the applicant and the third party.
9. Estimation of benefits that would flow to India/communities arising out of the third party transfer of accessed biological resources and traditional knowledge.
10. Proposed mechanism and arrangements for benefit sharing arising out of the proposed third party transfer.
11. Any other relevant information.

Declaration

I/We declare the Information provided in the application form is true and correct and

I/we shall be responsible for any incorrect/wrong information.

Signed

Name

Title

Place

Date

FORM V

(See rule-23(1)

Form of Memorandum of Appeal

BEFORE THE ________________ MINISTRY OF ENVIRONMENT AND FORESTS, NEW DELHI

OR

NATIONAL BIODIVERSITY AUTHORITY

(as the case may be)

(Memorandum of appeal under Section 50 of the Biological Diversity Act, 2002)

Appeal No. .... **of 200**

....

....Appellant(s)

Vs.

....

....Respondent(s)

(here mention the designation of the Authority/Board, as the case may be)

The appellant begs to prefer this Memorandum of Appeal against the order dated ________________ passed by the Respondent on the following facts and grounds.

1. FACTS:

(Here briefly mention the facts of the case):

2. GROUND:

(Here mention the grounds on which the appeal is made):

i.)

ii.)

iii.)

3. RELIEF SOUGHT:

i.)

ii.)

iii.)

4. PRAYER:

(a) In the light of what is stated above, the appellant respectfully prays that the order/ decision of the respondent be quashed/set-aside

(b) The policy/guidelines/regulation framed by the Respondent be quashed/modified/ annulled to the extent ___________________

(c) ______________________________________

Signature of the appellant with Seal

Place: ________________ Address:

Dated: ________________

VERIFICATION

I, the appellant do hereby declare that what is stated above is true to the best of my information and belief.

Verified on __________ day of ________________

Signature of the appellant with Seal

Address

Signature of the Authorized representative of the appellant

Enclosures: 1. Authenticated copy of the order/direction/policy decision against which the the appeal has been preferred.

FORM VI

(See rule 28)

BEFORE THE ______________________ MINISTRY OF ENVIRONMENT AND FORESTS, NEW DELHI

OR

NATIONAL BIODIVERSITY AUTHORITY

(as the case may be)

Appeal No. ________________ of 200

Between:

____

______________________________ .....Appellant(s)

Vs.

____

______________________________ .....Respondent(s)

NOTICE

Please take note that the above appeal filed by the appellant, against the order/direction/policy decision (give details) is fixed for hearing on ____________________

at _______________________.

The copies of the appeal memorandum and other annexure filed along with the appeal are sent herewith for your reference.

Please note that if you fail to appear on the said date or other subsequent date of hearing of the appeal, the appeal would be disposed of finally by placing you ex-parte.

Authorized signatory on behalf of the Appellate Authority (Seal)

Date: ________________

Place: ________________

FORM VII

FORM OF NOTICE

(See rule 24(1))

By Registered Post/Acknowledgement due

From,

Sri ______________

____

____

To,

____

____

____

Sub: NOTICE UNDER SECTION 61(b) OF THE BIOLOGICAL DIVERSITY ACT, 2002.

1. Whereas an offence under the Biological Diversity Act, 2002 has been committed/is being committed by ____________________.

2. I/We hereby give notice of 30 days under Section 61(b) of the Biological Diversity Act, 2002 of my/our intention to file a complaint in the Court against __________ for violation of the provisions of the Biological Diversity Act, 2002.

3. In support of my/our notice, I am/we are enclosing herewith the following documents as evidence of proof.

Place: ______________

Dated: ______________

Signature

Explanation

1) In case the notice to be given in the name of a company, documentary evidence authorizing the person to sign the notice on behalf of the company shall be enclosed to the notice.
2) Give the name and address of the alleged offender. In case of using biological resource/knowledge/research/bio-survey and bio utilization the intellectual property right/patent, without the approval of the Authority, the details thereof and the commercial utilization if any, may be furnished.
3) Documentary evidence shall include photograph, technical report, etc. for enabling enquiry into the alleged violation/offence. [No. J-22018/57/2002-CSC (BC)]

References

Anon. 1995. Legal issues in plant germplasm collecting. *In* L. Guarino, V. Ramanatha Rao, R. Reid. (eds.). Collecting Plant Genetic Diversity : Technical Guidelines. CAB International. pp 13-29.

http://www.biodiv.org/default.aspx

http://www.envfor.nic.in/divisions/biodiv/act/bio_div_act_2002.pdf

http://www.fao.org/ag/cgrfa/itpgr.htm

Section - II

Seed Enterprises

20

Seed Industry in India

Seed is the most important component for productive agriculture. Significant advances have been made in production of food crops in India during the last four decades. The quantum jump in the food production primarily accounted for by increased productivity and production in rice and wheat was the result of introduction of improved varieties and their seeds often nicknamed as Green Revolution. The food production which was nearly 50 million tones during 1950 – 51 reached to a level of 210 million tones during 2003 – 04. During the same period the productivity increased about 3 folds (Table 20.1).

Table 20.1. Trends in total food production in India

Years	Area (million ha)	Production (million tonnes)	Productivity (kg/ha)
1950-1951	97.30	50.80	522
1960-1961	115.60	82.00	710
1970-1971	124.30	108.40	872
1980-1981	126.70	130.00	1023
1990-1991	127.80	176.00	1380
1994-1995	123.60	191.10	1547
1997-1998	123.85	199.93	1601
1998-1999	125.17	203.04	1627
1999-2000	123.10	209.00	1704
2000-2001	121.05	196.81	1626
2001-2002	121.91	212.03	1739
2002-2003	113.13	182.57	1614
2003-04	124.20	210.80	1697
2004-05	124.00	206.50	1655

The role of seed sector has been substantial towards the increasing production and productivity. As a matter of fact the substantial upward trend in production of food grains in India occurred in 1990 – 91 and since then its continued upward trend stagnated at about a level of 210 millions tones at present. Table 20.2 on status of quality seed availability and distribution in India shows an upward distinct peak in seed distribution during 1990 – 91.

Table 20.2. Status of quality seed availability and distribution in India *(in lac q)*

Years	Quantity Distributed	Availability
1953-1954	1.83	NA
1960-1961	1.46	NA
1970-1971	5.16	NA
1980-1981	25.00	NA
1990-1991	54.10	NA
1995-1996	68.80	NA
1996-1997	70.00	NA
1997-1998	78.79	99.54
1998-1999	84.97	103.38
1999-2000	87.98	104.95
2000-2001	86.27	111.09
2001-2002	91.80	114.03
2002-2003	93.00	103.96
2003-2004	NA	124.00

The role of quality seed in increasing production and productivity is highlighted in Table 20.3 where increasing distribution of quality seed is parallel to increasing production and productivity of food grains in India. It is thus obvious that the quality seed will continue to be a vital component of agriculture for decades to come.

Table 20.3. Role of quality seed in production and productivity

Years	Area (million ha)	Production (million tonnes)	Productivity (kg/ha)	Distribution of seed (lac q)
1970-1971	124.30	108.40	872	5.16
1980-1981	126.70	130.00	1023	25.00
1990-1991	127.80	176.00	1380	54.10
1994-1995	123.85	191.10	1547	65.00
1999-2000	123.10	209.00	1704	87.98
2001-2002	121.91	212.03	1739	91.80

Seed Replacement Rate (SSR) of Major Crops

Seed replacement rate is defined as the production of actual quality seed distributed to actual seed required for planting the target area. The seed replacement rate of major crops in India for the year 1996 - 97 and 1997 - 98 is given in Table 20.4.

Table 20.4. Seed replacement rate of major crops in India in 1996 – 97

Crop	Area sown (million ha)	Sowing rate (kg/ha)	Seed requirement (lac q)	Seed distribution (lac q)	Seed supply % of requirement
Wheat	25.9	100	2590	232	8.95
Rice	43.4	30	1303	166	12.7
Sorghum	11.4	12	137	27	19.7
Pearl millet	10.0	4	40	17	42.5
Maize	6.3	20	125	19	15.2
Chick pea	6.8	75	514	14	2.7
Pigeon pea	3.5	20	70	6	8.6
Other pulses (Lentil, Moong, etc.)	12.1	20	242	22	9.1
Groundnut	7.6	150	1140	73	6.4
Rape seed/ Mustard	6.5	5	33	10	30.3
Soybean	5.4	75	405	31	7.7
Sunflower	1.9	10	19	6	31.6
Cotton	9.1	20	182	29	15.9
Jute and Mesta	1.1	10	11	3	27.3

Seed replacement rate of major crops during 1997-98

Crop	Area (million ha)	Seed rate (kg/ha)	Seed required (lac q)	Actual Seed distributed (lac q)	SRR%
Wheat	26.70	100	267.0	24.4	9.15
Rice	43.45	30	130.3	20.2	15.51
Maize	6.32	20	12.6	2.3	18.67
Sorghum	10.80	12	13.0	2.8	21.76
Pearl millet	9.67	4	3.9	1.6	41.79
Chickpea	7.56	75	56.7	1.0	1.87
Pigeonpea	3.36	20	6.7	0.6	8.77

Contd...

Table 20.4. Contd...

Groundnut	7.09	150	106.4	6.8	6.40
Rapeseed Mustard	7.04	5	3.5	1.0	30.00
Sunflower	1.74	10	1.8	0.7	41.67
Soybean	5.99	75	44.9	3.6	8.20
Cotton	8.87	20	17.7	2.8	16.21
Jute	1.10	10	1.1	0.3	30.90
Potato	1.21	3000	363.0	6.8	1.88
Total			1028.6	7.5	7.3

The data in Table 20.4 shows that average seed replacement rate has been 7.28 to 9.60 %. Now this indicates that there is tremendous scope to produce quality seed in greater amounts to increase the seed replacement rate. Even increasing the seed replacement rate from the existing level of around 10 % to 20 % will amount to doubling the current seed production which is at a level close to 8,00,000 tonnes (0.8 mt). In the major field crops like wheat and rice, the seed replacement rate is in the range of 9 - 12 %. A few crops where the SRR is more than 30 % are pearl millet and sunflower. In potato and chick pea seed replacement rate is hardly 2 %.

The status of area, seed requirement and percentage of hybrid seed in major vegetable crops in India is given in Tables 20.5 and 20.6 based on Arora (1998).

Table 20.5. Area and seed requirement for major vegetable crops in 1997 – 98

Crop	Area (000 ha)	Sowing seed rate (kg/ha)	Seed requirement (tons)
Beans	238	50.00	11880
Bhindi	372	12.50	4645
Brinjal	479	00.50	240
Cabbage	242	00.50	121
Cauliflower	304	00.75	228
Chilli	574	00.50	287
Gourds	409	05.00	2046
Melons	169	02.50	423
Onions	388	12.50	4849
Peas	227	75.00	17037
Tomato	486	00.35	169
Total	3888	160.10	42925

Source : Asian Seed, September 1998 issue : 'Vegetable seed sector' by Shri R.S. Arora

Table 20.6. Estimated area under hybrid vegetables in 1997 – 98 in India

Crop	Total Area (000 ha)	Area under hybrid vegetable (000 ha)	Share of hybrids %
Tomato	486	153	31.5
Bhindi (Okra)	371	20	5.4
Cabbage	242	79	31.4
Brinjal (egg plant)	479	85	17.8
Other vegetables			
Cauliflower	304	10	3.3
Chillies	573	14	2.4
Gourds	409	10	2.4
Melons	169	7	4.0

Source : Asian Seed, September 1998 issue : 'Vegetable seed sector' by Shri R.S. Arora

The area estimates of major vegetables given in Tables 20.5 and 20.6 are by and large unchanged during the present times (2004 - 05), although there has been substantial improvement in total production and productivity of vegetable crops. India produces about 90 million tones of vegetables from an acreage of 6 million ha with the productivity of 15 t/ha. In addition potato production is to the tune of 28 million tones from 1.3 million ha area with the productivity over 21 t/ha.

Indian Seed Industry

The detail review on Indian seed industry including its history, current status and future has been presented by Godwal (2003). According to this review the Indian seed industry is currently valued around Rs 2500 crores ($ 500 million) and is proposed to be around 3750 crores ($ 750 million) by 2002. There are about 150 organized seed companies in India today. Several companies have Government of India (DSIR) recognized research and development departments and have produced and released a large number of varieties and hybrids in several crops. The contribution of private research in terms of value is steadily increasing. The share of research hybrids in total turnover of crops like pearl millet, sorghum-sudan grass, sunflower, maize, sorghum and cotton was about 70% in 1997–98 compared to 46% in 1990–91. Private sector R&D's real investment in research has quadrupled between 1986 and 1998. Subsidiaries and joint ventures with multinational companies account for 30% of all private seed industry research A study made over nine private seed companies indicates that the amount spent on R&D ranged from 0.78% (0.49 crores) to 15.08% (22.62 crores) based on Companies Annual Reports 1998–99. Some of the companies initiated the work on development of transgenic crops. In March 2002 the first transgenic hybrid cotton seed was allowed for commercial cultivation in the farmers' field.

Seeds form the fundamental and crucial input for sustained growth in farm production, often stimulating the use of new methods, machinery and yield-enhancing agro-inputs. The role of the seed sector is not only to ensure adequacy in seed quality but also to ensure varietal diversity. Today, the Indian seed programme boasts one of the biggest seed markets in the world, with annual sales at around US $920 million. Of this, domestic market accounts for US $900 million and sales in the global market account for the remaining US $20 million.

The New Policy on Seed Development (NPSD), established in 1988 with the objective of augmenting productivity and output quality, stimulated major growth in the industry as it attracted a lot of investment in seed business from major domestic seed companies. Given the growth of the seed sector in recent years, India has the potential to become the foremost player in the seed export business in the developing world with prospective markets in Asia, Africa and South America.

Public Sector Seed Agencies

Like many agriculturally developed Asian nations, India has sizeable public and private sector seed businesses. Giant public sector players include the National Seeds Corporation (NSC), the State Farm Corporation of India (SFCI) and the thirteen State Seed Corporations (SSCs). NSC was the first public sector organization, established in 1963, and remained virtually the only agency for seed production for around 13 years. Its role extended to several developmental programmes including training, quality control and extension activities in seeds. This was followed by the setting up of the SSCs under two consecutive plan periods, supported by the World Bank, and these largely adopted the role of the NSC in the Indian States. These corporations engage principally in production and marketing of seeds of high yielding and hybrid varieties developed by the public sector.

Private Sector Seed Companies

Although private seed companies such as Pocha and Sutton have been established since the pre-independence era, accelerated growth of the private sector began only after the introduction of the new seed policy in 1988 which ushered in a liberal business climate. Currently there are over 200 private seed companies, together with a few multinational companies, and these tend to focus on low volume, high value crops with the principal effort being placed on creating hybrids for oilseeds, maize, cotton and vegetable crops. The private sector accounts for 70% of the market in terms of market turnover whereas the public sector has the greater share in terms of volume sales.

Global Initiatives

India today has a critical mass and level of growth that it could use not only to cater to the growing domestic requirement but also to make a concerted effort for global trade under provisions of GATT and WTO. Furthermore, India is endowed with second largest area of farmland, and the largest area of irrigated land, in the world and, with its huge germplasm diversity, its seed industry is well placed to serve both domestic and international markets.

Seed, Seed Industry and Seed Production Framework

Seed is a critical and basic input for attaining sustained growth in agricultural production. Seed is carrier of new technology for crop production. Distribution of assured quality seed is necessary for attaining higher crop yields. Policy initiatives taken by the Government of India during 1960's and 1970's for generating quality seed production and distribution of improved plant varieties developed by the scientists is one of the reasons for the country's self - sufficiency in food grain. The Indian seed industry has shown impressive growth and should continue to provide further potential for growth in agricultural production. The role of the seed industry is not only to produce adequate quantity of quality seeds but also to achieve varietal diversity.

The Indian seed programme largely adheres to the limited generation system for seed multiplication in a phased manner. The system recognizes three generations, viz., breeder, foundation and certified seeds and provides adequate safeguards for quality assurance in the seed multiplication chain to maintain the purity of a variety as it flows from the breeder to the farmer. The Indian seed programme involves the participation of Central and State Governments, the research system of the Indian Council of Agricultural Research (ICAR) and the State Agriculture Universities (SAUs), public sector, private sector and cooperative sector institutions. The seed sector in the country today comprises two national level Seed Corporations, i.e. the National Seeds Corporation (NSC) and the State Farms Corporation of India (SFCI), 13 State Seed Corporations (SSCs) and more than 200 prominent private seed companies. For quality control and certification, there are 20 State Seed Certification Agencies (SSCAs) and 99 State Seed Testing Laboratories. Though the private sector has started to play a significant role in the production and distribution of seeds over the last 7 to 8 years, the organized seed sector, particularly for food crops and cereals, continues to be dominated by the public sector.

The seed policy for the country provides a framework within which the full potential for the development of the seed sector can be realized. Government of India introduced a New Policy on Seed Development (NPSD) in October, 1988 with the objective of making the best planting material available anywhere in the world to Indian farmers for increased productivity and production. The policy provides for liberal imports (subject to certain parameters and regulations) to achieve this objective as well as to encourage the growth of export in seeds. The policy has provided a significant boost to Indian agriculture.

India has acceded to the Agreement on Trade Related Intellectual Property Rights (TRIPs) under the WTO Agreement under which it has agreed to enact a legislation for plant variety protection within a period of five years. This has already been done through Protection of Plant Varieties and Farmers' Rights Act-2001 enacted by the Govt. of India. The enactment of such legislation will have far-reaching consequences for the development of the seed industry. As seed policy has to provide a framework for the development of new varieties, it is intimately linked to issues related to conservation of agro bio-diversity flowing from the Convention on Bio-Diversity. This Convention articulates the sovereign right of nations on their genetic resources and provides for the recognition to the role of farming communities in the conservation and development of agro bio-diversity.

There is close link between the organized seed producers and the public sector research establishments in the country, with Seeds Division of the Department of Agriculture & Cooperation providing a vital link and focal point. After a variety has been notified in the country or for a State/group of States, seeds of the variety can be certified and made available to farmers. Assessment of demand of the requirement of seed is made at State level with the State Agriculture Department taking the lead and the State Seed Corporations providing back-up. Based on the assessment of demand, the production of foundation seed is organized by the State Departments and State Seed Corporations apart from NSC and SFCI. The requirement of breeder seed for producing the estimated quality of foundation seed is made available to the Seeds Division in advance in accordance with the notified schedule for this purpose. The Seeds Division consolidates the indents from both public and private organizations and places the indents on the ICAR which in turn informs the breeders of the variety or hybrid, including the sponsored breeders, to produce the required quantity. After production of the breeder seed, the quantity produced is intimated to the DAC at the end of the season and allocation of identified quantities is made to the indenters. Private seed producers are encouraged by being given proportionate allocation.Various seed development schemes being implemented by Seeds Division ensure increased availability of quality and certified seeds to farmers.

Criteria for Crop Selection by Private Companies

- Existence of hybridization systems
- Easy availability of germplasm, segregating or even finished lines
- Low yields or susceptibility to major diseases or pests in existing public or other company hybrids
- Regional adaptability or specific industry needs
- Newly introduced crop, hence higher chances of success
- Scope for value addition through chemical treatment or genetic engineering
- The impact of private research is apparent from the increase in area under proprietary hybrids

- Private R&D's real investment in research has quadrupled between 1986 and 1998
- Subsidiaries or joint ventures with MNCs account for 30 % of all private seed industry research in 1998
- Both domestic and MNCs have initiated greater focus on hybrids including more recently on hybrid wheat
- Once "Protection of Plant Varieties and Farmers' Rights Act-2001" comes in force, private sector would commence investment even in variety research

Structure of Seed Supply System

The structure of seed supply system as prevalent in India is given in Table 20.7.

Table 20.7. Seed supply system in India

Supporting organization	Activity
❒ Ministry of Agriculture (MoA)	❒ Overall direction and coordination
❒ Dept. of Agriculture and Cooperation (MoA)	❒ Administration and coordination on national basis
❒ Seed Division (MoA)	❒ All seed related administrative matters
❒ Agricultural Universities, ICAR institutes and Seed Companies	❒ Plant breeding and variety development
❒ AICVIP identification for release	❒ Variety development, testing and
❒ Central Seed Committee (CSC) affairs	❒ Seed related legislative and regulatory
❒ Central Seed Certification Board standards	❒ Certification procedures and seed
❒ Central Sub-committee on Crop Standards, Notification and Release of varities	❒ Evaluation of release proposals and recommendation for release and notification of varieties
❒ Seed Technology Research Centers under National Seeds Programme	❒ Research on various aspects of seed technology
❒ Agricultural Universities, ICAR Institutes, SFCI and Seed Companies	❒ Breeder and foundation seed production
❒ NSC, SFCI, SSCs and Private Seed Companies	❒ Foundation and commercial (certified/truthfully labeled) seed production
❒ Contract Seed Growers	❒ Production of certified/labeled seed

Contd...

Table 20.4. Contd...

❒ State Seed Certification Agencies harvesting, processing and packing for certification	❒ Field inspection, monitoring of
❒ Seed Testing Laboratories	❒ Evaluation of seed quality
❒ Seed company Depots, Distributors, Sub- distributors, Retailers	❒ Sale of seed to consumer farmers
❒ State Department of Ag, Designated Seed inspectors	❒ Monitoring seed quality in market

Legislation and Policy Issues

- ❒ Seed Act 1966
- ❒ Seed Rules 1968
- ❒ Seed Control Order 1983
- ❒ Essential Commodities Act 1955
- ❒ Package Commodities Order 1975
- ❒ Standards of Weights and Measures Act 1976
- ❒ Consumer Protection Act 1986
- ❒ Plant, Fruits and Seeds Order 1989
- ❒ Urban Land Ceiling Act 1976
- ❒ Plat Variety Protection and Farmers' Rights Act 2001

National Seeds Corporation (NSC) in the Service of the Seed Industry of India

- ❒ The NSC Limited is a Government of India undertaking under the administrative control of the Ministry of Agriculture, Government of India
- ❒ It was set up in March, 1963 and started functioning from July, 1963

Objectives of NSC

- ❒ To carry on in India the production, processing, drying, storing, distribution and transportation of agricultural seeds
- ❒ To enter into agreement with individuals, co-operative societies, and government agencies in the growing, processing, drying, storing, distribution, transplanting and selling of agricultural seeds
- ❒ To undertake by inspection and any other means, seed quality control measures in all the facets of seed business carried on behalf of or in co-operation with the concerned organizations
- ❒ To store and stockpile reserve supply of any seed needed for improvement of agriculture in India

Services Provided by NSC

- Foundation seed production
- Certified seed production
- Seed certification
- Seed quality control
- Seed processing, handling and packaging
- Seed marketing
- Information communication and public relations
- Sales promotion and advertising
- Project formulation and project management
- Training
- Consultancy services in all the above disciplines

NSC handles quality foundation and certified seeds of nearly 230 varieties of 70 crops.

NSC Certified Seeds for Farm Prosperity

- Latest and best NSC seeds made available are produced under scientific supervision
- Produced under suitable agro-climatic conditions in selected seed production areas
- Quality control at all stages of production under the care of experts
- Treatment with appropriate chemicals assuring protection in storage and during emergence in the field
- Healthy and vigorous NSC seeds are guaranteed for high germination and field standards
- Scientific processing ensures removal of weed seeds other crop seeds, under sized seeds, non-viable seeds and inert matter
- NSC seeds suitable for sowing under varying climatic conditions
- Sold in handy sealed packets or bags with tags containing all relevant information
- NSC seeds available from thousands of retail outlets at short bullock-cart trip away from every village in India

NSCs MIcro Objectives

- Monitoring availability of breeder, foundation and certified seeds of all India varieties and ensuring timely availability thereof
- Setting up adequate processing facility to ensure timely processing of seeds

- Setting up of conditioned as well as proper staring capacity to ensure proper staring of raw seed, processed seed and reserve stock
- Understanding field inspections, periodical quality control checks as well as continuous testing of seed to ensure quality of seeds
- Appointing co-operatives as far as possible to distribute seeds throughout the country
- Making available seeds to deficit states in time and in sufficient quantities
- Imparting training to personnel engaged in various activities of seed industry

Organization

National Seeds Corporation (A Government of India Undertaking), Beej Bhawan, Pusa Complex, New Delhi – 110 012)

- HQ
- Production Division
- Marketing Division
- R&D and Quality Control Division
- Training section
- Accounts and Finance Division
- Engineering Division
- NSP Coordination Cell (now wound up after expiry of the project)
- Administrative Division
- Publicity and Public Relations Division
- Publications, Publicity and Advertising
- Exhibition, Extension and Mailing Unit

Regional Offices, Units, Sub-Units and Farms of NSC

- **Ahmedabad Region**

  Sub-units
  - Ahmedabad
  - Vadodara
  - Rajkot
  - Panch Mahal
- **Bangalore Region**

  Sub-units
  - Bangalore
  - Belgaum
  - Bellary
  - Chickbalapur

- Devangere
- Dharwar
- Hassan
- Mysore

❒ **Bhopal Region**

Sub-units

- Bhopal
- Chhindwara
- Gwalior
- Indore
- Tikamgarh
- Raipur

❒ **Calcutta Region**

Sub-units

- Kharagpur (Midnapur)
- Malda
- Siliguri
- Burdwan

❒ **Bangalore Region**

Sub-units

- Bhubaneshwar
- Sambalpur

❒ **Chandigarh Region**

Chandigarh

Sub-units

Haryana

- Gurgaon
- Hissar
- Jind
- Karnal
- Faridabad
- Panipat

Punjab

- Amritsar
- Bhatinda
- Jallandhar

- Ludhiana
- Patiala

❒ **Delhi Region**
Delhi

❒ **Guwahati Region**
Guwahati

❒ **Hyderabad Region**
Hyderabad
Sub-units
- Kurnool
- Secunderabad
- Warangal

State Seed Corporations

1. Punjab
2. Haryana
3. Maharashtra
4. Andhra Pradesh
5. UPS and TDC (now UA and TDC)
6. Karnataka
7. Bihar
8. Rajasthan
9. Orissa
10. Madhya Pradesh
11. Gujarat
12. Assam
13. West Bengal
14. UP

State Seed Certification Agencies (SSCA)

1. Maharashtra
2. Andhra Pradesh
3. Karnataka
4. Orissa
5. West Bengal
6. Haryana
7. Rajasthan

8. Madhya Pradesh
9. Bihar
10. Punjab
11. Uttar Pradesh
12. Gujarat
13. Himanchal Pradesh
14. Tamil Nadu
15. Delhi
16. Uttaranchal
17. Assam
18. Jammu and Kashmir } Department of Agriculture
19. Kerala

The Market Status of Seed in India

The public sector component for seed production and marketing comprises corporations namely National Seeds Corporation (NSC) and State Farm Corporation of India (SFCI) and 14 State Seed Corporations. The private sector comprises around 150 seed companies which include national and multi-national companies and other seed producing/marketing companies. The seed industry has made the remarkable progress beginning with a modest area in 1962 – 63 to over 5 lac ha by 1995 – 1996. The quantity of seed produced and sold during the corresponding period has gone up by five times from 1.4 lac ton to 7.0 lac tons (DOAC, 1989).

The seed production scenario in 1990-91 and 1998-99 is shown in Table 20.8 (Gardwal, 2003).

Table 20.8. Indian seed market (quantity in tones)

Year	Saved seed	Bought seed	Total
1998-99	8953523	864899	9818422
1990-91	5125299	591200	5716499
Quantity increase over 1990-91	3828224	273699	4101923

Seed market scenario for the year 1990-91 and 1998-99 is given in Table 20.9.

Table 20.9. Indian seed market for the year 1990-91 and 1998-99 (Rs. in crores)

Category	1990-91	1998-99
Proprietary hybrids	95.00	636.00
Public-bred hybrids	124.65	245.00
Varieties	460.15	1368.55
Grand total	679.80	2249.55

Table 20.9 shows that in terms of value, the seed of open pollinated varieties was the largest followed in order by public hybrids and proprietary hybrids (hybrids produced by private sector seed companies).

The total bought seed in 1990-91 was estimated at around 5.91 lac tons valued at Rs. 679.80 crores. Seeds of the proprietary hybrids formed a significant portion of the total seed market. The total market for purchased seed was valued at Rs. 2249 crores. The composition of Indian seed market in terms of quantity for the year 1998-99 and 1990-91 is given in Table 20.10.

Table 20.10. Composition of Indian seed market - 1998-99 and 1990-91 (quantity in tons)

Category	Proprietary hybrid	Public hybrid	Variety	Total
Cereals	43287 (15600)	30208 (52171)	30208 (297329)	570336 (365100)
Oilseeds	5233 (2200)	2720 (1800)	142298 (76300)	150251 (80300)
Fibres	2475 (1500)	5776 (5700)	25280 (17400)	33531 (24600)
Pulses	-	-	41900	41900
Vegetables	-	-	1662 (17000)	1981 (17000)
Potato (seed)	-	-	66900 (70000)	66900 (70000)
Others	-	-	- (34200)	- (34200)
Grand total	51314 (19300)	38704 (59671)	774881 (512229)	864899 (591200)

The composition of Indian seed market in terms of rupees is given in Table 20.11.

Table 20.11. Composition of Indian seed market (value of marketed seeds over different categories of crops in 1998-99 and 1990-91 - Rs. in crores)

Category	Proprietary hybrid	Public hybrid	Variety	Total
Cereals	254 (26)	90.00 (63.15)	560.00 (153.55)	904.00 (242.70)
Oilseeds	109 (26)	20.00 (4.50)	312.60 (145.60)	441.60 (176.10)
Fibres	138 (33)	135.00 (57.00)	84.20 (24.00)	157.20 (114.00)
Vegetables	135 (10)	-	217.35 (65.00)	352.35 (75.00)

Contd...

Table 20.11. Contd...

Pulses	-	-	127.40 (33.00)	127.40 (33.00)
Potato seed	-	-	67.00	67.00
Grand total	636 (95)	245.00 (124.65)	1368.55 (460.15)	2249.55 (679.80)

Table 20.12 gives a break up of saved seeds and bought seeds in five cereal crops namely jowar, bajra, maize, paddy and wheat.

Table 20.12. Composition of Indian seed market (quantity of the seeds marketed over different categories of crops in 1998-99 and 1990-91 - quantity in tons)

Category	Saved seed	Bought seed	Total
Cereals	4208162 (3592806)	570336 (365100)	4778498 (3957906)
Oilseeds	1165552 (1313680)	150251 (80300)	1315803 (1393980)
Fibres	16709 (21602)	33531 (24600)	50240 (46202)
Vegetables	-	1981.11	1981.11
Potato (seeds)	3563100 (171680)	66900 (70000)	3630000 (241680)
Pulses	-	41900	41900
Grand total	8953523 (5125299)	864899 (591200)	9818422 (5716499)

A perusal of the Tables shows that there is significant increase in the sale of proprietary hybrids in case of jowar, bajra and maize.

Future Prospects of Seed Industry in India

Indian agriculture accounting for about 30 % national GDP, is a major component of Indian economy. It is the source livelihood for over 70 % of its population. India with total area of 328.80 million ha having gross cropped area of 187 m ha and cultivated area of 142 m ha ranks second only to USA in the sheer size of agriculture on account of its large, arable land area, sizable irrigated area, rich agro-biodiversity, diverse agro-climate and one of the largest agricultural research systems (NARS). India has all the ingredients to become a global power in agriculture. The remarkable growth registered in agricultural production in particular during the last 30 years has made the country self sufficient in food production which is currently at the level of over 210 million tons and is proposed to be raised to 260 million tons by the year 2025. Today the Indian seed programme is one of the biggest seed market in the world with annual sales at around US $ 920 million. Out of this, US $ 900 million accounts for domestic market and US $ 20 million for export market. Indian seed market is set to increase several folds in the near future as more farmers adopt commercial seeds to boost yield.

References

Arora, R.S. 1998. Vegetable seed sector. Asian Seed : September issue.

DOAC, 1989. Report of the expert group on seed. Department of Agriculture and Cooperation, Ministry of Agriculture GOI, New Delhi. pp 1-2.

Godwal, V.R. 2003. The Indian seed industry : Its history, current status and future. Current Science. 84 (3) : 399 – 406.

21

Global Seed Industry

Issue

2004-2005 saw an upsurge in seed industry takeovers and a shake-up in rankings. Today, the top 10 companies control half of the world's commercial seed sales. With a total worldwide market of approximately US$21,000 million [$21 billion] per annum, the commercial seed industry is relatively small compared to the global pesticide market ($35,400 million), and it is positively puny compared to pharmaceutical sales ($466,000 million). But corporate control and ownership of seeds – the first link in the food chain – has far-reaching implications for global food security. This chapter examines seed industry consolidation and other recent trends in the commercial seed industry.

Impact

With control of seeds and agricultural research held in fewer hands, the world's food supply is increasingly vulnerable to the whims of market manoeuvers. Corporations make decisions to support the bottom line and increase shareholder returns – not to insure food security. Ultimately, seed industry monopoly also means fewer choices for farmers. A new study by the US Department of Agriculture examines the impact of seed industry concentration on agri-biotech research. The study concludes that reduced competition is associated with reduced R&D. Despite seed industry claims to the contrary, concentration in the seed industry is resulting in less innovations – not more.

Players

A fistful of transnational firms, the Gene Giants, dominates global seed sales. Monsanto, Dupont, Syngenta – all among the world's top-ranking pesticide firms – lead the pack.

Policy

Seed industry concentration is already high on the agenda of civil society and farmers' organizations that are working to support and maintain peasants and farmer-controlled seed systems and against policies and technologies that seek to further privatize seeds. The implications of seed industry consolidation for food security and biodiversity must also be urgently addressed by governments at the FAO Conference in November and by the UN Convention on Bio-diversity in 2006.

World's Top 10 Seed Companies

The top 10 companies are listed in Table 21.1.

Table 21.1. World's top ten companies

Company	2004 seed sales (USD millions)
1. Monsanto (US) + Seminis (acquired by Monsanto in January 2005)	$ 2277 + $ 526
2. Dupont/Pioneer (US)	$ 2600
3. Syngenta (Switzerland)	$ 1,239
4. Group Limagrain (France)	$ 1,044
5. KWS AG (Germany)	$ 622
6. Land O' Lakes (US)	$ 538
7. Sakata (Japan)	$ 416
8. Bayer Crop Science (Germany)	$ 387
9. Taikii (Japan)	$ 366
10. DLF-Trifolium (Denmark)	$ 320

Value of The Global Seed Market

Expert opinions vary - a lot. According to the International Seed Federation, the estimated size of the market for seed and other planting material in 56 selected countries in 2005 was US$25,243 million. ETC Group believes that the figure overstates the value, particularly because ISF's figures include "other planting material" - which is not defined. July 2005 report by Phillips McDougall, UK-based agribusiness industry analysts, puts the value of the commercial seed market at US$19,000 million - and estimates that the top 10 companies control 51% of the total market. Given the enormous disparity in the figures, ETC Group estimates that the commercial seed market falls somewhere in between. Using an estimated market value of $21,000 million for commercial seed sales worldwide, ETC Group conservatively estimates that the top 10 companies control 49% of the global seed market. 2005 saw an upsurge in seed industry mergers and a change in rankings. The perennial kingpin, Dupont's Pioneer Hi-Bred International, was dethroned from the top spot when Monsanto acquired Seminis in January 2005 for $1,400 million. With the acquisition of Seminis, Monsanto became the world's leading seed company - and the world's biggest commercial vegetable seed company. Given the fast pace of mergers and acquisitions, it is expected that the top 10 rankings will flip-flop again soon. Despite controversy and lack of public acceptance - genetically modified (GM) seeds are gaining market share. According to Phillips McDougall, genetically modified seeds now account for one quarter of the total value of the commercial seed market worldwide. The market for biotech seed traits (herbicide tolerance and insect resistance) has shot up from $280 million in 1996 to $4,700 million in 2004 - a 17-fold increase over the past nine years. In 2004, Pioneer/Dupont earned 50% of its seed revenues from varieties that included a genetically modified trait. For 2005 Dupont offered the following products for the US market.

Pioneer/Dupont Product Offerings, USA, 2005

Seed product	Total no. of seed varieties	No. with biotech traits	Per cent
Maize/corn	298	150	50 %
Soybean	107	89	83 %
Canola	9	4	44 %

The growth in GM seed market share is remarkable, especially given that genetically modified seeds have been accepted in relatively few countries and continue to be steeped in controversy all over the world. When a few giant firms make up the cost of characters in the commercial GM seed market, it isn't necessary to have a superior product to win market share.

Seed Industry Concentration

Whether we're talking about sneakers, washing machines, beer or cell phones - corporate concentration is ubiquitous. In 2005, Adidas tied the knot with Reebock, Procter & Gamble gobbled Gillette, Molson guzzled Coors, SBC acquired ATT, Verizon merged with MCI and Maytag is wooing Whirlpool. But seeds are fundamentally different from razors and running shoes. When ownership of seeds - the first link in the food chain - is tightly held by a fistful of transnational firms - the world's food supply becomes vulnerable to the whims of market maneuvers. Corporate boards make decisions to support the bottom line and increase shareholder returns - not necessarily to insure food security.

Monsanto – The World's Largest Seed Company – At a Glance

In 2004, Monsanto's biotech seeds and/or trait technology accounted for 88% of the total GM crop area worldwide. According to Monsanto, its biotech trait acreage covered 175.7 million acres in 2004 - roughly the size of Zambia. (one acre = .4047 hectare). Monsanto's share of GM crops are as follows:

- **GM Soybeans :** Monsanto accounted for 91% of the worldwide GM soybean area in 2004. (119.5 million acres GM soybeans worldwide; Monsanto GM soybean 109 million acres)

- **GM Maize** : Monsanto accounted for 97% of the worldwide GM maize area in 2004. (47.7 million acres GM maize worldwide; Monsanto GM maize 46.4 million acres)
- **GM Cotton** : Monsanto accounted for 63.5% of the worldwide GM cotton area in 2004. (22.2 million acres GM cotton worldwide; Monsanto GM cotton 14.1 million acres)
- **GM Canola** : Monsanto accounted for 59% of the worldwide GM canola area in 2004. (10.6 million acres GM canola worldwide; Monsanto GM canola 6.3 million acres)

  Note: These statistics are based on **industry** sources: Monsanto and ISAAA.
- **Global Seed Market Share: Corn/Maize** : Monsanto controls 41% **global** market share

  Soybeans : Monsanto controls 25% **global** market share
- **Cotton :** In April **2005** Monsanto acquired **Emergent Genetics** for $300 million. Emergent is the third largest cotton **seed** company in both India and the US, with approximately 12% of the US cotton **seed** market and 10% of the Indian hybrid cotton **seed** market.

Monsanto Takes the Garden Path

With the January **2005** acquisition of Seminis for $1,400 million [1.4 billion], Monsanto takes a dominant position in the fast-growing vegetable **seed** market – a previously untapped **seed** segment for Monsanto. Under a variety of brand names, Seminis supplies over 3,500 **seed** varieties to fruit and vegetable growers in 150 countries. The Seminis acquisition includes the following brands:

- Royal Sluis
- Petoseed
- Bruinsma
- Asgrow Vegetable Seeds

For Monsanto, "vegetable **seed** is the next logical strategic move" because it's a "high value, high growth segment in agriculture." According to Monsanto, **seed** and trait gross profit as percent of sales is higher for vegetables (64%) than for soybeans (63%) or for corn (57%). Monsanto now assumes a leading market share in the **global** vegetable **seed** market, where they were virtually invisible before:

Beans : Monsanto controls 31% of the global seed market

Cucumbers : Monsanto controls 38% of the global seed market

Hot Pepper : Monsanto controls 34% of the global seed market

Sweet Pepper : Monsanto controls 29% of the global seed market

Tomato : Monsanto controls 23% of the global seed market

Onions : Monsanto controls 25% of the global seed market

A new study by the US Department of Agriculture looks at how seed industry concentration affects research." In the US, private sector spending on crop variety R&D increased 14-fold between 1960 and 1996, while public expenditures stagnated. It's widely acknowledged that intellectual property laws (plant breeders' rights and patents) that give companies exclusive monopoly rights over plant varieties spurred seed industry concentration. Looking at biotech maize, cotton and soybeans, USDA researchers found that research intensity slowed as seed markets became more concentrated. 'Those companies that survived seed industry consolidation appear to be sponsoring less research relative to the size of their individual markets than when more companies were involved. Also fewer companies developing crops and marketing seeds may translate into fewer varieties."

The USDA study also notes that public research on crop variety R&D has a simulative effect on private biotech research. The increasing publicly funded research for plant breeding may not only sustain the high rates of public return to public research, but could also promote additional private research. In the case of agbiotech in the USA, reduced competition is associated with reduced R&D. Despite seed industry claims to the contrary, concentration in the seed industry has resulted in less innovations. Ultimately, a highly concentrated seed market means less choice for farmers, not more.

Milestone or Miasma?

According to zealous proponents of biotech, "sometime during the early days of May 2005, a farmer somewhere in the world" planted the 1 billionth acre of biotech crops (400 million hectares)." For supporters of GM crops, tracking the total number of acres planted with genetically modified seeds is like McDonald's posting the number of hamburgers served under the golden arches.

The global proliferation of hamburgers sold by the world's mega fast-food enterprises is an appropriate comparison to the spread of the Gene Giants' biotech seeds. There are enormous social costs imposed by the ubiquity of these corporate products - costs that are not usually recognized or addressed until after the product/technology is widely deployed.

Take GM contamination, for example – unwanted gene transfer via cross-pollination from GM crops to conventional or organic crops nearby. In the early days of biotech it was discussed as a remote possibility, it soon became a reality, then a nuisance, and now a crisis (for some). With the rapid expansion of GM crop area, farmers are finding it increasingly difficult to produce purely non-GM varieties. According to Canadian researchers writing in Ecological Economics earlier this year, "Loss of, or limited ability to produce certified non-GM crops has the potential to impose significant production and consumption externalities on producers, consumers and other downstream users." They note that "the opportunity costs to farmers could rise dramatically as the number of GM-free producing areas declines, especially if disgruntled consumers choose simply to avoid food product lines with GM content."

Today, scientists are debating over how to measure GM contamination and what it means. Regulators and scientists are trying to determine isolation distances to thwart pollen dispersal and cross- pollination. At least 28 governments plus the European Union are grappling with requirements for the labeling of GM foods, and how to set tolerance levels for GM content (in non-labeled foods). Consumers in some markets are rapidly losing the ability to choose non-GM foods, or must pay more to avoid biotech foods. Farmers who choose not to grow GM crops are faced with loss of markets if their crops are contaminated, or face lawsuits if unauthorized patented genes are found on their property.

How will society deal with social costs imposed by GM crops and GM contamination? Who will decide? Who will pay? The Canadian researchers offer two troubling suggestions. Either a tax on GM growers could be used to compensate non-GM growers for any loss of income due to co-mingling or, alternately, non-GM growers could pay GM growers to restrict their planting. Both proposals appear unacceptable because neither puts the burden on the source of the contamination. There are signs that policy decisions are being made to accommodate the Gene Giants, farmers and consumers and to transfer the costs and burdens of GM contamination to farmers and consumers.

In the United States, state governments are passing laws that make it illegal for local governments to ban or restrict genetically modified crops. Determined not to allow local prohibitions of GM crops like those approved by three counties in California, the industry has pushed at least 14 states to pass preemption bills that are designed to neutralise citizen concerns about GM crops.

Another example is that the seed industry and allied governments are aggressively promoting Terminator seed technology as a viable method to stop gene flow. Delta & Pine Land, the US-based company that is actively developing genetic seed sterilization, makes the outrageous and scientifically unsupportable claim that Terminator "provides the biosafety advantage of preventing even the remote possibility of transgene movement".

New seed laws are being adopted in many countries that aim to restrict the rights of farmers to control and use their seeds. An in-depth report by GRAIN examines the imposition of new, repressive seed laws that are replacing older seed legislation in many countries. According to GRAIN, "the main purpose of these seed laws is to provide better protection of private seed varieties developed by companies and sideline farmers'own seeds completely."

GM Seeds and Pesticides

Since the early days of biotech, the industry has promised that genetically modified seeds would reduce the use of pesticides in agriculture. A 2004 study by

Charles M. Benbrook looks at GM crops and pesticide use in the US from 1993-2004 (the US accounts for 60% of the total GM crop area worldwide). The study concludes that overall the use of agrochemicals on GM crop acreage has risen about 4.1% since 1996. The findings contradict the industry's oft-repeated claim that biotech crops reduce pesticide use in agriculture.

According to Benbrook GE [engineered] corn, soybeans and cotton have led to a 122 million pound increase in pesticide use since 1996. While Bt crops have reduced insecticide use by about 15.6 million pounds over this period, HT [tolerant] crops have increased herbicide use 138 million pounds. Bt crops have reduced insecticide use on corn and cotton about 5 percent, while HT technology has increased herbicide use about 5 percent across the three major crops. But since so much more herbicide is used on corn, soybeans and cotton, compared to the volume of insecticide applied to corn and cotton, overall pesticide use has risen about 4.1 percent on acres planted to GE varieties. – Charles M. Benbrook, "Genetically Engineered Crops and Pesticide Use in the United States: The First Nine Years." Biotech InfoNet, Technical Paper Number 7, October 2004.

Farmers are now applying greater amounts of herbicides on genetically modified herbicide tolerant crops because some weeds have developed resistance in the face of heavy reliance on herbicide tolerant crops. Benbrook finds that "reliance on a single herbicide, glyphosate, as the primary method for managing weeds on millions of acres planted to HT varieties" is the primary factor that requires farmers "to apply more herbicides per acre to achieve the same level of weed control."

Glyphosate, the world's most commonly used agricultural chemical, is typically considered less hazardous than other chemical weed killers. But new studies on glyphosate, and Monsanto's proprietary formulation, RoundUp, raise serious concerns about the safety of the chemical for human health and the environment. About three-quarters of the worldwide area devoted to GM crops last year was planted with crops engineered to tolerate spraying of glyphosate.

Seed industry consolidation means less competition, resulting in fewer choices for farmers and greater vulnerability for local farming communities and global food security. Choices are further reduced with the spread of GM contamination encroaching on conventional and organic crops.

Seed Industry Concentration-2005

ETC Group's list include many of the world's largest seed companies and their acquisitions and /or subsidiaries. It is not a comprehensive list of all seed companies, but includes many of the top 20 firms that sell commercial field, vegetable and garden seeds. The list is as given in Table 21.2.

Table 21.2. Seed company subsidiaries/acquisitions

Seed Company	Subsidiaries/Acquisitions	Comments
Royal Barenbrug Group (Netherlands)	• Barenbrug Belgium • Barenbrug China • Barenbrug France • Barenbrug Holland BV • Barenbrug Luxembourg • Barenbrug Polska • Barenbrug South East • Barenbrug UK • Barenbrug USA • Barenbrug Production • Heritage Seeds Pty (Australia) • Modern Forage Systems Inc • New Zealand Agriseeds • Palaversich y Cia (Argentina)	Specializes in forage crops and turf grass. 20 subsidiaries in 12 countries. Annual sales: €160 million in 2004.
BASF (Germany	• SunGene (Germany) • Metanomics • ExSeed Genetics LLC	BASF, the world's largest chemical company, holds 40% equity stake in Svalof Weibull (see below).
Bayer (Germany) Subsidiary: Bayer Crop Science	• Aventis Crop Science (6/02) • AgrEvo • Plant Genetic Systems • Nunhems BV • Nunza BV • Sunseeds • Cannon Roth • Pioneer Vegetable Genetics • Dessert Seed • Leen de Mos (Neth. & Spain) • Castle Seed • Keystone Seed • Genex (Australia) • AgrEvo Cotton Seed Intl. (Australia) • Biogenetic Technologies	In 2002 Bayer purchased Aventis Crop Science (formed by the 1999 merger of Hoechst and Rhone Poulenc). Nunhems is one of top 5 vegetable seed companies in the world. Bayer Crop Science 2004 annual seed sales: €311 million

Contd..

Table 21.2. Contd..

	• Sementes Ribeiral (Brazil) • Mitla Pesquisa Agricola (Brazil) • Sementes Fartura (Brazil) • Granja 4 Irtmaos (Brazil) • Associated Farmers Delinting • Gustafson (3/04)	
DLF Trifolium ((Denmark)	• DLF International Seeds (USA) • DLF-TRIFOLIUM Ltd. (UK) • Hladke Zlvotice s.r.o (Czek Rep.) • Top Green (France) • Prodana Seeds • DLF Group China • Danespo Holding A/S (50%) • DLF Seeds Ltd. (NZ) • DLF-TRIFOLIUM A/S, Moscow • DLF-TRIFOLIUM Deutschland • Cebeco Seeds Group (The Netherlands) • Cebeco Saaten GMBH • Cebeco Seeds S.R.O. • Cebeco-Verstenit GMBH & Co. KG • Cebeco Zaden B.V. • La Maison Des Gazons S.A. • N.V. Zaden Van Engelen S.A. • Oliver Seeds Ltd. Proco Sem S. A. • Oliver Seeds Ltd. • Proco Sem S.A. • Seed innovations Ltd. • Wiboltt Fro A/S	Worlds largest breeder of cool season clover and grass seed. Purchased Cebeco Seeds Group in 2003 (grains and flax). Annual turnover US$320 million (for year ending 31 May 2004).
Delta and Pine Land, USA	• Ellis Brothers Seed • Arizona Processing • Mississippi Seed Co. • Hartz Cotton	World's largest cotton seed company. Developer of Terminator Technology. Annual sales, 2004:

Contd..

Table 21.2. Contd..

	• Sure Grow Seeds • D & PL South Africa, Inc. • D & PL Semillas Ltda (Costa Rica) • Deltapine Australia Pty.Ltd. • Turk Deltapine, Inc. (Turkey) • Deltapine India Seed Private Ltd. • D & M International, LLC: • D & PL China Pte Ltd. • Hebei Ji Dai Cottonseed Technology • Company Ltd. • CDM Mandiyu S.R.L. (Argentina) • MDM Sementes De Algodao, Ltda. (Brazil) • DeltaMax Cotton, LLC (50%)	US$315 million
Dow Chemical Co. (USA) Subsidiary: Dow Agrosciences	• Mycogen • Agrigenetics • Cargill Hybrid Seeds • United Agriseeds • Morgan Seeds (Argentina) • Kelten & Lynks • Delta & Pine Land (corn & sorghum only) • Dinamilho Carol Productos (Brazil) • Hibridos Colorado Ltda. (Brazil) • FT Biogenetica de Milho (Brazil) • Phytogen (w/J.G. Boswell) • Empresa Brasileira de Sementes (Brazil)	In 1998, Mycogen, a biotech company, became a wholly-owned subsidiary of Dow Agrosciences. The company specializes in maize, sunflowers, soybeans, canola and alfalfa.

Contd..

Table 21.2. Contd..

Dupont (USA)	• Pioneer Hi-Bred Intl. (USA) • Pioneer Argentina S.A. • Pioneer Hi-Bred Australia Ply Ltd • Pioneer Hi-Bred Services GmbH (Austria) • Pioneer Hi-Bred Northern Europe (Belgium) • Pioneer Sementes Ltda. (Brazil) • Pioneer Semena Bulgaria • Pioneer Hi-Bred Limited (Canada) • Semillas Pioneer Chile Ltda. • Shandong Denghai-PIONEER Seeds (China) • DuPont de Colombia S.A. • Pioneer Sjeme d.o.o. (Croatia) • Pioneer Hi-Bred Services (Czech Rep.) • Misr Pioneer Seed Company (Egypt) • Pioneer Hi-Bred Seeds (Ethiopia) • Pioneer Semences SAS (France) • Pioneer Hi-Bred N. Europe (Germany • Pioneer Hi-Bred Hellas (Greece) • Pioneer Hi-Bred Magyarorszdg Kft. (Hungary) • PHI Seeds Ltd. (India) • PT DuPont Indonesia • Pioneer Hi-Bred Italia • Pioneer Hi-Bred Japan • Farmchem Seedlinks Limited (Kenya) • Chemicals & Marketing Co. (Malawi)	Dupont's Pioneer Hi-Bred lost its longstanding title as world's largest seed company in early 2005 when Monsanto purchased Seminis. Dupont now ranks number 2, with 2004 seed sales of $2,600 million. Pioneer develops hybrids of corn, sorghum, sunflower and canola, and varieties of soybean, alfalfa and wheat for worldwide markets. The company has subsidiaries "on every crop-producing continent of the world."

Contd..

Table 21.2. Contd..

	• PHI Mexico SA de CV • Pioneer Hi-Bred N. Europe (Neth.) • Genetic Technologies, Ltd. (New Zealand) • Pioneer Pakistan Seed Ltd • Melo & Cia, CA. (Panama) • Pioneer Hi-Bred Philippines • Pioneer Hi-Bred Services GmbH (Poland) • Pioneer Hi-Bred Sementes de Portugal • Pioneer Hi-Bred Puerto Rico • Pioneer Hi-Bred Seeds Agro (Romania) • Pioneer Semena Holding GmbH (Russia) • Pioneer Hi-Bred Services (Serbia & Montenegro) • Pioneer Hi-Bred Slovensko (Slovakia) • Pioneer Hi-Bred Services (Slovenia) • Pioneer Hi-Bred RSA (South Africa) • South Korea O.M.C. • Pioneer Hi-Bred Spain SL • Bytrade Tanzania Limited • Pioneer Hi-Bred (Thailand) Co. • Pioneer Tohumculuk (Turkey) • Pioneer Nasinnya Ukraine, LLC • Pioneer Hi-Bred N. Europe (UK) • Agar Cross Uruguaya S.A • Semillas Pioneer de Venezuela • Farmchem Services Ltd. (Zambia) • Pioneer Hi-Bred Zimbabwe	

Contd..

Table 21.2. Contd..

KWS AG (Germany)	• AgReliant (joint venture with Limagrain) • AgroMais • APZ • Betaseed • CPB Twyford • KWS ARGENTINA • KWS AUSTRIA SAAT GMBH • KWS BENELUX • KWS CHILE • KWS FRANCE • KWS ITALIA • KWS KLOSTERGUT • WIEBRECHTSHAUSEN • KWS MAIS FRANCE • KWS MAIS GMBH • KWS OSIVA s.r.o. • KWS POLSKA • KWS RAGT HYBRID KFT • KWS RUS • KWS SAAT AG • KWS SCANDINAVIA AB • KWS Semena Bulgaria EOOD • KWS Semena d.o.o. • KWS Semena s.r.o. • KWS SEME YU • KWS SEMILLAS IBERICA • KWS Sjeme d.o.o. • S.C. KWS Seminte S.R.L. • KWS TURK • KWS Ukraine T.O.W. • Lochow-Petkus GmbH • Lochow-Petkus Polska • MOMONT • Pan Tohum • PLANTA • Razes Hybrides • SAKA-RAGIS	KWS specializes in sugar beets, maize, cereals. Generates over 70% of its revenue outside Germany. 4 largest maize breeder in the US. With annual sales of US$622 million in 2004, KWS is world's fifth largest seed company.

Contd..

Table 21.2. Contd..

	• Semena AG • ZKW • Producers Hybrid	
Landec Corp. (USA)	• Landec Ag Inc. (USA) • Heartland Seed	US maize seed company. With the acquisition of Heartland Seed in 2005, the company's annual seed revenues are approximately $34 million.
Groupe Limagrain (France)	• Vilmorin Clause & Cie • Advanta BV (European field crop division) • Force Limagrain (France) • Limagrain (Bulgaria) • Limagrain Cental Europe (France) • Limagrain Ceska Rep (Czech Rep.) • Limagrain Genetics (France) • Limagrain (Ilalia) • Limagrain Magyaroszag (Hungary) • Limagrain Moldova • Limagrain Nedertand • Limagrain Nickerson GrnbH (Germany) • Limagrain Polska (Poland) • Limagrain Romania • Limagrain Slovensko • Limagrain Verneuil Holding • Mais Angevin Nickerson (France) • Nickerson UK • Nickerson Intl. Research (France) • Nickerson Sur (Spain) • Soltis (France)	Limagrain is an independent cooperative and the European leader for maize and wheat seeds. Its subsidiary, Vilmorin Clause & Cie, is the world leader for home garden seeds (commercial). Limagrain's 2004 turnover for field seeds, vegetable seeds and garden seeds was approximately US$1,044 million.

Contd..

Table 21.2. Contd..

	• Alliance Semillas (Chile) • CHMT (South Africa) • Clause Tezier ltalia • Clause Tezier do Brasil • Clause UK • Clause-Tezier Iberica (Spain) • Clause Tezier (Ftance) • Clause • CNOS Vilmorin (Poland) • Ferry-Morse (US) • Flora-Fey (Germany) • Flora-Fey (Austria) • Harris Moran • Henderson • Kyowa • Marco Polo • Niagra • AgReliant Genetics (Joint venture w/KWS) • AgReliant Generics US (Joint venture w/KWS) • Hazera Genetics • Producers Hybrids (acquired by a AgReliant)	
Monsanto (USA)	• Seminis • Emergent Genetics • American Seeds Inc. • Channel Bio Corp. • Crow's Hybrid Corn • Midwest Seed Genetics • Wilson Seeds • NC+Hybrids • Advanta Canola Seeds • Interstate Canola Seeds • Asgrow (soybean & corn) • Holden's Foundation • Jacob Hartz • Hybritech • Calgene	Monsanto became the world's largest seed company in 2005 when it purchased Seminis - the world's largest vegetable seed business for $ 1.4 billion. Though traditionally a chemical company, Monsanto now earns more money from seeds and biotech traits than from its pesticide business. 2004 annual seed sales (including Seminis): US$2,803 million.

Contd..

Table 21.2. Contd..

	• Agracetus • Plant Genetics Inc. • Ameri-Can Pedigreed • Monsoy (Brazil) • First Line Seeds (Canada) • Plant Breeding Intl. (UK) • Agroceres (Brazil) • Cargill's intl. seed division • Dekalb Genetics (USA) • Custom Farm Seed • Sensako (South Africa)	
Nidera Corporation (The Netherlands)	• Nidera Semillas (Argentina) • Nidera Sementes (Brazil)	Nidera Seeds is a division within the privately-owned Nidera Corporation, a grain, oil seeds and oil processor company. Nidera started in the seed industry in 1988 with a program to improve sunflowers, and has expanded in crops such as corn, soybean, grain sorghum, forage sorghum, alfalfa and canola and wheat. In 2005. Nidera acquired Bayer's corn & soybean interests in Brazil. Annual turnover: approx. US$80 million
Panner Group (South Africa)	• Pau Seeds USA (formerly owned by Bayer) • Pannar Genetics, Inc. • Kaystan Seed • Pannar Seeds, Inc. (US) • Kombat (South Africa) • Starke Ayres (South Africa) • Mascor (South Africa)	Founded in 1958, privately-held Pannar Group is the largest seed firm in Africa, with subsidiary companies in several African countries, the EU, US and Argentina. Specializes in maize hybrids, sunflower,

Contd..

Table 21.2. Contd..

	• Pannar Seed Kenya • Pannar Seed Lda (Mozambique) • Pannar Seed Z (Zambia) • Pannar Seed BV (The Netherlands)	grains, pasture seed. The company does not publish or release turnover or profit figures.
Saaten – Union GmbH Ltd. (Germany)	• Subsidiary companies in UK, Poland, France, Romania. • Hybrinova (Dupont's hybrid wheat business) • Monsanto's hybrid wheat business	Specializes in hybrid wheat. 2004/05 annual sales for the German-based operations Euro 134 million. The entire group had 04/O5 turnover of approximately Euro 155 million in 2004/05 (does not include sugar beet activity).
Sakata (Japan)	• SakataUK • Sakata Ornamentals UK • Sakata Holland • Sakata Ornamentals Europe (Denmark) • Frisa Planter (Denmark) • Sakata Polska (Poland) • Sakata Korea Co. • Sakata Seed (Suzhou) China • Sakata Siam Seed (Thailand) • Sakata Seed Oceania • Sakata Seed Corporation (India) • Sakata Vegetables Europe (France) • Sakata Middle East (Jordan) • MayFord Seeds (South Africa) • SakataVegenetics (South Africa) • Sakata Seed Iberica (Spain)	Founded in 1913, Sakata specializes in vegetable and flower seeds and ornamentals. The company and subsidiaries worldwide. Annual seed sales: $416 million (for year ending May 31, 2004).

Contd..

Table 21.2. Contd..

	• Alf Christianson Seed (USA) • Sakata Seed America (USA) • Sakata Seed de Mexico • Sakata Seed de Guatemala • Sakata Centroamerica (Costa Rica) • Sakata Seed Sudamerica (Brazil) • Sakata Seed Chile • Sakata Ornamentals (Chile)	
Seminis – majority owned by Fox Paine, a buyout firm (sold in 2005)	• Asgrow Seed Co. • Petoseed • Royal Sluis • Hungnong Seed Co. (S. Korea) • Ang Seed Co. (S. Korea) • Sementes Agroceres (vegetable seed division) • Barham Seed	Monsanto (See above) acquired Seminis, the world's largest producers of vegetable and fruit seeds, in January 2005 for $1.4 billion.
Svalof Weibull AB (Sweden)	• Danisko Seeds • SW Seed Canada • Newfield Seeds (Canada) • Riding Valley Agro (Canada) • Promark Seed (Canada) • Priorty Lab Servtces (Canada) • Wheat City Seed (Canada)	SW is 60% owned by Lantmannen (the Swedish Farmers Supply and Crop Marketing Association) and 40% owned by BASF. The company specializes in cereals, oilseeds, forage crops. Annual sales: €116 million (2004)
Syngenta	• Advanta BV (North American corn and soybean business – Garst brand) • Petoseed • Bruinsma • Northrup King (NK) • Asgrow Vegetable Seeds • Funk Seed Intl. • Rogers Bros. • Zaadunie BV (Neth.)	Syngenta was formed in 2000 when Novartis merged with AstraZeneca's agribusiness. In 2004, 52% of Syngenta's seed sales came from field crops; 48% from vegetables and flowers. Annual seed sales: US$1,239 million

Contd..

Table 21.2. Contd..

	• McNair Seed • Cokers Pedigreed • Fredonia • Hilleshog • Agritrading • CC Benoist • Maisadour Semences • Eridania Beghin-Soy • Golden Harvest (6/04) • Dia-Engei (Japan) 2/04 • CHS Research LLC (04) • GA21 (technology) (04)	
Takii and Co., Ltd. (Japan)	• American Takii, Inc. • CTT Seed Co. (Thailand) • Qingdao Huang Long (China) • T.W. Company (Hong Kong) • Takii Chile • Takii Europe (Netherlands) • Takii France • Takii Korea Co. • Pahuja Takii Seed (India) • Takii do Brasil	Founded 1835 in Japan. Leading vegetable and flower seed company. Annual seed sales: $366 million (for year ending April 2004).
Land O Lakes (USA)	• Croplan Genetics • Hytest Seeds • Agriliance (joint venture with CHS, Inc.) • ABI Alfalfa • Seed Research of Oregon • Pickseed Companies Group • Seeds Ohio • Forage Genetics Inc.	Land O' Lakes is a giant agribusiness cooperative with total sales of $7,700 million. The company's 2004 seed sales were $538 million, specializing in alfalfa, maize, soybeans and in forage and turf grasses. About one-half of the seed sold by Land O' Lakes is purchased from Monsanto & Syngenta and then sold to coops.

Reference

http://www.etcgroup.org/documents/Comm90GlobalSeed.pdf.

22

Profile of Major Seed Companies in India

Indian seed market is one of the largest in the world and entire supply is through almost locally produced seed. The value of currently marketed seed (excluding the major share of seed saved and used by the farmers) is about USD1 billion (Rs.4500/crore). Of this, roughly a third includes the traded seed making the seed business worth about Rs. 1500 crore. This figure is likely to touch Rs. 4500 crore (USD 1 billion) crore mark by 2010. In comparison to about USD 1 billion seed business in India, the worldwide market of seed is worth about USD 21 billion. The global pesticide market is worth USD 35 billion. These values, however, are much smaller in comparison to global sale of pharmaceuticals worth USD 466 billion.

ON global basis 10 companies control half of the seed market. The top 3 leading transnational firms dominate the seed business. There are :

- Monsanto ($ 2.8 bn)
- Dupont ($ 2.6 bn)
- Syngenta ($ 1.3 bn)

Indian seed industry used to be dominated by public sector seed companies. However following the easing of government regulations and implementation of a new seed policy in 1988, private sector seed companies have started playing major role in seed research and marketing. Further, Government's decision to embrace biotechnology as a means of achieving food security has attracted several biotechnology focused multinational seed companies in India. The turnover by volume has reportedly reached a ratio of 60 : 40 between the private and public sector seed companies.

Private Sector Seed Companies in India

The private sector now includes about 150 seed companies in India. They are classified as :

- Large companies (annual sales turnover exceeding rupees 20 crores)
- Medium companies (annual sales turnover between rupees 2 and 20 cores)
- Small unorganized companies (annual sales turnover less than rupees 2 crores)

The profiles of major seed companies operating in India is given below :

Monsanto Seed Company

Monsanto is a leading provider of agricultural products and technology based solutions. Monsanto uses unparalleled innovations in plant biotechnology, genomics and breeding to improve productivity and to reduce the costs of farming. It produces leading seed brands, including DEKALB and Asgrow, and develops biotechnology traits that integrate insect control and weed control into the seed itself. It makes Roundup, the world's best selling herbicide, and other herbicides, which can be combined with seeds and traits to offer farmers integrated solutions. It manages business into segments, namely, Seeds and Genomics, and Agricultural Productivity.

Seeds and Genomics

The Seeds and Genomics segment consists of global business in seeds and related biotechnology traits, and technology platforms based on plant genomics, the science of studying and using the genes in living plants.

Agricultural Productivity

Agricultural Productivity segment consists of Roundup and other herbicides, lawn-and-garden herbicide business and animal agriculture business.

World Headquarters is Located at

Monsanto Company
800 North Lindbergh Boulevard
St. Louis, MO-63167

The Activities of Monsanto are as follows :

Seeds and Genomics

Monsanto serves farmers with high-quality, brand-name seeds, such as DEKALB and Asgrow. It also uses a broad, high-quality collection of genetic material called germplasm to develop new varieties that are licenced through Holden's/Corn States business. Monsanto develops and produces seeds with top performance and quality. Monsanto's branded and licensed seeds hold leading positions in key corn and soybean seed markets. Their plant breeding, including genomics based marker-assisted breeding (MAB), drives strong yield gains in seeds. The Seeds and Genomics segment generated 42 percent of sales and 53 percent of gross profit in fiscal year 2004. In 2004, Monsanto's brands of corn seed gained 1.4 points in U.S. market share. MAB continued to produce yield gains in corn at double the conventional breeding rate. Monsanto has the most biotechnology products in the market in three crop focus areas: corn, cotton and oilseeds, such as soybeans and canola. MAB is expected to have an increasing effect on yield gains in commercial corn and soybean seeds. Monsanto intends to continue to support strong seed brands and to license germplasm and traits to seed partners.

Biotechnology traits, such as herbicide tolerance in Roundup Ready soybeans and insect protection in YieldGard corn, enable farmers to produce crops more efficiently. Monsanto traits help farmers reduce their tillage and their pesticide use. Having set the industry standard for insect-protection and herbicide-tolerance traits, Monsanto holds the competitive advantage in delivering improved, second-generation traits, as well as seeds with two or more traits. Their strong germplasm base in both branded and licensed businesses allows them to launch trait products in the varieties and hybrids farmers want most. In 2004, they received final Japanese import approval for YieldGard Plus. This approval sets the stage for the fiscal year 2005 launch of the industry's first triple-stacked product, YieldGard Plus with Roundup Ready Corn 2. Monsanto is likely to launch YieldGard Plus throughout the U.S. Corn Belt in 2005. Their plan for mid-term growth involves both second-generation and stacked-trait products (including the first triple-stack product) and the expansion of existing products in approved markets. The pipeline includes output traits that produce healthier foods and other consumer benefits.

Agricultural Productivity

Roundup Herbicide

Roundup is a nonselective herbicide widely used because of its efficacy and environmental protection. It is the herbicide of choice for conservation tillage and for use over the top of RoundupReady crops, which are improved through biotechnology to tolerate Roundup. Monsanto is focused on maintaining the market leadership and brand position of Roundup. They are responding to intense competition by continuing to develop improved formulations and by reducing marketing infrastructure commensurate with price and share declines. Agricultural Productivity accounted for 58 percent of sales and 47 percent of gross profit in fiscal year 2004. The Roundup agricultural herbicide, the world's bestselling herbicide, complements Monsanto's biotechnology products as part of RoundupReady weed control system. The post patent performance of Roundup in the United States is expected to follow the global pattern, where Monsanto has maintained market leadership in the face of intense competition.

Other Agricultural Productivity

Selective herbicides, such as Harness Xtra, Machete, and Maverick, control specific weeds in corn, rice and wheat. Animal agriculture products improve dairy cow productivity and swine genetics. Monsanto also produces lawn-and-garden herbicide products for the residential market. Posilac bovine somatotropin is the largest-selling product in its field. It makes a significant contribution to gross profit. Monsanto's selective herbicides complement their ability to offer integrated solutions, particularly in corn production. Most crop protection chemicals face declining markets and competitive pressures. The new bulk powder production facility for Posilac was approved by the U.S. Food and Drug Administration. New retail outlets and national television advertising helped drive lawn-and-garden sales higher. Monsanto believes that new product formulations and more efficient manufacturing capacity can support the growth of Posilac. Monsanto expects the lawn-and-garden herbicide business to remain a strong cash generator and to support Monsanto's brand equity in the market place.

Monsanto has worldwide locations as given below:

Argentina, Australia, Austria, Belgium, Brazil, Bulgaria, Canada, Chile, China, Colombia, Croatia, Czech Republic, Denmark, Ecuador, France, Germany, Greece, Hong Kong, Hungary, India, Indonesia, Italy, Japan, Kenya, Korea, Malawi, Malaysia, Mexico, Pakistan, Philippines, Poland, Romania, Russian Federation, Senegal, Singapore, Slovakia, South Africa, Spain, Taiwan, Tanzania, Thailand, Turkey, Ukraine, United Kingdom, United States, Vietnam

Monsanto in India

From a humble beginning nearly a century ago, Monsanto is today a global leader in providing agricultural solutions the world over. Monsanto in India has its corporate office at Bombay and adopts top quality, cost effective and integrated approaches for improving farm productivity to produce better quality foods, strictly adhering to stringent standards for eco-care and safety. Monsanto Company has many revolutionary products and technological breakthroughs in the agricultural sciences to its credit.

Monsanto's work in India began over 50 years ago, soon after independence. Machete, the first rice herbicide was launched in India in 1975, and still continues to be a leading weed control tool for soybean farmers in western India. Roundup is providing plantation growers with a leading-edge weed control solution since 1991. Roundup has gone on to become an important weed control product for farmers, large and small, throughout the country. Leader, a new generation herbicide, was launched in 1998 for post-emergent weed control in wheat.

Monsanto is a committed participant in the seed industry. High-yield crop varieties and hybrid crops such as corn, sunflower and cotton are important crops that the company is committed to developing.

Monsanto, through its field operations, knowledge of agricultural production and constant search for new agricultural solutions and technologies, seeks to develop, communicate and provide sustainable solutions to Indian farmers.

As a responsible corporate citizen, Monsanto runs some community outreach initiatives. "Monsanto Vidyarthi" is a programme for educating children. Other projects include training farmers with small holdings on agricultural techniques under the banner of "Vanarai" which improves their crop productivity and preserves the environment.

The financial position of the company on global basis can be visualized from Table 22.1.

Table 22.1. Financial highlights of Monsanto-2005 (in USD million)

Years ended August 31 (in millions except per share amounts)	**2003**	**2004**	**2005**	**Change (2004 to 2005)**
Operating results				
Net sales	$ 4924	$ 5423	$ 6294	16 %
EBIT	$ 158	$ 445	$ 347	(22) %
Net income	$ 68	$ 267	$ 255	(4) %
Diluted earnings per share	$ 0.26	$ 0.99	$ 0.94	(5) %
Other selected data				
Free cash flow	$ 646	$ 999	$ 70	(93) %
Capital expenditures	206	$ 210	$ 281	(3) %
Depreciation	453	$ 452	$ 488	8 %
Diluted shares outstanding	261.8	269.2	272.7	1 %

Monsanto has acquired Seminis (the biggest vegetable seed and fruit plant material supplier company in the world for $ 1.4 billion in cash in March 2005 and assumed that a performance based payment of upto $ 125 million payable by the end of fiscal year 2007. The Monsanto India has corporate office in Bombay and research centre in Bangalore. The annual turnover of Monsanto India is about rupees 900 crores and the net profit is about rupees 60 crores.

Important Hybrids of Monsanto India

ASGROW -Sunflower Hybrids

SH 322

- High oil yield
- Good yield potential
- Good disease tolerance
- Early medium maturity and wide adaptability

SH 41

- High yield and oil content
- Full season hybrid for high yield environment

SH 177

- Highly compact seed setting
- Uniform head size
- Seed with high oil content
- Medium maturity with medium height and good uniformty

SH 416

- High oil and high yield
- Compact seed set
- Downy mildew and virus tolerant

SH 88

- Midium early maturity hybrid
- Compact seed set
- Suited to Punjab, Haryana and UP for planting after potato crop
- Released by Punjab Agril. University, Ludhiana

Maize Hybrids

ASGROW 501

- Early maturity
- Attractive orange yellow grain colour

ASGROW 900 M

- High yield potential
- Hybrid with wide adaptation
- Good grain colour with semi-flint kernels
- One of the largest selling hybrid in India

DEKALB -Maize Hybrids

HISHELL

- High shelling percentage
- Golden orange grain colour
- Good keeping quality of grain
- Performance good under moisture stress condition

ALLROUNDER

- Wide adaptation
- Drought tolerance
- Cold tolerance
- Orange yellow grain colour

PRABAL

- ❒ High yield and stable performer
- ❒ Good tip filling
- ❒ Sturdy look and healthy plants
- ❒ Drought tolerance

DOUBLE

- ❒ Can be used for green cob or grain purpose
- ❒ Good cob length and good filling
- ❒ Stays green and gives good fodder yield

Cotton

Bollgard

Mahyco Monsanto Biotech Ltd. developed first genetically modified cotton to be grown in India. Its three varieties MECH 162, MECH 12 and MECH 184 were permitted for commercial cultivation in March, 2002.

- ❒ Strength to fight bollworms within the plant
- ❒ Reduces insecticide use
- ❒ Higher yield
- ❒ Safe to beneficial insects
- ❒ Fits in integrated pest management

Syngenta

Syngenta is a world-leading agribusiness company committed to sustainable agriculture through innovative research and technology. The company is a leader in crop protection, and ranks third in the high-value commercial seeds market. Sales in 2005 were approximately $8.1 billion. Syngenta employs more than 19,000 people in over 90 countries.

Drug giants Novartis and AstraZeneca, the European pharmaceutical companies spun off their respective agrochemical and seed business in 2000, creating Syngenta, the world's largest agrochemical company (ahead of Bayer Crop Science). The company produces crop protection products (insecticides, herbicides, fungicides), field crop seeds (soybeans), vegetable seeds (corn, beans, tomatoes), and flowers. Syngenta and Myriad Genetics have mapped the rice genome, which could make rice crops more resistant to disease and lead to the genetic mapping of other cereals like wheat and corn.

Financial Summary

This is given in Table 22.2.

Table 22.2. Financial summary of Syngenta

Sales	Excluding restructuring, impairment and discontinued operations		
	2005 $ m	2004 $ m	2003 $ m
Sales	8104	7269	6525
Net income	779	623	340
Earnings per share	$ 7.67	$ 7.19	$ 3.34
Earnings per share before one-off tax credit	$ 7.67	$ 5.87	$ 3.34
Recommended dividend per share (CHF)	3.30	2.70	1.70

Syngenta

Syngenta Field Crops include corn, soybean, sugar beet, cereals, sunflower and oilseed rape. Syngenta sells field crops through its major brands NK@, GARST@, GOLDEN HARVEST@ and HILLESHOG@. The company has a strong position across the field crops market. Syngenta is the world's leading supplier of sunflower seed, and the third largest of corn and soybeans.

Syngenta provides growers with more than 1,000 different field crop varieties. In 2005, the company's drive for innovation and tailored solutions generated over 200 new products, each developed to meet specific needs, they enabled growers all over the world to raise yields further.

The company's ability to market high-quality germplasm continues to provide an important competitive edge, for example in the US corn belt. In the USA, Syngenta successfully integrated the GARST@ and GOLDEN HARVEST@ brands acquired during 2004. The company is now enhancing its Seeds offer to US growers through an expanded choice of biotechnology traits.

Syngenta seed varieties offer benefits to growers, processors and consumers. The company's development work focuses on improved agronomic properties as well as applications for food, animal feed and renewable fuels.

Sales of corn and soybean grew across all regions. GARST® and GOLDEN HARVEST® made a successful first year contribution in the USA and market share was maintained; growth excluding these acquisitions was 20 per cent. Syngenta is the world leader in flowers and the second largest provider of vegetable seeds.

The company sells vegetables under its S&G and ROGERS brands. The range includes tomatoes, lettuce, cucumbers, cabbage, sweet corn, peppers, melons, watermelons and beans. Syngenta has an integrated global technology network focusing on 14 major crops. The company's development program produces over 140 new varieties each year. Syngenta now fully owns the fresh produce company, Dulcinea, in the USA. In Europe, recent successful launches of branded products include the Solinda'M watermelon, Toscanella'M and Kumato'M tomatoes.

S&G@ Flowers continues to build on its global market leadership in seeds, cuttings and young plants. A wide choice of flowers with strong visual appeal is crucial to continuing success in satisfying consumer demand. Excellence in the supply chain contributes to further growth. After a slow start to the year due to poor weather in Europe, vegetables sales showed a marked recovery in the second half, with S&G@ reinforcing its strength in tomatoes. Sales in the emerging markets of Latin America and Asia Pacific continue to expand rapidly. In fresh produce, sales of watermelon and cantaloupe has strong growth in the USA.

New Products Through Plant Breeding

The richness and diversity of Syngenta's germplasm resources, combined with molecular breeding, has accelerated the rate of innovation in Seeds. In Field Crops, R&D focuses on growers primary aim, higher quality and yield. Raising the essential nutritional components, starch, oil or protein further improves value for food and feed, and biofuel applications. Marker-assisted breeding is helping to increase disease resistance, enhance cooking, baking and brewing quality for processors and consumers. It is also used to improve plants' agronomic properties such as size and germination.

In Vegetable Seeds, Syngenta develops new products to provide consumers with consistent high quality, improved appearance, taste and texture. Powerful analytical science has been expanding knowledge of taste, flavor and nutrition. Combined with advanced breeding technology, this is accelerating the introduction of novel varieties.

In Flowers, Syngenta breeding programmes offer gardeners and consumers an exciting flow of richer, more vibrant colors, better flowering and longer visual attraction, as well as providing growers and retailers with ever- improving convenience. For customers, this means a greater choice of improved crop varieties and hybrids, with hundreds of new Seeds products launched each year.

Input Traits

Input traits, which control crop pests, continue to gain importance, especially in corn and soybean in the USA and Latin America. Syngenta is on track to deliver a full suite of stacked traits in corn, glyphosate tolerance and resistance to both leaf and soil insects, building on the quality and breadth of its germplasm from the combination of GARST®, GOLDEN HARVEST® and NK®. These stacks will be introduced over the next few years.

Development of the novel insect control trait for cotton, VIPCOT™, through a partnership with the premier cotton seed company in the USA, has continued to progress towards launch. This trait has reached the final stage of selection of the technology options for later commercialization.

Output Traits

While input traits such as insect resistance provide important benefits for the grower, output traits add value for processors and ultimately consumers. Work on plant produced pharmaceutical compounds progressed in 2005 to the stage of clinical trials. However in view of the extended time lines and associated costs of full commercialization it has been decided to seek third parties to take these projects through development to commercial launch. Resources released in R&D will be redirected to their business.

A major use of corn is for animal feed, and more recently for ethanol production. Traits that bring benefits in these applications are expected to provide new value creation in the future. Syngenta has products in development in both areas. Incorporaiton of the enzyme phytase into the grain itself provides additional benefits to adding enzyme at the pelleting stage of animal feed production. Similarly, corn expressing the enzyme amylase is being developed for more efficient ethanol production, as it speeds up starch conversion into sugar.

Pioneer

Pioneer Hi-Bred International, Inc., a DuPont company, is the world's leading developer and supplier of advanced plant genetics to farmers worldwide. With headquarters in Des Moines, Iowa, Pioneer develops, produces and markets a full line of top-quality seeds and forage and grain additives and provides services to customers in nearly 70 countries.

DuPont is a science company, delivering science-based solutions that make a difference in people's lives in food and nutrition, health care, apparel, home and construction, electronics and transportation.

Worldwide Locations Listing

Operating on every crop-producing continent of the world, Pioneer supplies superior seed to the world's farmers. The important locations of this company are as follows:

Belgium (BE) 32
Pioneer Hi-Bred Northern Europe
Sales Division GmbH
Belgisch Filiaal / Filiale Belge
Uitbreidingstraat 84 bus 3
2600 Berchem

Belize (BZ) 501
Midwest Seeds & Agro Supplies Co. Ltd.
20-02L Center Road
Spanish Lookout
Cayo District

Bolivia (BO) 591
Semillas Villacruz S.R.L.
Gilberto Molina No. 3190
Barrio Chacarilla
Zona del Hospital Japonés
Santa Cruz de la Sierra

Brazil (BR) 55
Pioneer Sementes Ltda.
Santa Cruz do Sul
BR 471 Km 49, Distrito Industrial
Caixa Postal 1009
96810-970
Santa Cruz do Sul, RS

Bulgaria (BG) 359
Pioneer Semena Bulgaria
Skobelev Street, Entr. B-62
1606 Sofia

Canada (CA) 1
Pioneer Hi-Bred Limited
Chatham-Administrative Offices
2 Highway West, PO Box 730
Chatham, ON N7M5L1

Chile (CL) 56
Semillas Pioneer Chile Ltda.
Avda. El Bosque Norte 500, Oficina 1102,
Las Condes,
Santiago

China (CN) 86
Shandong Denghai-PIONEER Seeds Co.
Unit 1101, China World Tower 2
No. 1 Jianguomen Wai Avenue
Beijing 100004

Colombia (CO) 57
DuPont de Colombia S.A.
Calle 114 # 9-01 Torre A Piso 14 Of. 1401
Santa Bárbara
Bogotá

Costa Rica (CR) 506
Agrícola Piscis SA
1 Km. Sur de la entrada principal del Parque Industrial del Cartago

Croatia (HR) 385
Pioneer Sjeme d.o.o.
Culinecka cesta 2b
10 040 Zagreb

Czech Republic (CZ) 42
Pioneer Hi-Bred Services GmbH
obchodní zastoupení pro ER
J. Opletala 1279
690 59 Boeclav

Dominican Republic (DO) 809
Productora de Semillas Dominicanas
Av. Antonio Guzmán
Km. 6
La Herradura
Santiago

Ecuador (EC)
Egypt (EG) 20
Misr Pioneer Seed Company
9B Army Forces Buildings
In Front of Workers' University
Nasar City, Cairo 11371

El Salvador (SV) 503
Duwest
Calle el Progreso y Boulevard Venezuela
Colonia Roma
Frente al Hospital de Antel
San Salvador

Ethiopia (ET) 251
Pioneer Hi-Bred Seeds (Ethiopia) P.L.C
Bole Road
P.O Box 1134
Addis Ababa

France (FR) 33
Pioneer Semences SAS
Chemin de L'Enseigure
31840 AUSSONNE

Germany (DE) 49
Pioneer Hi-Bred Northern Europe
Sales Division GmbH
Apensener Straße 198
21614 Buxtehude

Greece (GR) 30
Pioneer Hi-Bred Hellas S.A.
15, Fleming Str.
151 23 Maroussi
Athens

Guatemala (GT) 502
Duwest
Via 4,6 - 65
Zona 4
Guatemala

Honduras (HN) 504
Duwest
1ra. Avenida 15 y 16
Calle Sur Oeste
Barrio La Guardia
San Pedro Sula, C.A.

Hungary (HU) 36
Pioneer Hi-Bred Magyarország Kft.
2040 Budaörs
Neumann J. u. 1

India (IN) 91
PHI SEEDS LTD
3rd Floor, Babukhans' Millennium Centre
Raj Bhavan Road, Somajiguda
Hyderabad 500,082

Indonesia (ID) 62
PT DuPont Indonesia
Menara Mulia Building, 5th Fl.
Jl. Jend.l Gatot Subroto Kav. 9-11
Jakarta 12930

Italy (IT) 39
Pioneer Hi-Bred Italia Srl
Via Pari Opportunita' n.2
26030 GADESCO PIEVE DELMONA (CR)

Jamaica (JM) 876
109 Marcus Garvey Drive
Kingston 11
Jamaica, West Indies

Japan (JP) 81
Pioneer Hi-Bred Japan Co., Ltd.
Minato-ku Toranomon 3 - 7 - 10
Landic Toranomon Bldg. 7f
105 - 0001 Tokyo

Kenya (KE) 254
Farmchem Seedlinks Limited
Wundanyi/Ndume Rd, off Lunga Lunga Rd, Industrial Area
P.O. Box 18407 - 00500
Nairobi

Malawi (MW) 265
Chemicals & Marketing Co. Limited
Kidney Crescent Road
P O Box 1230,
Blantyre

Mexico (MX)
PHI Mexico SA de CV
Carr. Guadalajara-Morelia Km 21 #8601-A
Tlajomulco de Zúniga, Jalisco
C.P. 45645

Netherlands (NL) 31
Pioneer Hi-Bred Northern Europe
Sales Division GmbH
Nederlandse Vestiging
Verlengde Poolseweg 34 - 46
4818 CL Breda

New Zealand (NZ) 64
Genetic Technologies, Ltd.
Level 37, Vero Centre
48 Shortland Street
P.O. Box 105303
Auckland 1030

Nicargua (NI) 505
Duwest
Plaza 3F, Segundo Piso
Modulo B-5
Km. 5
Carretera Norte
Managua

Pakistan (PK) 92
Pioneer Pakistan Seed Ltd
2-G/4, Canal Bank, Johar Town
Lahore

Panama (PA) 507
Grupo Melo
Vía Espana Rio Abajo 2313
Panamá
Republica de Panamá

Peru (PE) 51
Representaciones Agrícolas de Miguel Villanueva E.I.R LTDA
AV. J.J. ELIAS No. 450
ICA - PERU

Philippines (PH) 63
Pioneer Hi-Bred Philippines, Inc.
24F Antel Global Corporate Center
Dona Julia Vargas Avenue
Ortigas Center, Pasig City 1600

Poland (PL) 48
Pioneer Hi-Bred Services GmbH
Przedstawicielstwo w Polsce
Swadzim, ul. Poznanska 16
62-080 Tarnowo Podgórne

Portugal (PT) 351
Pioneer Hi-Bred Sementes de Portugal, SA
Campo Pequeno, no 48 - 6o Esq.
Edifício Taurus
1000-081 Lisbon
Portugal

Puerto Rico (PR) 1
Pioneer Hi-Bred Puerto Rico, Inc.
Shipping:
P.O. Box 22
Aguirre 00704-0022
Business Address:
Carr. #3, Km. 1549
Salinas 00751

Romania (RO) 40
Pioneer Hi-Bred Seeds Agro SRL
D.N. 2, km. 19,7, O.P. Afuma
jud. llfov, Cod pootal 077010, Romania

Russia (RU) 7
Pioneer Semena Holding GmbH
Representative Office in Moscow
13/2, B. Palashevskiy pereulok
103104 Moscow

Serbia & Montenegro (CS) 891
Pioneer Hi-Bred Services GmbH
Predstavnistvo Novi Sad
10 Laze Teleckog
2100 Novi Sad
Serbia & Montenegro

Slovakia (SK) 421
Pioneer Hi-Bred Slovensko s.r.o.
Oravska 13
90301 Senec
Slovensko

Slovenia (SI) 386
Pioneer Semena Holding GmbH
Parndorf
PSH poslovno svetovanje
Podružnica Murska Sobota
Markišavska cesta 10
9000 Murska Sobota

South Africa (ZA) 27
Pioneer Hi-Bred RSA (Pty) Ltd
4th Floor, Outspan House
1006 Lenchen Avenue North
P.O. Box 8010
Centurion 0046
Republic of South Africa

South Korea (KR) 82
O.M.C.
Dacho Bldg. Room 201
1341-2 Seocho-Dong
Seocho-Ku
Seoul

Spain (ES) 34
Pioneer Hi-Bred Spain SL
Av. del Reino Unido s/n
Edificio ADYTEC 2a planta
41012 Sevilla

Switzerland (CH) 41
Pioneer Hi-Bred Switzerland S.A.
DuPont Agriculture and Nutrition
via Cantonale / Galleria 3A
CH-6928 Manno, Switzerland

Tanzania (TZ) 255
Bytrade Tanzania Limited
213 Maweni Street, Upanga
P.O. Box 3491
Dar-es-Salaam, TZ

Thailand (TH) 66
Pioneer Hi-Bred (Thailand) Co., Ltd.
9th Floor, Olympia Thai Tower
444 Ratchadapisek Road
Samsennok, Huay Kwang
Bangkok 10310 Thailand

Trinidad and Tobago (TT)
Sheppard Enterprises Ltd.
24 Hillside Ave. Cascade
Port-au-Spain

Turkey (TR) 90
Pioneer Tohumculuk A.S.
Büyükdere Cad. No:122
Özsezen o Merkezi A Blok
Kat:2 aioli 34394 stanbul

Ukraine (UA) 380
Pioneer Nasinnya Ukraine, LLC
152, Zabolotnogo Street
03143 Kiev

United Kingdom (GB) 44
Pioneer Hi-Bred Northern Europe
Service Division GmbH
United Kingdom Branch
Central Boulevard
Blythe Valley Business Park
GB-Solihull, B90 8AG

United States (US) 1
Corporate Headquarters
7100 NW 62nd Avenue
PO BOX 1014
Johnston, IA 50131-1014

Uruguay (UY) 598
Agar Cross Uruguaya S.A
Cno. Gral. Máximo Santos 4900
12400 - Montevideo

Venezuela (VE) 58
Semillas PIONEER de Venezuela
Zona Industrial Carabobo final 9o Transversal
frente a la planta DuPont
Valencia
estado Carabobo

Zambia (ZM) 260
Farmchem Services Limited
9A Sheki Sheki Road
Emmasdale Industrial Area
P.O.Box 30654
Lusaka

Zimbabwe (ZW) 263
Pioneer Hi-Bred Zimbabwe (pvt) ltd
Mutual Gardens
100 The Chase West
Emerald Hill
Harare

Pioneer India

Pioneer India began research and parent seed operations in 1970. A network of efficient distributorships now provide proprietary Pioneer@ brand corn and sunflower hybrids, grain sorghum, pearl millet, rice and mustard seed to farmers across the country. In 1992 Pioneer's India production, sales, and marketing operations became part of a 50-50 joint venture with the Southern Petroleum Industries Corporation (SPIC) as SPIC PHI Seeds Ltd. In January 2001 Pioneer purchased SPIC's share of the joint venture and now wholly owns the operation.

Product Overview

Alfalfa

Pioneer offers growers over 15 different Pioneer@ brand alfalfa varieties. Each year, alfalfa researchers harvest more than 30,000 test plots worldwide to measure product and parental performance in various environments. Technology traits are also available to help protect against insect pressures. Pioneer sales professional can help select the right variety(ies) for farmers.

Canola

At Pioneer, new standards are set for canola growers for future canola products, bringing farmers improvements of yields and traits, but in hybrid packages. To assure this steady flow of products, Pioneer researchers plant more than 65,000 test plots each year in seven countries. With either the Roundup Ready gene or Clearfield trait, both Pioneer@ brand canola hybrids and varieties offer consistent yields, harvestability, agronomic traits and herbicide management options. Local sales professional can help select the right hybrid or variety for the farmers.

Corn

Pioneer offers corn growers the industry's broadest lineup, with more than 240 unique and different Pioneer@ brand hybrids. Localized research results in products with top yield potential and necessary agronomic traits for all the major corn growing areas of the country. Technology traits are also available to help protect against insect and weed pressure. Nearby Pioneer sales professional can help select the right hybrids for farming operation.

Crop Protection Products

DuPont offers a complete line of crop protection products for row crops, cereals, specialty crops, fruits and vegetables. No two fields are alike and one size doesn't fit all. That's why DuPont offers a broad portfolio of products to meet farmers' needs.

Inoculants

Pioneer offers inoculants for corn and alfalfa silage, alfalfa hay, and high moisture grain. With a library of more than 4,000 strains of bacteria, Pioneer microbial researchers develop inoculants with carefully selected, unique, patented strains of bacteria.

Sorghum

Pioneer offers more than 14 sorghum hybrids in choice of red, yellow, and bronze, with seed treatment options to meet specific needs. Localized product research results with top yield potential, drought tolerance and the industry's only proprietary genetics specific to agronomic growing conditions that also withstand tough pest and disease challenges.

Soybeans

With over 50 new varieties introduced since year 2000, Pioneer varieties feature outstanding yields, with strong emergence, disease, pest and herbicide resistance, solid standability and smooth harvesting to meet the field by field needs of growers. Innovative bulk options, such as the PROBOX@ seed handling system, offer improved handling convenience.

Sunflower

At Pioneer, developing products with stable yield performance is a key goal. Researchers worldwide are dramatically increasing the number of hybrid combinations tested for yield. Each year, more than 130,000 test plots are planted for yield potential. Improving agronomic traits such as emergence, root and stalk strength, seedling vigor, and fast dry down also are emphasized. Local Pioneer sales professional can be colntacted about the top sunflower hybrids and market opportunities in particular area.

Wheat

With more than 12 soft red winter and soft white winter varieties, Pioneer@ brand wheat offer growers the convenience of bagged wheat with the agronomic package that fit their needs. An extensive testing network is used in North America and Europe, as more than 50,000 yield plots are planted each year. In Europe, the focus is on the development of varieties that meet the rigorous performance and end-use qualities of key target markets. With an eye on the future, hybrid development will provide a vehicle for delivering value-added traits to the marketplace.

Seminis

Seminis is an United States (US) company whose main activity is the production of hybrid vegetable and fruit seeds. The corporation has its administrative and research and development (R&D) headquarters in California, where much of its basic research activities are concentrated. Although Seminis is an American company, its European division is largely independent. The company built part of its technological and marketing capabilities through its acquisition of one of the largest European vegetable seed companies—Royal Sluis. It has 15 research locations in Europe and its European headquarters are located at Enkuizen in the Netherlands. Its biotechnological research activities are predominantly located in the Netherlands, France, and California. However, there are additional biotechnology research stations in other European countries, such as in Spain and Italy, which are also very important.

Seminis majority is owned by Savia, which is part of the Pulsar group of companies located in Mexico. Pulsar operates in the agro-technology, insurance, and packaging industries. Savia is basically a large conglomerate with strong interests along the entire food chain, which are realized by two companies—one of which is Seminis (which is concentrated in seed production) and the other is Bionova (which is concentrated in fruit and vegetable production, distribution, and R&D).

In 2000, Seminis' annual sales exceeded $US 474 million and R&D expenses were about $US 58 million. In that year, Seminis had 3,800 employees, of which 765 worked on R&D activities. Seminis is today the world leader in hybrid seeds for fruit and vegetables, with production capabilities in 30 countries and R&D facilities in 34 countries around the world. Now it has been acquired by Monsanto in USD 1.4 billion.

History, Markets and Products

Seminis was born in 1995 from the merger between the vegetable divisions of three seed companies—Asgrow (US), Petoseed (US), and Royal Sluis (the Netherlands). After the creation of Seminis, the seed and coating technologies of these three divisions were all placed under the INCOTEC name. In July 1998, Seminis then acquired 70% of Hungnong Seed, Korea's largest seed and vegetable firm, and Choong Ang Seed (AGROW, 1999). Hungnong and Choong Ang are leading companies in the oriental vegetable market. In November 1998, Seminis purchased the vegetable division of Sementes Agroceres, a Brazilian company that produces and distributes vegetable seeds throughout Brazil. In 1998, Seminis acquired the distribution rights to LSL PlantScience tomato varieties. LSL PlantScience is a world market leader in tomatoes developed for long shelf life. In September 1999, Seminis acquired Barham Seeds, a company dedicated to the research and development of seedless watermelon varieties.

As a result of these acquisitions Seminis has the following brands: Asgrow, Petoseed, Royal Sluis, Bruinsma, California, Genecorp, Hungnong, Choong Ang, and Horticeres. Asgrow, Petoseed, and Royal Sluis are full-line brands (with a broad catalogue of products which operate at a global scale); the other brands are regional or specialty brands.

INCOTEC provides coating and other seed preparation technologies to Seminis' subsidiaries, as well as to other seed companies. The majority of the more fundamental research of INCOTEC takes place in the Netherlands. Through INCOTEC, Seminis is directly linked to the pesticides industry—as some of INCOTEC's technologies incorporate pesticides in seed coatings.

In 1998, Seminis had the following share of the vegetable seed market: 26% of the global market; 39% of the US market; 24% of the European market (Seminis, 2000); and 15% of the Asian market (Joon-Hun, 1998). Seminis' main markets are North America and Europe. In 2000, 35.8% of Seminis' sales took place in North America, 29.8% in Europe, 17.7% in Asia, 8.6% in South America, and 8.1% in Middle East and Africa. Seminis is highly diversified in its products, customers, and regions in which it operates, "no one customer or product represents more than 2 percent of their sales." (Seminis, 2000). In Europe, Italy and Spain are the most important regional markets. In 2000, 29.6% of Seminis' European sales took place in Italy and 20% in Spain.

Innovation Strategy

The main aim of Seminis' innovation strategy is to strengthen its leadership position in the global vegetable seed market. To reach such an objective, Seminis has completed nine acquisitions of seed companies to date and has reinforced its presence in several regional markets. Seminis was established in Europe through the acquisition of two Dutch companies—Bruinsma (part of Asgrow) and Royal Sluis—both with a long tradition in plant breeding. The main products of Seminis are listed in Table 22.3.

Table 22.3. Main products and primary markets of Seminis' brands.

Brand	Main Products	Primary Markets
Asgrow	Tomato, carrot, onion, pepper, bean, pea, corn, melon, cucumber, brassica.	Worldwide
Petoseed	Tomato, pepper, eggplant, brassica, onion, melon, cucumber, lettuce.	Worldwide
Royal Sluis	Broccoli, cabbage, carrots, cauliflower, leeks, spinach, bean, tomato, lettuce.	Europe, Asia
Bruinsma	Greenhouse tomato, pepper, cucumber.	Europe
California	Cucumber, melon, squash, tomato.	Middle East and North Africa
Choong Ang	Chinese cabbage, hot peppers, oriental melon, radish, watermelon.	South Korea
Hungnong	Broccoli, cabbage, Chinese cabbage, hot peppers, oriental radishes.	Asia
Genecorp	Lettuce.	US
Horticeres	Beans, lettuce, okra, tomato, tropical cauliflower.	Brazil

Seminis – At a Glance

- Subsidiary of Mexico based conglomerate Savia
- 1999 seed revenues : US $ 531 million
- World's largest vegetable seed company
- World's fifth ranking seed company
- Controls 40 % of US vegetable seed market
- Presence in 120 countries : 70 research stations in 19 countries and production sites in 32 countries.
- Acquired by Monsanto in March 2005 in US $ 1.4 billion

Nunhems

Nunza has been the holding company for an international group of enterprises active worldwide in research breeding, production and marketing of vegetable seed for professional horticulture business. Nunza Head Office is located in

Netherlands. The brand name Asia/Pacific, Africa and Europe is Nunhems whereas Sunseeds as the brand for the America's. The principal products for Asia are : watermelon, hot pepper, eggplant, melons, cauliflower, onion, tomatoes, okra, asparagus, cabbage, sweet pepper, cucumber, carrot, squash and gherkins. Today, Nunza is among the worlds four leading vegetable seed companies. The companies are located in USA, Chile, Mexico, Brazil, The Netherlands, Spain, France, Italy, Germany, Hungary, Turkey, UK, Poland, Ukrain, Middle East, China and India.

Nunhems seeds India is fully integrated vegetable seed company involved in a wide range of crops. Its corporate office is in Gurgaon. The registered office is based at Hyderabad. The research stations of this company are located at Bangalore, Gurgaon and Kullu. The products of this company are as given in Tables 22.4-22.17.

Shriram Bioseed Genetics Limited

DCM Shriram Consolidated Ltd. (DSCL) is a Rs. 1550 crore (approx. US $ 300 m) public limited, multi-location conglomerate based in New Delhi, India. Shriram Bioseed Genetics India Limited has become a subsidiary of the Company whose primary business are :

- Agri-Business (Urea, fertilizer, Sugar, Farm inputs marketing such as DAP, Pesticides, Seeds, Agri retailing - Haryali Kisan Bazaar).
- Plastics (PVC and PVC compounds)
- Chemicals (Chlor-Alkali)

Based at of Hyderabad, the Shriram Bioseed Genetics India Ltd. Unit of DSCL is one of the emerging leaders in hybrid seeds, especially in corn hybrids. Other seed varieties include - cotton, bajra, jowar, etc.

Product Description

Corn

DSCL's Shriram BioSeed subsidiary offers the following range of products for corn hybrid seeds :

Bioseed - 9681 (Yellow Orange), Bioseed - 9637 (Orange Yellow), Bioseed - 9636 (Yellow Orange), Bioseed - 9690 (Yellow), Bioseed - 9621 (White), Bioseed - 9718 (White)

Bajra

DSCL's Shriram Bioseed subsidiary offers the following range of products for Bajra seeds :

Bioseed - 8421, Bioseed - 8434, Bioseed - 8469, Bioseed - 8448, Bioseed - 8493

Table 22.4. Morphological characteristics of eggplant hybrids of Nunhems

Variety	Plant habit	Days to first harvest	Av. Fruit weight (g)	Fruit shape	Fruit colour	Calyx colour	Overall features
BE 707	Tall, erect and upright	65-70	300-400	Oval round	Deep purple	Green	Thornless, solitary bearing, partial resistance to mites
Sandhya (257)	Erect, bushy	45-50	250 -300	Round	Glossy deep purple	Green	Thornless, solitary bearing, very early
Abhishek (256)	Tall, erect and upright	55-60	250-300	Round	Deep purple	Green	Thornless, solitary bearing, partial resistance to sucking pests
Black Boy (175)	Tall, erect and upright	50-55	200-250	Oval round	Blackish purple	Green	Thornless, solitary bearing, partial resistance to sucking pests
Yamini (177)	Erect and upright	45-50	100- 150	Oval	Glossy blackish purple	Green	Thornless, long shelf life, solitary bearing, partial resistance to sucking pests
Kareena (1002)	Bushy and vigorous	50-55	80-100	Oval round	Glossy blackish purple	Green	Partial resistance to sucking pests and high temperature
Jaya (1001)	Bushy and spreading	55-60	40-50	Round	Purple	Green	Cluster bearing (3-4), thornless, firm fruits
Kanha (091)	Erect and bushy	50-55	50-60	Oval round	Glossy deep purple	Green	Cluster (3-4) bearing, partial resistance to sucking pests
BE 708	Spreading and bushy	55-60	200-250	Oblong	Deep purple	Purplish Green	Thornless high yielder
BE 706	Tall Erect and Bushy	50-55	100-130	Long	Deep purple	Green	Thornless, solitary bearing, straight fruit, higher yielder
Shilpa (743)	Erect and Bushy	50-55	100-125	Long thin	Blackish purple	Purplish Green	Partial resistance to sucking pest and resistant to bacterial wilt
Chaya (3715)	Erect and Bushy	55-60	70-80	Half long	Shining blackish purple	Green	Cluster bearing (2-3), resistance to bacterial wilt
BE 925	Upright and Erect	60-65	80-100	Long	Glossy green	Green	Solitary bearing, partial resistance to sucking pests
Mohini	Erect	55-60	75-80	Oval	Glossy pink with white stripes	Green	Cluster bearing, thornless
Super Mohini	Prostrate and Bushy	50-55	75-80	Oval to round	Glossy pink with white stripes	Green	Cluster bearing, partial resistance to sucking pests and rain fall
Kamini	Erect and bushy	60-65	65-75	Oval	Pink with white stripes	Green	Thorny and cluster bearing
Harshita (1346)	Semi Erect	50-60	70-100	Oval to round	Green with white stripes	Green	Thornless, cluster bearing and partial resistance to bacterial wilt
Pinki (1364)	Erect and upright	50-55	50-60	Long	Pink	Purplish Green	Cluster bearing, resistance to B. wilt

Table 22.5. Morphological characteristics of hot pepper hybrids of Nunhems

Variety	Plant vigour	Maturity	Fruiting habit	Fruit length (cm)	Fruit diameter (cm)	Skin thickness	Fruit colour	Pungency	Uses	Pepper type	Other features
New Vardan (693)	Strong and semi erect	Early	Pendent	12.1	0.88	Thin	Light green	Pungent	Fresh and drying	Cayenne	Heat tolerant
Ujala (2600)	Strong	Early	Pendent	12.5	1.10	Thin	Light green	Very pungent	Fresh and drying	Cayenne	Partial resistant to PVY, mites and stress
Sona (MH-2)	Meidum	Early	Pendent	8.0	0.80	Thin	Light green	Very pungent	Fresh and drying	Cayenne	Resistant to PVY
Mashal (169)	Strong and tall	Early	Pendent	15.0	1.20	Medium	Light green	Mild pungent	Fresh and drying	Cayenne	-
Red Star	Strong	Early	Pendent	14.0	1.00	Thin	Light green	Mild pungent	Fresh and drying	Cayenne	Resistant to CVMV
Sankranti (695)	Strong	Early	Pendent	13.0	1.20	Medium	Light green	Mild pungent	Fresh and drying	Cayenne	Resistant to CVMV
Dushhera	Strong	Medium early	Pendent	11.5	1.20	Medium	Green	Pungent	Fresh and drying	Cayenne	Good for processing and storage
Devanur Deluxe	Strong	Medium early	Pendent	12.0	1.30	Medium	Green	Very pungent	Fresh and drying	Cayenne	Resistant to CVMV and stress
Soldier (MH-1)	Strong	Medium early	Pendent	10.0	1.00	Thin	Green	Very pungent	Fresh and drying	Cayenne	Resistant to CVMV
Deepali (010)	Strong	Medium early	Pendent	8.0	0.90	Thin	Green	Very pungent	Fresh and drying	Cayenne	G-4 type hybrid
Green Wonder	Strong	Medium early	Pendent	15.0	1.70	Thin	Green	Mild pungent	Fresh and drying	Anaheim	Partial resistant to high temperature
Rudra (1689)	Medium	Early	Pendent	12.0	1.40	Thin	Light green	Mild pungent	Fresh and drying	Anaheim	-
Kranti (341)	Strong	Early	Pendent	13.5	1.40	Thin	Light green	Mild pungent	Fresh and drying	Anaheim	-
Veeru	Strong	Early	Pendent	12.0	2.00	Thin	Green	Mild pungent	Fresh and drying	Anaheim	-
Shikha (347)	Strong	Early	Pendent	8.0	0.80	Thin	Green	Very pungent	Fresh and drying	Cayenne	Cluster bearing
Anjula (2K002)	Medium	Early	Pendent	6.0	1.20	Medium	Dark green	Very pungent	Processing	Jalopeno	-
Sindoor (2K212)	Strong	Early	Pendent	13.0	1.20	Medium	Green	Mild pungent	Fresh and drying	Cayenne	High yield, continuous growth and partial resistant to mites

Table 22.6. Morphological characteristics of tomato hybrids of Nunhems

Variety	Market type	Days to 1st harvest	Av. Fruit wt. (g)	Fruit shape	Fruit firmness	Mature fruit colour	Disease resistance	Other features
Hybrid determinate								
Raja	FM	60-65	70-80	High round	Firm	Deep red	-	High yield and early maturity
S-7712	FM	65-70	80-90	Oval to oblong	Very firm	Red	Partial resistance to high temperature	Wide adaptability, partial resistance to LB

Contd...

Table 22.6. Contd...

Hybrid large determinate								
S-230	FM	75-75	100-110	Heart	Firm	Bright red	Fusarium 1 & 2	Good shipping quality
S-176	FM	75-80	100-110	Heart to oval	Firm	Bright red	Early blight, V F_1 and F_2	Good shipping quality
S-7704	FM	75-80	80-90	Oblong	Very firm	Red	V, F_1 and tolerant to BW	Good foliage cover
Hybrid indeterminate								
S-7730	FM	60-65	80-90	High round	Firm	Red	-	Early maturity
S-7610	FM	60-65	70-80	Deep oblate	Firm	Deep red	V, F_1 and F_2, high level partial resistance to BW	-
Avtar (7711)	FM	70-75	90-100	High round	Firm	Red	High level of partial resistant to BW	Good hilly areas
Tomato leaf curl virus – large determinate								
NP 7725	FM	65-70	90-100	Round	Firm	Deep red	TLCV	Early maturity
Processing								
6109	Processing and FM	60-65	90	Blocky	Firm	Deep red	-	Attractive fruit colour and high brix (5.5)
Mini tomato – indeterminate								
Anant	Mini tomato	45-50	12-13	Round	Medium firm	Red	-	Sweet taste, high Brix (7.5)

Table 22.7. Morphological characteristics of Nunhems Watermelon hybrids

Variety	Type	Rind colour	Fruit shape	Fruit weight (kg)	Flesh colour	Brix sugar content	Other features
Madhuri (64)	Jublee	Light green with dark green stripes	Oblong	10-12	Red	12 %	Granular texture, early maturity
Madhubala (80)	Jublee	Light green with dark green stripes	Oblong	10-12	Red	12 %	Granular texture, medium maturity, wide adaptability
Khushboo (10)	Jublee	Blackish green stripes on green background	Oblong	10-12	Deep red	12-13 %	Granular texture, wide adaptability, medium maturity, excellent shipping quality

Contd...

Table 22.7. Contd...

Mumtaz (31)	Jublee	Light green with green stripes	Oblong	12-13	Red	12 %	Medium maturity, god field holding and excellent for shipping
Red honey	Jublee	Light green with dark green stripes	Round	4-5	Red	11 %	Small attractive fruits
Surbhi	Sugar Baby	Dark green	Round	7-8	Red	11-12 %	Granular texture, high productivity
Maya	Sugar Baby	Dark green	Oblong	7-8	Red	12-13 %	Granular texture, good shipping quality

Table 22.8. Morphological characteristics of okra hybrids of Nunhems

Variety	Plant height	Days to 1st harvest	Fruit colour	Fruit angle	Fruit length (cm)	Fruit width (cm)	Other features
S-40	Tall	45-50	Green	Pentagon	12-12.5	1.4	High level of resistance to YVMV
Uphar	Tall	42-45	Green	Pentagon	12.5	1.4	Early maturity, wide adaptability, high level of resistance to YVMV
S-008	Tall	48-52	Green	Pentagon	13.4	1.5	Suitable for South India
Makhmali	Medium tall	45-50	Glossy deep green	Pentagon	11-12.4	1.5	Attractive fruit colour, high level of resistance to YVMV and suitable for all okra growing areas
Tulsi	Medium tall	45-46	Glossy dark green	Pentagon	12-12.5	1.4	Attractive fruit colour, high yield and high level of resistance to YVMV
Shaina	Tall	48-50	Glossy dark green	Pentagon	12-13	1.4	High level of resistance to YVMV, suitable for all okra growing areas
S-90	Medium tall	45-48	Glossy dark green	Pentagon	10-11	1.8-2	High yield and best suitable for fresh expon and processing
Ladys Pride (NP 1103)	Medium tall	45-46	Glossy dark green	Pentagon	11-12	2.04	High level of resistance to hot and humid weather and yellow vein mosaic virus

Table 22.9. Morphological characteristics of cabbage hybrids of Nunhems

Characteristics	Bahar	Mansi	Bharat (45)	Pragati	Unnati	RIa (105)	Greenbali
Leaf type	Medium spreading	Less spreading	Medium spreading	Less spreading	Spreading	Medium spreading	Less spreading
Leaf colour	Bluish green	Bluish green	Bluish green	Green	Green	Bluish green	Green
Head shape	Round	Round to oval	Round	Round	Flat	Round	Round

Contd...

Table 22.9. Contd...

Head compactness	Very compact	Very compact	Compact	Compact	Compact	Compact	Compact
Head weight (kg)	1.5-2.0	0.9-1.25	1.25-1.5	0.9-1.0	2.5-3.0	1.5-2.0	0.6-0.7
Days to first harvest	65-75	60-65	55-60	55-60	80-85	70	55-60
Reaction to bursting	Partial resistance	Partial resistance	Partial resistance	Partial resistance	Partial resistance	Partial resistance	Partial resistance
Reaction to diseases and DBM	-	Intermediate level to black rot, soft rot and heat	Intermediate level to black rot, soft rot and heat	Partial resistant to fusarium yellows and cold	Intermediate level to black rot, soft rot and heat	Black rot, yellows and soft rot tolerant	-

Table 22.10. Morphological characteristics of cauliflower hybrids of Nunhems

Variety	Plant height	Maturity	Head av. wt. (g)	Head shape	Head colour	Season	Other features
Himgouri	Semi upright	55-60	400-800	Semi dome	Wjite	June - 15th Aug.	Heat tolerant and compact head
Sarita (NP1000)	Semi upright to . upright	60-65	800	Dome/semi dome	Pure white	15th June - 15th Aug.	Good self covering leaves, uniform plant stand and wide adaptability
Parvti	Open to semi upright	55-60	600-800	Semi-dome to flat	Creamy white	15th June - 15th Aug.	Vigorous plant with compact curd quality
Shweta	Semi upright	65-70	600-800	Dome	White	July - August	Compact white curds
Tushar	Upright	55-60	800-1000	Semi dome	Snow white	May - September (Western plains)	Excellent curd quality, wide adaptability
Gaurav (87002)	Upright	60-65	700-800	Dome	Snow white	May - July (Oct. - Nov.) (Southern plains)	Uniform plant type and attractive curd quality
Himpushpa (NP 1801)	Upright	65-70	800-1000	Dome-semi dome	White	Aug. - Sept. (Northern and western plains)	Compact white curds, good market value

Table 22.11. Morphological characteristics of Sweet Pepper of Nunhems

Variety	Maturity	Plant vigour	Fruit colour	Fruit firmness	Thickness	Other features
Anupam	Early	Green, red	Green, red	Very firm	Medium thick	More fruiting, tolerant to heat and pungent
Alankar (6602)	Medium early	Deep green, red	Deep green, red	Excellent	Thick	Blocky fruits (9 x 8 cm), high level of partial ressitance to TMV and PVY
Harit (6703)	Medium early	Glossy green, red	Glossy green, red	Very firm	Thick	Blocky fruits (9x8 cm), thick walled and processing type
Nutan	Medium early	Glossy green, yellow	Glossy green, yellow	Very firm	Thick	High yielder, blocky fruits, 11-12 x 8-9 cm

Table 22.12. Morphological characteristics of pickling cucumber of Nunhems

Variety	Fruit shape	Fruit colour	LB ratio	Fruits/kg (Grade-1)	Fruit surface	Internal quality	Fruit firmness	Shelf life
Alax	Slightly tapered	Medium green to dark green	3.1:1	200-250	Spines	Excellent	Very firm	Good
Stimora	Cylindrical	Dark green	3:1	200-250	Smooth	Excellent	Very firm	Good

Table 22.13. Morphological characteristics of melon hybrids of Nunhems

Characteristics	6281	Castella	Draks
Maturity	Mid-late	Medium early	Early
Shape	Round	Round to oval	Round
Skin colour before maturity	Green	Green	Green
Skin colour at full maturity	Yellow	Grey green	Yellow
Netting	Semi-netted	Semi-netted	Semi-netted
Sutures	Yes	Yes	Yes
Av. Fruit weight	800-900 g	1200-1400 g	1200-1400 g
Sugar content	High	High	High
Flesh colour	Orange	Deep orange	Orange
Aroma	Fragrant and aromatic	Externally fragrant and internally aromatic	Fragrant and aromatic
Shipping ability	Very good	Very good	Very good
Disease reaction	Resistant to F.o.m. races, 0,1,partial resistant to powdery mildew	Resistant to F.o.m. races, 0,1,partial resistant to powdery mildew	Resistant to F.o.m. races, 0,1, partial resistant to powdery mildew

Table 22.14. Morphological characteristics of onion hybrids of Nunhems

Characteristics	Rojo	9730	9539
Plant vigour	Very good	Good	Very good
Foliage length	45 - 50 cm	40 - 45 cm	50 - 60 cm
Foliage colour	Green to dark green	Green to dark green	Green to dark green
Bulb shape	Round	Globe	Oval round
Bulb size	Medium large	Medium	Medium large
Exterior colour	Dark red	Dark red	Pink
Pungency	Less	Less	Medium
Bulb uniformity	Good	Good	Good
Bulb firmness	Firm	Very firm	Firm
Neck	Medium	Narrow	Medium

Table 22.15. Morphological characteristics of ridge gourd hybrids of Nunhems

Variety	Vine length (meter)	Plant vigour	Fruit colour	Fruit length (cm)	Fruit weight (g)	Fruits per vine	First harvest (days)	Duration of crop (days)
Green Gold	3.5	Good	Green	45-50	300	20-22	48-50	120
Narmada	3.5	Very good	Green	50-55	200-210	20-25	50-55	120
Lata	3.0	Very good	Bright green	45-50	225-230	22-25	45-48	115-120

Table 22.16. Morphological characteristics of bitter gourd hybrids

Variety	Plant habit	Maturity	Shape	Fruit size (cm)	Other features
Summer Green	Medium vigorous	55-60 days	Spindle	12 x 3.2	Extra early, good for stuffing
Chaman	Vigorous	60 days	Straight with spines	15 x 3.2	Prolific, crispy, good for frying and stuffing
Kiran	Vigorous	60 days	Straight long with spines	20 x 2.8	Productive, crispy and good for frying

Table 22.17. Morphological characteristics of bottle gourd hybrids of Nunhems

Characteristics	Gauri (99001)	Nargis
Plant vigour	Good	Good
Days to 1st harvest	42-45 days after sowing	50 days after sowing
Fruit shape	Long cylindrical	Long cylindrical
Fruit colour	Light green	Light green
Av. Fruit length	30 – 32 cm	30 – 35 cm
Av. Fruit girth	6 – 7 cm	6 – 7 cm
Av. Fruit weight	800 – 1000 g	1000 – 1200 g
Av. Number of fruits/plant	16 – 20	20 – 25
Special features	Early and high yielder	High yield potential

Jowar

DSCL's Shriram Bioseed subsidiary offers the following range of products for jowar hybrid seeds :

Bioseed - 7508, Bioseed - 7707

Cotton

Bioseed - 6569, Bioseed - 6527, Bioseed - 6188, Gabbar, Bioseed - 6669 (North), Bioseed - 6317, Bioseed - 6488 (North)

SSG (Sorghum Sudan Grass)

Bioseed - 766 (white), Bioseed - 777 (pink)

Paddy

Bioseed - 401 (Sonali), Bioseed - 403 (Shabnam)

Sunflower

Bioseed - 5457, Bioseed - 5468, Bioseed - 5445

Pro-Agro Seeds

The development of Plant Biotechnology (also referred to as 'Green Biotechnology') started in the early 1980's. The last decade has seen a dramatic restructuring of the seed / agro-chemindustry to secure market access. Through the acquisition of Aventis Crop Science, Bayer has entered into the field of Biotechnology & Seeds.

In India, the activities of biotechnology are undertaken through Proagro Seed Company Private Limited. Proagro Seeds, a company of the Bayer Crop Science group, is one of the leading seed companies in India and is engaged in breeding, development, production and marketing of high quality hybrid field crop seeds. To cater to the specific needs of farmers, the Company offers a wide product range covering corn, cotton, rice, pearl millet, grain sorghum, sunflower and mustard. The company has an excellent reputation for high yielding field crop seeds as well as dedicated and experienced personnel.

Product Development through Genetic Research

The company has Biotech laboratories in Gurgaon and Hyderabad where genetic markers, DNA fingerprinting and other tested techniques are used to provide support to the company's plant breeding efforts.

Seed Production, Conditioning and Distribution

Proagro has established seed production centers in 4 states and plans to develop production centers in other states as well. The Company has seed conditioning plants in Hyderabad , Gurgaon and Banglore. A state-of-the - art seed conditioning plant has been commissioned at Toopran in Andhra Pradesh. Company warehouses are located throughout the country, including a central warehouse at Toopran with 10,000 mt capacity. Proagro has a well-established network of distributors across the country who enable timely reach of seeds to farmers in remote areas.

Quality Assurance

Proagro ensures that only the highest quality seed reaches the farmers. Quality is assured by testing the seeds for germination, seed moisture, physical and genetic purity. Extensive tests are carried out in laboratory and fields grow–outs to ensure high quality standards.

With several of Proagro's products notified for cultivation by the Government of India, the company's product portfolio is designed to help farmers increase productivity, ensure high yielding crops and provide innovative solutions to the agricultural community. The company has received acclaim from several quarters for its initiatives and in recognition of its efforts to provide high quality products, the Government of India honored Proagro with the National Award for Excellence in R&D efforts in 1990. The company's centers for carrying out in-house Research and Development are also well- recognized by the Government of India. The Company's seed-testing laboratory at Hyderabad has also received accreditation by the International Seed Testing Association.

Advanta Seeds

Advanta Seeds is a leader in the research and development of canola and sunflower seed technology. They have a rich history of performance in Canada and are leaders in seed innovation, being the first company to introduce hybrid Canola to the Canadian market. Their goal is to accept no limits on what can be achieved through breeding efforts. They want to provide producers with the best seed in the world, by combining values and experience in plant breeding with the latest in modern science and technology. The result is an impressive portfolio of seed technology families that include: Hyola hybrid canola; Hysyn Polish synthetic canola; HyLite Open Pollinated Canola; and Interstate sunflower hybrids.

Advanta Seeds is growing in Canada. They have sales, research, and production agronomists located across the country to support broad distributor network and growing list of research and production locations. Their head office for research, marketing and administration is in Winnipeg, Manitoba with canola seed production based in southern Alberta, Cranbrook, British Columbia and New Brunswick. Their canola breeding program is based in Manitoba with research and demonstration trials located across the Prairies.

Advanta India Limited- Mission

To be leaders in developing market-focused products of superior value based on world class technology which can enhance agricultural productivity and farmer prosperity, while growing a sound and profitable enterprise for the stakeholders by an entrepreneurial organisation committed to total quality.

A Brief Profile

Advanta India Ltd. (formerly ITC Zeneca Ltd.) is the Indian joint venture company of Advanta B.V. which is among top five seed companies of the world. Advanta B.V. was formed by the merger of Zeneca Seeds U.K. and Vanderhave Seeds of Netherlands.

Since the launch of new seed policy on seed development in 1988, several research based international seed companies established their presence in India. ITC ZENECA LTD. was established in India as a 50:50 joint venture between ITC and ZENECA U.K. with the prime objective of conducting research, development, production and marketing of superior high yielding hybrid seeds of crops of national importance.

Since its establishment ITC Zeneca Ltd. has progressed with its vision of growing for a better future with its objective of plant breeding in India for the Indian farmers while networking with Zeneca technology worldwide.

Due to restructuring of promoters/shareholder businesses ITC Zeneca has changed its name to Advanta India Ltd from October 1st, 1998 but continues to maintain the objective of latest technology promising prosperity. Their web site provides details on the products of this technology, which is playing its role to increase Indian agricultural productivity and enhancing rural prosperity.

The prosperity of India's agriculture based economy is dependent on use of progressive techniques by the Indian farmers. The use of hybrid seeds, which is the most cost effective way of improving productivity, is one such critical areas. Backed by sound infrastructure which includes research stations in various agro-climate zones, two state of art seed conditioning plants and national sales and distribution network, Advanta India Ltd. is committed to serve this critical requirement of Indian agriculture with an assurance of total quality and service. Now it has been merged with India Phosphorus Ltd.

Products

Advanta has products in the following crops in India :

- Sunflower
- Maize
- Bajra
- Grain sorghum
- Cotton

- Paddy
- Mustard
- Sorghum Sudan Grass

Information about these products are:

Sunflower PAC – 36

Largest selling hybrid in Indian sunflower seed market specially suitable to areas with low moisture.

High yield
Widely acceptable
High oil (40 - 42 %)
Moisture stress tolerant
Medium duration (90 days)
Medium height
Notified by Govt. of India

Sunflower PAC – 8699

Extremely high yielding full season hybrid with very high oil content specially suitable to irrigated regions of southern India and early Rabi sowings in Bihar.

Very high Yield
Very high oil (42 - 44 %)
Dark black grain
Full season (110 days)
Medium tall, suitable to irrigated/assured rainfall conditions
Notified by Govt. of India

Sunflower PAC-308

A high yielding medium maturing hybrid suitable to irrigated regions

High yield
High oil (40 - 42 %)
Medium duration
Medium tall
Specially suitable to irrigated conditions
Identified by Indian Council of Agriculture Research

Sunflower PAC – 302

An early maturing hybrid especially suitable to irrigated regions

Early maturity (80 days)
High yield
High oil (40 - 42 %)
Black grain
Wide acceptability
Medium tall
Specially suitable to irrigated conditions/assured rainfall

Maize PAC – 701

Early maturing, high yielding hybrid with orange flint grins specially suitable for green cob and early maturity requirement

Early maturity (80 - 85 days)
High yield
Orange flint grain
Uniform size green cobs
Relativily tolerant to blight
Identified by Indian Council of Agricultural Research

Maize PAC – 9714

Widely adaptable medium duration hybrid with high yield, orange flint grains and stay green character

Widely adaptable
High yield
Medium maturity (110 - 115 days)
Orange flint grain, fetches better price
Stay green character
Notified by Government of India

Maize PAC – 9112

Extremely uniform, widely adaptable medium duration hybrid with high yield and orange flint grains

Extremely uniform
High yield
Medium maturity (110 days)
Wide adaptability
Orange flint grains, fetches better price

Maize PAC – 9735

Extremely high yielding, full season hybrid with fair amount of moisture stress tolerance and orange yellow semi-flint grains

Extremely high yield
Wide adaptability
Medium maturity (110 - 115 days)
Semi-flint orange yellow grains
Fairly tolerant to moisture stress

Baby Corn PAC 792

A very high yielding baby corn with golden yellow ears

High yield
Golden yellow coloure baby corn
Less rejection
Excellent quality

Bajra PAC – 931

A medium maturity high yielding hybrid with wide adaptability specially suitable for relatively heavier soils

Medium maturity (80 – 85 days)
Very high yield
Wide adaptability
Compact long candle shaped ears
Bold grains
Medium height, tolerant to downy mildews

Bajra PAC – 938

An early maturity high yielding hybrid specially suitable for relatively lighter soils and drier conditions

High yield
Early maturity (70 – 75 days)
Relatively suitable for lighter soils
Fairly tolerant to drier conditions
Tolerant to downy mildews

Grain Sorghum PAC – 501

A widely adaptable high yielding hybrid with pearly white quality and bold grains

High yield
Widely adaptable
Early maturing (95 - 100 days)
Compact long candle shaped ears
Bold seeded semi loose panicle

Grain Sorghum PAC – 505

A very early hybrid with high yield and pearly white grains
Very early maturity (90 - 95 days)
Escapes prolonged terminal moisture stress
High yield (difference is more if the crop gets exposed to terminal moisture stress)
Semi-loose panicle
Pearly white grains

Grain Sorghum PAC – 537

A very high yielding medium maturity hybrid with fairly good grain quality

Very high yield
Wide adaptability
Medium maturity (110 days)
Semi compact panicle
Bold grain

Cotton PAC - 101

Early maturing hybrid with wide adaptability, high yield and fine quality lint

Early maturity (140 - 150 days)
Wide adaptability
Synchronous maturity
High yield
Yellow coloure petals with spot
Good quality fine lint

Cotton PAC – 103

Early maturing hybrid with excellent lint quality

Early maturity
High lint yield
Tolerant to *Heliothis*
Very big, good boll size
Quality lint

Cotton PAC-131

Medium maturity, high yielding hybrid with excellent lint quality

High yield
Full season maturity (160 - 170 days)
Wider adaptability
Tolerant to sucking pests
Excellent lint quality

Cotton PAC - 132

A widely adaptable stable performing medium maturity hybrid with good lint quality

High adaptability
Medium maturity (155 - 160 days)
High yield
Good rejuvenation capacity
Good fibre quality

Cotton PAC – 133

Extremely high yielding full season hybrid with excellent lint quality

High yield
Full season maturity (165 days)
Wider adaptability
Excellent lint quality

Paddy PAC-801

Very high yielding adaptable hybrid suitable to par boiled regions

Very high yield
Early maturity
Highly adaptable
Less chaff
High tillering
Suitable to par-boiled areas

Mustard Hyola PAC-401

A double zero mustard hybrid with extremely high yield, resistant to white rust and tolerant to frost

High yield
Very high oil (42 %)

Resistant to white rust

Tolerant to frost

Low erucic acid (< 2 % - healthy oil)

Low Glucosinolates (< 30μM/g – export quality cake)

Sorghum Sudan Grass

SSG JUMBO

An extremely high yielding SSG suitable to all pars of the country

Extremely high forage yield

Thin stems

High palatability

Low HCN (< 30 mg/100 g DM)

Late first cut

Notified by Govt. of India

Corporate and Registered Office

ADVANTA INDIA Ltd.

405, 4th Floor, 'A' Wing, Carlton Towers

No. 1, Airport Road, Bangalore – 560 008

Website : www.advantaindia.com

MAHYCO (Maharashtra Hybrid Seed Company)

Mahyco was founded in 1964, by Dr. Badrinarayan R. Barwale (Winner of the World Food Prize-1998). 'Mahyco' is a pioneer and leader in the Indian Seed Industry and is India's largest private sector seed company. Since its inception it, has been engaged in plant genetic research and production of quality hybrid seed for the farming community of India in hybrid sorghum, hybrid pearl millet, hybrid sunflower and now hybrid wheat. First company in the world to successfully commercialize Hybrid Cotton based on GMS / CMS. In 2002 became the first Indian company to be allowed to commercially grow and market Bt cotton- India's first GM crop.

Currently, engaged in the research, production, processing and marketing of approximately 200 hybrids in 30 crop species including cereals, oilseeds, fibre and vegetables, Mahyco is now all set to launch products developed through the process of biotechnology. It has made a breakthrough in the development of hybrids in rice and wheat. This will go a long way in addressing the need of feeding the growing world population.

Maharashtra Hybrid Seed Company (MAHYCO) produces and markets a broad range of quality seeds developed with the latest scientific advances in biotechnology, plant breeding and crop production in India. The headquarters is located at Jalna, about 60 km from Aurangabad in the state of Maharashtra.

MAHYCO is on the verge of greater achievements in the years to come. MAHYCO is poised to launch and commercialize breakthrough products like the bollworm resistant cotton, hybrid wheat and hybrid rice and a host of new products that will provide increased productivity and prosperity to the farmers. MAHYCO is also making a major thrust in exports and has the confidence to build a truly global organization in the coming years.

Infrastructure

More than 100,0000 grower farmers, 33 production centers, 8 state-of-art processing units with in-house quality assurance facilities, a million quintals of seed processing capacity at 14 processing plants and more than 25000 quintals of dehumidified storage for sensitive seed material are highlights of Mahyco's truly huge infrastructure. The production and the processing network is supported by a well developed all India marketing network consisting of more than 5000 sales outlets.

ISO Certification

An ISO 9001 company with largest multi-locational ISO certification Member of ISTA (International Seed Testing Association) under Article 1V (C) of the ISTA certificate, MAHYCO has the ISO 9001 certification award. It is the largest multilocation and certification seed industry in India and probably one of the largest in the world covering 59 locations with a strong R & D infrastructure developed near Jalna (Maharashtra).

Commitment to Research and Development

R&D facility is equipped with Zellweger Uster HVI 900A fiber testing machine, Perkin Elmer Autosystem gas chromatography (GC) machine and a Nuclear Magnetic Resonance (NMR) from Newport Instrument. These instruments are extensively being used for various biochemical analyses of seeds and seed products such as oil from castor, cotton, etc. Substantial investment in R&D has been made by Mahyco to be able to produce superior newer products and maintain leadership. It is the first seed company worldwide to successfully commercialize Fl hybrid cotton based on GMS/CMS system. It is also the first private enterprise in India to produce & market hybrids in sorghum, pearl millet, wheat and sunflower.

Mahyco Research Center is built on 110 acres with buildings for Administrative Offices, Research Laboratories, Greenhouses, etc. near Jalna (Maharashtra). Mahyco Life Sciences Research Centre (MLSRC), at Dawalwadi near Jalna is one of Asia's most advanced seed industry R&D establishments and is well equipped with state of art labs for seed health, molecular biology, cytogenetics, pathology, entomology, molecular virology and plant transformation. The germplasm preserved is one of the most extensive and diverse in the land. Best multi- disciplinary talent, trained nationally and internationally work in the centre. The company carries out its research at 6 centres all over the country. In addition, the multilocational crops performance testing is presently carried at 16 different centers in the country.

Sequencing Lab

Equipped with Perkin Elmer ABI Prism 377, automated DNA sequencer, up to 96 sequences can be made in one run. Up to 800 base pairs in a DNA sequence can be resolved in each sequence. Virus sequences, gene sequence analysis, novel gene sequences, quality control of new constructs are undertaken.

Bio-Technology Lab

Gene discovery projects aim for useful agronomic traits such as drought resistance and insect resistance antibody mediated and gene silencing strategies for pathogen control, alternative male sterility system.

Plant Transformations

Transformations and molecular analysis of transgenic plants are undertaken. Routine genetic transformations of many different crops including paddy, sorghum, tomato, brinjal, okra, etc. are in use.

Molecular Virology Lab

There is an inhouse facility for detection, identification and molecular characterization of plant viruses (geminiviruses, TOSPO, ILAR and Poty groups) of economic importance. Virus resistance screening and identification of prominent genotypes for product development are taken into account. Diagnosis of plant virus diseases by using electron microscopy and serological properties is done.

Plant Disease Clinic

Isolation and identification of pathogen from infected samples is routinely done for detection and recognition of plant diseases. Screening for disease resistance in seeds, etc. is also done.

Applied Molecular Genetics & Molecular Biology Lab

To study the plant pathogen interactions at the molecular level and apply the results in developing durable and/or enhanced disease resistance, microbial genomics programme is in progress to identify novel and agronomically useful genes or gene products from micro flora native to India and apply such identified gene(s) or gene products in developing new tools and technologies for crop improvement. Developing molecular markers for enhancing breeding programmes and to characterize genetically Mahyco's germplasm is on priority. The lab is having fermentation and associated facilities to develop indigenous, cost effective fermentation technologies to produce agriculturally important microbial metabolites.

Greenhouses

6500 sq. meter of climatically controlled Greenhouses have been built to specifications of Govt. of India, Biotechnology Department. Experimental trials and breeding programmes with transgenic cotton, sorghum, rice, brinjal, tomato, etc. are undertaken.

Seed Testing Lab

This labis capable of analyzing isozymes, seed health, germination, physico-chemical characteristics, qualities such as taste, flavor and acceptability of grains. Experimental trials and breeding programmes with transgenic cotton, sorghum, rice, brinjal, tomato, etc. are undertaken. Mahyco products are known for their excellent quality. The production, processing and packaging of seeds are done as per stringent quality assurance norms. Mahyco's seed testing lab (QC) has been granted the status of Member Laboratory of the International Seed Testing Association (ISTA), Zurich, Switzerland since 999. Seed testing for genetic quality is necessary to certify a genetically pure seed. Seeds that do not contain any other seeds of the same crop or other species are highly desirable. Among other tests, isozyme analysis is an important tool to determine genetic purity in seed lots. Mahyco employs this technique for rapid identification of undesirable inbred seeds in Fl seed production programmes. It carries out isozyme analysis in seed quality testing. Sorghum, bajra, maize, sunflower, paddy, tomato, cantaloupe and chilli are some of the crops that are routinely subjected to this test.

DNA Sequencing

MAHYCO - LSRC is equipped with state of the art Perkin Elmer ABI Prism 377 automated DNA sequencer. Up to 96 sequencing reactions can be performed in one run. In each reaction up to 800 base pairs in a DNA sequence can be resolved. Projects relating to virus sequences, plant and yeast gene analysis, etc., are being routinely carried out here.

Plant Transformation

Mahyco Plant transformation facility is capable for genetic transformation of various crops such as rice, wheat, sorghum, tomato, brinjal, etc. It is fully equipped with state of the art equipments necessary for the work. The scientists working in the facility are fully capable of transformation and molecular analyses of transgenic plants.

Electron Microscopy

MAHYCO LSRC's virology division is an in-house facility for detection, identification and diagnosis of plant viruses. It is equipped with Philips CM100 Transmission electron microscope, LKB ultra microtome and Poloron E 6000 carbon coating unit besides a well-trained team of scientists.

Plant Disease Diagnostics

Detection and recognition of plant disease in the field, isolation and identification of pathogen from the infected samples in the laboratory and screening for disease resistance in cereals, vegetables, oilseeds and pulses are the major activities of this facility. It also helps farmers by way of analyzing diseased samples collected from field and suggesting proper control measures as a free service. This laboratory has expertise, in-house facility and well-trained staff to screen various economically important plant diseases.

Mahyco Seeds Ltd.

Hybrid sorghum
Hybrid pearl millet
Hybrid paddy
Hybrid maize
Hybrid wheat
Hybrid cotton
Hybrid sunflower
Hybrid castor
Mustard

Mahyco Vegetable Seeds Ltd.

Hybrid okra
Hybrid egg plant
Hybrid tomato
Hybrid pepper
Hybrid gourd
Hybrid cabbage
Hybrid cauliflower
Hybrid beetroot, radish
Hybrid watermelon

HYBRID PADDY (*Oryza Sativa*)

White Jewel-1

White Jewel- 1 is medium maturity hybrid, tolerant to blast and brown leaf blight. It has excellent tillering habit with more grains per panicle. Grains are medium slender, bold, and translucent with mild aroma.

White Jewel-2

White Jewel- 2 is an early maturing hybrid, tolerant to blast, good tillering habit with more grains per panicle. Long slender, bold attractive grains with mild aroma.

White Jewel-3

White Jewel- 3 is medium maturity hybrid, resistant to blast, has excellent tillering habit, and more grains per panicle and medium slender, medium size translucent grains.

HYBRID WHEAT (*Triticum aestivum* L.)

New Gold

New Gold is medium maturity hybrid with profuse tillering habit; medium bold and amber color shiny grains having excellent bread making quality. It has yet to make its impact.

HYBRID COTTON (*Gossypium arboreum*)

Magic AAC-1

Magic AAC- 1 is highly tolerant to drought and sucking pest, with good seed cotton retention.

Magic AAC-2

Magic AAC- 2 is highly tolerant to drought and sucking pest, with good seed cotton retention and good rejuvenation capacity, medium to big boll size.

(*Gossypium hirsutum*)

Bounty HHC-1 (Gossypium hirsutum)

Bounty HHC- 1 is an early to medium maturity hybrid, resistant to CLCV.

Bounty HHC-2

Bounty HHC- 2 is a hybrid with big boll size, good stay green character, excellent fiber properties, tolerant to CLCV.

Bounty HHC-3

Bounty HHC- 3 is drought tolerant hybrid having dense leaves.

Bounty HHC-4

Bounty HHC- 4 has profuse bearing, performs well under irrigated as well as non-irrigated condition.

(G. hirsutum X G barbadense)

White Silk HBC-1

White Silk HBC- 1 is an open plant type with profuse bearing, extra long and fine cotton, good boll opening.

White Silk HBC-2

White Silk HBC- 2 is an open plant type with semi okra leaves and good boll opening.

HYBRID SUNFLOWER *(Helianthus annuus* L.)

Sunrise-1

Sunrise- 1 is black seeded, convex headed with good centre filling and tolerant to downey mildew.

Sunrise-2

Sunrise- 2 is a hybrid with brown seed having stripes and convex medium size head.

HYBRID SORGHUM (*Sorghum bicolor* (L) Moench.)

White Pearl-1

White Pearl- 1 is a hybrid with semi-compact ear head bearing creamy white round bold grains.

White Pearl-2

White Pearl- 2 is tolerant to shoot fly and aphid infestation resulting clean fodder having lustrous creamy white attractive semi-round grains.

White Pearl-3

White Pearl- 3 is also tolerant to shoot fly and aphid infestation resulting clean fodder having creamy white round medium grains with proven ratooning ability.

Green Grand-1

Green Grand- 1 is dark green, tall, has profuse tillering, TSS 14-16 brix, gives sweetness to palatable fodder, ideal for multiple cutting.

Green Grand- 2

Green Grand- 2 is taller than Green Grand -1.

HYBRID OKRA (*Abelomoschus esculentus* (L.) Moench.)

Green Power

Green Power is lush dark green in color and tolerant to yellow vein mosiac virus with semi-okra leaves and tender fruits. It exhibits delayed fibre formation.

Green Energy

Green Energy has dark, tender fruit and is tolerant to yellow vein mosaic virus.

Green Glory

Green Glory is dark green and longer than Green Energy and tolerant to yellow vein mosaic virus.

Green Jewel

Green-Jewel is dark green in color and tolerant to yellow vein mosaic virus, with semi-okra leaves and thin tender fruits. It exhibits delayed fibre formation.

HYBRID MUSTARD (*Brassica juncea* (Linn)

Sharda-1

Sharda- 1 is medium tall and semi compact type hybrid with blackish brown bold seeds, close bearing silique.

Sharda-2

Sharda- 2 is a medium tall and semi compact type hybrid like Sharda- 1 with shiny blackish brown bold seeds and sparse bearing silique.

HYBRID PEARL MILLET (*Pennisetum glaucum* (L.)

Moti-1

Moti- 1 is medium maturity hybrid with round bold whitish grey grains, long cylindrical, semi-compact ear head and tolerant to downey mildew.

Moti-2

Moti- 2 is an early maturing hybrid with round bold whitish grey grains. It has conical compact ear head and is tolerant to downy mildew and drought condition to some extent.

Moti-3

Moti- 3 is medium maturity hybrid having round bold whitish grey grains, conical ear head and is resistant to downey mildew.

HYBRID MAIZE (*Zea mays*)

Emperor-1

Emperor- 1 is two-cob tendency hybrid, with golden orange grain and excellent shelling %. It is tolerant to rain fed conditions, stem and root lodging.

Emperor-2

Emperor- 2 is late maturity hybrid. Cob is round conical and of big size, yellow, semi-dent, bold grains with good shelling %.

Emperor-3

Emperor- 3 is medium maturity hybrid with cylindrical big size cob having orange semi-flint, bold grains with good shelling %.

Emperor-4

Emperor- 4 is also two cob tendency hybrid like No.1 but hybrid has yellow, semi-flint, and bold grains with good shelling %.

CASTOR (*Ricinus communis* L.)

Black Diamond-1

Black Diamond- 1 is a hybrid with very long compact raceme and spiny capsule.

Black Diamond-2

Black Diamond- 2 is an early maturing hybrid having medium long semi-compact raceme with non-spiny capsules and high oil content.

Black Diamond-3

Black Diamond- 3 is an early maturing hybrid, with spiny capsule.

HYBRID BEETROOT (*Beta vulgaris* L.)

Lalraj

Lalraj is globe shaped with deep red, uniform, sweet and crispy root.

HYBRID RADISH (*Raphanus sativus* L.)

White Early

White Early is early maturing hybrid. The roots are smooth cylindrical with obtuse base.

White Wonder

White Wonder has long tapering white roots with crispy mild taste. It is medium-maturing hybrid.

HYBRID TOMATO (*Lycopersicon esculentum* Mill)

Abhiruchi-1

It has an appealing dark red blocky fruits of uniform size, it is good for fresh market/processing.

Abhiruchi-2

Abhiruchi- 2 has appealing dark red blocky fruits of uniform size, it is good for fresh market and processing.

Abhiruchi-3

Abhiruchi-3 has attractive dark red fruit, ideal for fresh market and is resistant to bacterial wilt.

Abhiruchi-4

Abhiruchi- 4 has oval and compact dark red uniform fruit with good keeping quality and transportability.

Spurthi-1

Spurthi- 1 has eye catching dark red oval fruit, good for fresh market/ processing and provides prolonged span of extended marketing.

HYBRID BOTTLE GOURD (*Lagenaria siceraria* (Mol.) standl.)

Green Major

Green Major has cylindrical, shiny green fruit with lining, retains its lustre and green color for a longer time, ideal for packing and long distance transport.

Green Sergeant

Green Sergeant has an attractive, cylindrical, absolutely straight fruit, ideally picked when 40 - 45 cm long (60 - 65 days old), ideal for packing and long distance transport.

Green Golu

Green Golu has green round fruit raised at pedicel end with average fruit weight of 400 - 500 g.

HYBRID PUMPKIN (*Cucurbita moschata* Duch)

Thunder Ball

Thunder Ball has round flat medium sized fruit. The average fruit weight is 3-3.5 kg.

Sumo

Sumo has big oblong fruit (7 to 8 kg) with solid flesh.

HYBRID SNAKE GOURD (*Trichosanthes cucumerina*)

Sequoia

Sequoia has uniform fruits with prolific bearing. The fruits are attractive white and medium long.

HYBRID ASH GOURD (*Benincasa hispida* (Thunb). Cogn.)

Power Giant

Power Giant has oblong, medium sized, white waxy fruit weighing 5 to 6 kg.

HYBRID SPONGE GOURD (*Luffa cylindrica* (L.) Roetn.)

Emerald

Emerald has shiny dark green slender fruits.

Green Splendour

Green Splendour has dark green colored, long fruits.

HYBRID RIDGE GOURD (*Luffa acutangula* L. Roxb.)

Ridge Queen

Ridge Queen has light green colored long fruit than Ridge King with prominent ridges.

Ridge Long

Ridge Long has eye catching shiny green colored long fruits with prominent ridges.

Ridge King

Ridge King is early hybrid with light green colour fruit with prominent ridges.

Bright Splendour

Bright Splendour has shining-light green slender medium long fruit.

Future Challenges for Indian Agriculture and Mahyco

Indian agriculture faces tremendous challenges of producing enough food for its growing population. This must be accompanied on same or declining land mass. There is a continuing need to increase food production in the developing countries and this increase has to come from increased yields from major crops grown on existing cultivable lands. Many approaches will have to be deployed simultaneously to double India's food production by 2020. Insects not only cause direct loss to the agricultural produce, but also indirectly due to their role as vectors of various plant pathogens. In addition to direct losses caused by insects, there are additional costs in the form of pesticides applied for pest control, currently valued at US $10 billion annually. In crops such as pearl millet, sorghum, pigeonpea, chickpea and groundnut grown under subsistence farming conditions in the developing countries, the losses due to various biotic and abiotic factors have been estimated to be over US $ 2 billion annually. Massive application of pesticides results in adverse effects on the beneficial organisms, leaves pesticide residues in the food causing environmental pollution. In addition, the target pests evolve and become resistant to the pesticide, requiring higher volume of spray for control. As a result, the chemical control of pests is under increasing pressure. There is a greater need to develop alternative or additional technologies, which would minimize pesticide use, and provide adequate crop protection for sustainable food, feed and fiber production in the future. Besides the insect pests, diseases caused by viruses, bacteria, fungi and other microorganisms are emerging and pose greater threats to Indian farming community.

A number of conventional and biotechnological methods that are now available make it possible to transfer, track, isolate and study specific genes for specific traits. With the advent of genetic transformation techniques, it is now possible to insert genes in plant that provide a unique trait. Technologies are now borderless, genes from bacteria can be transferred to plants or genes from unrelated plant species can also be transferred. For example, gene from bacterium such as *Bacillus thuringiensis* (Bt) has been identified and through plant transformation introduced in crops for pest control on a commercial scale. In an attempt to adopt biotechnological innovations for the improvement of Indian agriculture, Mahyco has invested considerably in its R&D during the last 5 five years. Mahyco has in-house research programme and collaboration with other companies. In a joint venture programme with Monsanto Co., U.S.A. genetically engineered cotton containing the 'Cry lAc gene from the bacterium Bt has been brought in the country and this gene introgressed with Mahyco's cotton hybrids. The Bt cotton hybrids which contain the 'Cry lAc' gene, make it resistant to the cotton bollworm pest that inflicts heavy damage to the crop all over India. As this is a genetically enhanced product, regulatory approval is required from Govt. of India. Only recently, Indian government has cleared large-scale field trials on genetically engineered cotton making it the first transgenic crop to be given such permission in India. The Genetic Engineering Approval Committee (GEAC) under the Indian Ministry of Environment and Forests (MOEF) gave the clearance.

With the adoption of biotechnology and state of the art sophisticated instrumentation facility in Mahyco, genetic transformation of various crop species such as rice, wheat, sorghum, brinjal, tomato, etc. is now a routine by Mahyco scientists at R&D) center. Mahyco is also involved in mining microbes and plants for agronomically useful genes from indigenous sources native to India to be introduced in crop plants.

Indo American Hybrid Seeds

IAHS is a pioneering institution and one of the most innovative enterprises of its kind in India. Since inception in 1965, IAHS has made remarkable advances in every sphere of Commercial Horticulture. IAHS is at present engaged in the production and marketing of hybrid seeds in vegetables, flowers, agronomic crops like sunflower, cotton and rice, ornamental plants, biotechnology products, construction of greenhouses and landscape architecture.

Infrastructure, Sales and Marketing

IAHS has a very strong product base with an excellent back up of R& D which has won the National Award for excellence. They have also repeatedly won the APEDA (Agricultural & Processed Food Products Development Authority) Export Award. IAHS is one of the largest hybrid vegetable seed producers and marketers in India with a nation wide strong network of dealers and distributors and regular export of hybrid vegetable seeds and flower seeds.

Product Range

The main items of hybrid vegetable seeds that have been exported are Bhindi (okra), bitter gourd, bottle gourd, brinjal (eggplant), capsicum, hot pepper, radish and tomato. These hybrids are characterised by their quality and high resistance to pests and diseases resulting in better quality crop and increased yields. In open pollinated vegetable seeds several varieties have been popularised such as Bhindi (okra), bitter gourd, bottle gourd, hot pepper, onion, radish, spinach, tomato, etc.

Quality

Due to the strong R& D backup IAHS has been able to maintain the highest level of international quality. Their products have yielded excellent results in both domestic and international markets. The sustained performance has enabled them to get a good market share both in India and abroad.

Vision

- Maintain national technological superiority and strive for global leadership
- Breed, produce and market consistently superior international quality planting material
- Provide other selected agri-horticultural products and services of the highest quality
- Contribute significantly to greening the environment and national food and nutritional security

Research and Development

IAHS has long history of in-house R&D, using conventional breeding to develop high yielding, disease tolerant hybrids of vegetables like amaranthus, bottle gourd, brinjal (eggplant), canola, coriander, cotton, cucumber, hot peppers, melons, okra (Bhindi), onion, radish, spinach bitter gourd, sunflower, sweet pepper, tomato. In recent past, they have developed high yielding hybrids of cotton, sunflower & rice as well. The genetic engineering laboratory of IAHS has major interest in developing transgenic vegetables with tolerance to various diseases and pests using molecular tools in an eco-friendly way. The laboratory is also using tools of molecular breeding like MAS to introduce polygenic desirable traits. Techniques like SDS - PAGE, IEF & DNA fingerprinting are routinely used to estimate the genetic purity of hybrid seeds. The virologists have successfully developed ELISA based diagnostic kits for early and efficient detection of various plant diseases. Haploids are also used to enhance the breeding programme and develop new varieties of various vegetable crops.

Biotechnology is the science which uses the tools of modem genetics to enhance the beneficial traits of plants by modifying fruits, vegetables, grains, and other food crops to produce nutritional benefits, those which are disease tolerant with higher yields and extended shelf lives in addition to other desirable attributes. These innovations include planting and growing techniques and testing procedures for purity and safety.

Biotechnology is very much poised to make a huge difference in the availability of nutritious foods for a booming world population while protecting the depleting natural resources. Primary consumer benefits will be better taste and nutrition and finer quality products.

Contract Production of Seeds

Production Under Controlled Conditions

IAHS has been the first to introduce Greenhouse Technology in India for commercial production of seeds and ornamental plants and also first to establish climate controlled fiberglass covered Greenhouses for commercial production of hybrid seeds and ornamental plants in India. The company takes up construction of such greenhouses on turnkey basis.

Production Under Open Field Conditions

The company has set-up an export oriented unit consisting of 140 greenhouses built with in-house technology for production and export of Fl hybrid capsicum seeds to U.S.A. The project consists of 5,04,000 sq. ft. of Greenhouse covered area which is supposed to be largest unit in Asia. These climate controlled greenhouses have provided employment to hundreds of rural youths and women. This unit has already started production and exported about Rs. 6 crores worth of hi-tech capsicum seeds to U.S.A.

Comprehensive and Intensive Customer Care

IAHS is known for its exemplary customer service-from rendering cultural advice to reliable on- time delivery of seeds and plants. Their sales staff and customer service team work together with production managers and specialists to ensure that the customers get the right information and product they need. Their customer range includes housewives, home owners, market growers and processors throughout the country and abroad. IAHS offers intensive short-duration training on modem production technology.

Major Developments and Awards

1965	IAHS founded by Dr. Manmohan Attavar
1966	First Greenhouse constructed and production commenced
1967	First batch of Fl hybrid flower seeds exported to USA
1968	Mr. N.K. Bhat joined as partner
1973	First Fl hybrid vegetables in India - capsicum & tomato released
1988	Commercial Tissue Culture Lab established, Oilseed Improvement Programme launched
1989	The first ever YVMV tolerant okra hybrids (Varsha, Vijaya, Vishal) in India released
1990	First two Fl sunflower hybrids released, first Genetic Engineering Laboratory for Agriculture in private sector inaugurated
1991	Export Award from APEDA. (1987, 1988, 1990, 1991, 1992, 1996, 1997,1998)
1994	Cotton hybrids, Sanju and Gauri released
1995	The first ever aromatic long grain rice hybrid (Indam 100-001) released "National Award for Excellence in R&D effort" instituted by the Department of Science and Industrial Research, Government of India.
1996	The first ever indigenous leaf curl virus tolerant tomato hybrids (Indam 9501, 9502, 9503 and 9504) released, Visveshwaraiah Industrial Trade Centre Export Award
1997	INDAM CANOLA a "Double Zero" variety of rapeseed (*B. napus*) released Asia's largest climate controlled Greenhouse complex for seed production of capsicum & tomato for exports commissioned
2000	ISTA membership granted to IAHS
1990	Dr. Manmohan Attavar awarded an honorary Doctorate in Science.
1991	Dr. Manmohan Attavar receives Rajyotsava Award for his contribution to Horticulture
1995	Dr. Manmohan Attavar receives the Dr. M. H. Marigowda National Award for outstanding contribution to Horticulture
	Dr. Manmohan Attavar receives the prestigious APEDA Award for "Development and Export Promotion of Commercial Floriculture and Seeds for life time achievement

	Dr. Manmohan Attavar receives the Golden Jubilee International Award instituted by the International Chrysanthemum Society, California, USA
1996	Dr. Manmohan Attavar elected to executive committee and as member of Vegetable Crops of Federation of International Seedsmen, Switzerland
1998	The Republic Day Civilian Honour of "Padma Shri" Govt of India bestowed on Dr. Manmohan Attavar Chairman & Managing Director

Seeds of Different Crops

Cauliflower

Cauliflower is a vegetable that grows well during cool-seasons and is a member of the cabbage family.

Open Pollinated Variety	Indam Early - Creamy curds weighing about 500 g, slightly heat tolerant, curd formation between 20- 25°c
Synthetics	Indam 88-52 - White curds, weighing up to 1 kg, recommended for June-July sowings in South India
Hybrids	Indam 19 - Early, 45-50 days maturity, creamy white curds weighing 500-700 g under curding temperature of 20-25°c
	Indam 2435 - Creamy curds, early, recommended for South Indian summer
	Indam 9803 - White curds, weighing 1.5-2.0 kg, maturing in 80-85 days.

Chilli

Chillies contain capsaicin which is responsible for the hot taste of the fruit. Chillies are rich in vitamins especially vitamins A and C. Chillies are a must in Indian cuisine to lend the hot taste and pungent flavour. The common red chilli dried is the most tangy of all spices.

Hybrid -Green	Indam 6 - Early, slender light green glossy fruits weighing about 5 g, highly pungent
	Indam 9 - Early, semi- dwarf plants producing slender slightly wrinkled light green fruits of Jwala type, weighing 6 g
	Indam 10 - Early, semi- dwarf plants producing slightly thick smooth light green fruits, weighing about 7 g
Hybrid -Green & Dry	Indam-5 - Medium thick, high yielding, dark green fruits turning dark red after ripening, medium pungent, tolerant to thrips and mites
	Indam -42 - Moderately thick and green glossy fruits turning dark red after ripening, slightly pungent, tolerant to thrips & mites

	Indam -20 – Thick & long fruits, high yielding, dark green fruits turning dark red after ripening, medium pungent. Indam -21 - Slender & long, dark green slightly wrinkled fruits turning dark red after ripening, medium pungent, open pollinated variety

Open Pollinated Variety	Indam Jwala - Slender, light green, medium pungent fruits weighing 4-5 g

Capsicum / Bell Pepper / Sweet Pepper

These have straight, woody stems and the fruits are green at first and change to red or yellow. The fruits present a pretty sight above the rich green foliage. Seeds germinate within 6 to 8 weeks when sown in warm, well irrigated soil.

Hybrids	Mahabharat - 3-4 lobed, blocky dark green fruits, having good shelf life Golden Summer - Relatively tall, spreading, with light green foliage with blocky fruits 3-4 lobed and golden yellow in colour Bharat - 3-4 lobed, blocky dark green fruits, tolerant to TMV, excellent shipping qualities Indam 3 - Dwarf, compact plants producing blocky dark green fruits, slightly heat tolerant Indam 5 – Medium tall plants producing blocky dark green fruits weighing about 200 g
Open Pollinated Variety	California Wonder - Blocky dark green fruits weighing about 100 g

Round Cluster

Hybrid/Open Pollinated Varieties/Green

Hybrid/Open Pollinated Varieties/Green	Round Cluster - Round to oval fruits weighing 50- 60 g Long Cluster - Extra early, long slender fruits weighing 30 – 35 g Mysore Green Long - Long slender fruits weighing 80-100 g, borne solitarily

Paddy

Rice is the staple food crop over most of Asia. With the advent of hybrid rice technology the yield plateau in rice has been broken. The adoption of hybrid seeds has made possible the optimum seed rate which in turn brought about reduction of seed cost and with other factors chipping in, eventually reducing cost of cultivation with increased profit. IAHS has hybrid varieties that enhance faster yields, produce long slender grains and which are adaptable to different soil conditions.

Aromatic-open pollinated variety	Indam 9855 - Early maturing variety with long & slender grain
Non-aromatic-early	Indam 200-003 - Early duration hybrids with 115-120 days, suitable for medium & upland areas during Kharif and Rabi seasons
Non-aromatic-mid early	Indam 200-Oil - Suitable for Kharif & Rabi seasons. 120-125 days duration, dwarf & grain comparable to non lodging Parimal type of North India
Non-aromatic-mid late	Indam 200-014 - Suitable for Kharif & Rabi season, dwarf & non lodging, Sona Mahasuri and Samba Mahasuri type grains
	Indam 200-12 - Suitable for Kharif& Rabi, dwarf, non lodging
Non-aromatic-salt tolerant	Indam 300-001 - Kharif hybrid, high yielding can be grown in soils of pH up to 9.5
Hybrid	Indam 100-001 - Medium duration hybrid maturing in 130 days, with long slender grains and white translucent kernels

Green Gram

Indam 10-10 - Dwarf, early maturing, bold grain variety

Sunflower

Sunflower is a tall, strong bright yellow flower that faces the sun, filled tightly with hundreds of seeds. Sunflower has been used symbolically to denote sun and energy throughout history. Sunflower seed oil is today widely used as cooking oil and is rich in vitamin E content. It is rich in polyunsaturates and linoleic oil and has a fairly low rancidity level compared to other oils. Sunflower seeds have good fibre content and are rich particularly in thiamine, pyridoxine, niacin and pantothenic acid.

Sunflower	Ravi - 80-85 days for maturity with 38-40% oil content Arun - 85-90 days for maturity with 36-38% oil content Aditya - 85-90 days for maturity, tolerant to rust and downy mildew and 38-40% oil content

Canola

The quality and height attained by the canola plants squarely depend on growing conditions. Each plant produces groups of yellow flowers which mature into small green pods that become brown on ripening. On harvest, each pod yields about 20 seeds that are crushed to extract the oil and to make meal or stored for future use. In addition to its edible uses, canola oil is used in a number of inedible applications such as cosmetics and printing inks. Canola meal is used as a feed for livestock, poultry and as a fertilizer.

Indam: Double zero variety with high oil content & good performance at higher temperature.

Amaranthus

These are tender annuals which grow straight with the height touching almost 8 ft. The leaves are long or oval, from 2 to 4 inches long and they start out as green or dark red and have pink centres or uniform green colour. The plant is usually cut back to encourage branching and delay flowering. This is one of the best heat-resistant and commonly used greens for cooking.

Indam 1 : Broad leaved green leaves with pink center

Indam Pink : Broad pink leaves, late bolting

Indam Green : Broad uniform green leaves

Indam Red : Strong, narrow, red leaves

Coriander

Coriander is an annual plant with multi lobed leaves. The volatile oil that imparts the fragrant smell to the plant is present in the leaves and also in the seeds. The leaves are used for seasoning and garnishing in Indian and Eastern cuisines. Seeds are used as a condiment, an ingredient of curry powder.

Indam 1 : Early, suitable for multiple harvests

Indam 2 : Dark green leaves with strong aroma, late bolting variety

Brinjal /Eggplant

Brinjal / eggplant fruits are consumed as vegetable that grows well in a long warm season for best yields. This vegetable requires careful attention for a good harvest. Ornamental varieties are also available lending colour to garden.

Hybrid-Purple/Blackish purple-Solitary	Suphal - Bell shaped fruits weighing 400-500 g Supriya - Round fruits weighing 250 – 300 g
	Indam 451 - Round dark purple fruits weighing 200-250 g
	Sourabh - Long slender dark purple fruits weighing 90-120 g
	Suchitra - Oval dark purple fruits weighing 125 – 150 g
Hybrid-Purple / Blackish Purple-Clustrous	Indam 1632 - Oval dark purple fruits weighing 80-90 g

Hybrid-Purple with white stripes	Indam - 2524 - Extra early, with elongated fruits weighing 55 -65 g
	Indam-35 - Oval fruits weighing 90-100 g
Hybrid-Green	Hybrid Green - Solitary long fruits weighing 80-100 g
	Gota-3 - Solitary oval fruits with thorns on calyx
	Gota-4 - Solitary big round fruits with thorn on calyx
Open Pollinated Varieties>Purple	Round Cluster - Round to oval fruits weighing 50-60 g Long cluster - Extra early, long slender fruits weighing 30-35 g
Open Pollinated Varieties>Green	Mysore Green Long - Long slender fruits weighing 80-100 g, borne solitarily

French Beans

French Beans is an annual vegetable with the pods picked just before the seeds begin to swell, at which time they are tender and palatable. The food value is high, comprising 20% proteins and at least 50% carbohydrates.

Indam 2 : Medium long, slender, dark green string- less pods

Cluster Beans

Guar gum extracted from cluster beans is used as a thickening agent and is also a good source of dietary fibre.

Indam CB-2 : Medium tall plants with big pods in clusters

Bitter Gourd

Bitter gourd is a slender, climbing annual with long stalked leaves. The fruit is warty-looking gourd, usually oblong and resembling the ordinary cucumber in shape. The gourd can be parboiled in salted water or rubbed with some salt to remove part of the bitterness. It is considered a fighter against diabetes. 50 ml raw bitter gourd juice daily improves blood glucose tolerance in non-insulin dependent diabetics.

Open Pollinated varieties	Indam Green Long – Green, spiny fruits weighing about 125 g
	Indam White weighing about 125g
Hybrid-Green	Idman -1124 Meidum green spiny fruits, with good keeping quality weighing 150 g
Hybrids-white	Indam 4625 – Spiny fruits, weighing 110 g with high bitterness

Squash

Squash, is a tender, warm-season vegetable that can be grown throughout the year. Healthy and well- maintained plants produce abundant yields. The peel is packed with nutrients so never peel squash.

President - Dark green smooth fruits maturing in 50 days

Clarimore - Yellowish green fruits, mature in 50 days

Ambassador - Dark green fruits coming to maturity in 50 days, longer harvests

Bottle Gourd

Bottle gourd is an annual vegetable, a climber with strong tendrils belonging to the Cucumber family. Leaves are heart to kidney-shaped. The fruits in the green stage are used as vegetables and the dried gourd serves many purposes as water-containers, etc.

Long Cylindrical	Indam 1720 - Pale green fruits weighing 1.2 kg maturing in 60-65 days
	Indam 204 - Pale green fruits, weighing 1.2 kg maturing in 55-60 days
	Indam 9720 (OP) - High yielding open pollinated variety weighing about 1 kg
	PSPL (OP)
Short Cylindrical	Indam 320 - Pale green fruits weighing about 750 g suitable for carton packing, very early

Broccoli

Broccoli is a hardy vegetable of the cabbage family that is high in vitamins A and D. It develops best during cool seasons of the year. New heat tolerant varieties allow broccoli to be produced in all but the hottest parts of the season. A member of the cabbage family and a close relative of cauliflower, broccoli packs more nutrients than any other vegetable.

Packman - Dark green heads, compact curds weighing about 300 g, maturing in about 60-65 days

Khol-Khol

The stem is the edible part and enlarges into a round knob immediately above the ground. It has to be cooked before the flesh becomes woody.

Indam Early White - Early, uniform round knobs weighing 200 g maturing in 55-60 days

Cabbage

Cabbage is a vegetable that grows well in fertile soils. There are different varieties of cabbage available in various shades of green, as well as red or purple. The shape could also vary from the standard round to flattened or pointed. Cabbage is regarded as a good source of vitamins.

Small Frame	Indam 296, Indam 1299, Indam Express – Early hybrids with round heads maturing in 35-60 days and weighing about 1-1.2 kg
Medium Frame	ST(NF)-01 ,ST(NF)- 02 - Early hybrid with globe shaped green heads weighing 1.2-1.5 kg, maturing in 65-70 days
Large Frame	Krishna - Large, globe shaped, bluish green heads weighing 2.5-3 kg, good field holding capacity

Beet Root

The top leaves of the plant are cooked as greens and the roots are cooked or even pickled. The rich crimson colour is a feast to any salad preparations where it lends colour as well as taste. Beet leaves are an excellent source of vitamin A and the roots are a good source of vitamin C. They are particularly rich in foliates and are also high in fibre, both soluble and insoluble.

Open pollinated varieties	Ruby Queen - Round uniform roots with crimson red core
	Indam Ruby Red - Round big roots with red core
Hybrids	High round big roots with deep red core

Onion

Onion is a cool-season vegetable that can be grown successfully throughout most parts of our country. The bulbing process of onions depends on day lengths. High temperatures and low humidity are advantageous during bulbing and curing. Onion is an undisputed ingredient in almost all of the Indian curries.

Indam DR-1 : Globe shaped, deep red medium sized kharif onion

Creole Red : Round, intense red, medium sized bulb

Cucumber

Cucumber is a tender, juicy warm-season vegetable that produces well with proper care and protection. They are the main ingredient for salads lending that crispy juicy and decorative touch.

Fresh Market	Indam 9531 - Medium long, medium green fruits, tolerant to downy mildew
	Priya - Medium long, cylindrical, dark green fruits suitable for slicing
	Indam 10 - Light green fruits tolerant to downy mildew
Processing	Calypso - Suitable for pickling and other products of Gherkin industry
Open pollinated variety	Poinsette - Dark green, medium long, cylindrical fruits

Cotton

Medium / Long Staple	Sanju - Suitable for rainfed cultivation with fibre length of 27.7 mm and 40-50 counts
	Indam 2194 - Suitable for rainfed cultivation with fibre length of 28.5 mm and 40-50 counts
	Indam 206 - Suitable for rainfed cultivation with fibre length of 29.5 mm and 40-50 counts
Extra Long Staple	Gown - Suitable for irrigated cultivation with fibre length of 35 mm and 80-100 counts

Spinach

Spinach is nutrient packed vegetable which finds use daily in most cuisines as it is rich source of Vitamin A, iron, calcium and protein. Spinach can be grown and multiple harvests can be carried out in the right soil and moisture condition.

Indam Spinach I : Dark green foliage, pink base and pigmentation on the stem

Indam Spinach 2 : Light green foliage, green stem, late bolting.

Methi

Fenugreek or Methi is an ancient spice, a brownish-yellow rhombic shape seed which is slightly bitter and aromatic. Dry light roasting can enhance the flavour of the seed, reduce the bitterness and is a must ingredient for Indian pickles. The fresh Methi leaves make a tasty vegetable and is an exotic green used in the North Indian cuisine. This slightly bitter herb is valuable for cooking as they enhance the food's quality most noteworthy being the fact that, they stimulate bile secretion thereby aiding in digestion.

Indam Methi -It has vigorous ash green leaves

Soya

Indam Soya is an early maturing variety with good flavour.

Pak Choi

Pak Choi grows best in a sunny climate all year through in a tropical country like ours. The plant grows well with generous watering and highly fertile soil.

Indam Pak Choi has soft & thick petioled light green leaves

Watermelon

Watermelons grow and thrive well in the warmer parts of the country. Watermelons are low in calories and very nutritious. The hardy green cover protects the rich red juicy flesh inside which has the properties of a powerful antioxidant. Watermelons have a cooling effect.

Open pollinated variety	Sugar Baby - Round, dark green fruits with deep red flesh
	Crimson Sweet - Round, dark green with light green stripes with deep red flesh
Hybrids	Madhu - Medium green, oval fruits with dark green stripes, 8-10 kg
	Pata Negra - Round, intense green fruits weighing 7-8 kg
	Indam 365 - Oblong, light green stripes on dark surface
	Indam 9307 - Oblong, green striped fruits weighing 8-9 kg
	Sweet Dew - Oblong, dark green with light green stripes, fruits weighing about 7-8 kg
	Olympia - Oval, light green stripes on dark green background, fruits weighing about 8 kg

Musk Melon

Muskmelon, also known as cantaloupe, is a tender, heat-loving vegetable. It derives its name from the musk smell that most varieties have. It is mainly a dessert fruit which is very wholesome and nutritious.

Swarna - Slightly elongated, thinly netted, yellowish orange fruits

Sona - Round, closely netted, orange cream coloured fruits with deep orange flesh

Ridge Gourd

Ridge Gourd belongs to the Cucurbit family like cucumber. It is a good source of carbohydrates, vitamin A, vitamin C and minerals. They are vine plants, with smooth, sharply angled 5-lobed leaves. Fruits vary in size, and colour varies to different shades of green. Ridge gourd grows well in a hot and humid climate in a variety of soils.

Hybrids	Indam 1222 - Slender dark green fruits with sharp ridges weighing 300 g
Open Pollinated Variety	Indam -1 - Medium green fruits with sharp ridges weighing about 175 g

Sponge Gourd

Sponge gourds thrive well in a sunny location in a well-drained soil base. Also known as Luffa, this is a vine plant that can climb up to 15 feet which grows fast in a warm climate. The fruit is edible only when it is immature. The matured fruit forms sponge-like fibres inside. The dried matured fruits are often used as scrubbers, cleaning pads or bath sponges and recently they have found uses in a variety of applications.

Indam 23 - Early maturing hybrid, producing dark green fruits of about 150 g

Tomato

Tomato is the most popular vegetable and ingredient which is a must for Indian curries to give the gravy, colour and taste. Tomatoes are easy to grow and each plant produces adequate harvest. Tomatoes are a store house of vitamin C, potassium, fibre and vitamin A. It is a very good appetizer with a refreshing taste.

Open pollinated varieties	Indam-13- Green shouldered deep oblate fruits for fresh market
	Indam-15 – Round fruits for both fresh market and processing
Hybrids-Determinate-Fresh Market	Meghana - Round fruits of 75 g
	Vaishali - Deep oblate fruits with slight heat tolerance
	Ruchi - Oblate, acidic fruits
	Indam 2106 - Round fruits of 85 g
	Indam 2108 - Round fruits of 95 g with good acid and sugar blend
Hybrids-Determinate-Processing	Indam 88-3 - Pear shaped fruits of 85 g in a concentrated red colour
	Rashmi Improved - Square round fruits of 90 g in a concentrated red
Hybrids-Determinate-Dual Purpose	Ramya - Square round fruits of 90 g, recommended for rainy season
	Rajani - Early hybrid, square round fruits
	Rupali - Square round fruits, fruits set at temperature of 35°c
	Rohini II - Square round fruits of 75 g, suitable for long distance shipment
	Akash - Square round fruits of 90 g, suitable for long distance shipment
	Utpan - Early hybrid with round fruits of 80 g in concentrated red

	Rohini - Square round fruits for post rainy season and long distance shipping
	Rashmi - Round fruits for cultivation during hot and dry weather
Hybrids-Indeterminate-Dual Purpose	Naveen - Square round fruits of 90 g with good shelf life
	Naveen 2000+ - Square round fruits of 85 g and wide adaptability
	Rakshita - Square round fruits of 90 g
Hybrids-Disease Tolerance-Leaf Curl Virus	Indam 9501 - Suitable for north Indian conditions
	Indam 9502- Oblate fruits with good heat tolerance
	Indam 9802 - South & Western India, oblate fruits of 80-85 g
	Indam 9714 - South India, round fruits of 65 g
Hybrids-Disease Tolerance-Bacterial Wilt	Indam 1116 - South India, round fruits of 90 g, suitable for fresh markets
	Indam 2107 - Round fruits of 90 g suitable for fresh market
	Rakshita - East India

Radish

Radish is a cool-season, fast-maturing, easy-to-grow vegetable. Radishes grow well in fertile soil with plenty of sunshine and moisture. The popular red radish is low in calories with a distinct flavour. The fleshy root can be eaten both raw as well as cooked. The vegetable contains vitamin B&C and is said to have a cooling effect.

Indam 4 - Long roots, cold season variety, edible leaves

Indam 7 - Chetki type, fleshy leaves, short, stout, cylindrical roots

Namdhari Seeds

Namdhari Seeds was founded by Mr. Uday Singh in 1985 at Urgahalli, Bidadi, Bangalore for contract production of vegetable seeds for export from India. In the due course of time a separate division was created for Research & Development of hybrid vegetable seeds, suitable for Indian agro-climatic conditions. Later on Namdhari Seeds Pvt. Ltd., was formed for domestic marketing of developed hybrid vegetable seeds. His Holiness Sri Satguru Jagjit Singh Ji's role as the mentor and guide of Namdhari Seeds Pvt. Ltd. in setting the success achievements is impeccable. Namdhari Seeds export high quality seeds to USA, Japan, Europe and other countries, on the basis of contract production. They are selling locally bred and developed hybrid vegetable seeds like tomato, watermelon, hot pepper, etc. in Indian market and neighboring countries. Urgahalli Estate was originally bought for a study farm, keeping the mild climatic conditions in mind. The farm measuring 185 acres was identified and selected by His Holiness Sri Satguru Jagjit Singh Ji. But as it was totally undeveloped and dry land area, it became difficult to grow many crops in this farm. A part of the farm was just suitable to grow sugarcane and ragi. Thereafter the farm development work started with a small quantity of OP seed production for local companies. However, that was not the only target of Namdhari Seeds. With the deep visionary thinking of Chairman cum Managing Director Mr. Uday Singh, the first contract order of US $ 6000 was bagged from a U.S company for production of F1 hybrid tomato seeds, and that was the beginning of their growth.

Tomato Dual Purpose and Processing Hybrids

NS 815, NS 812, NS 816, UTSAV, AMOGH, NS 2535, NS 29, NS 40, NS 681, NS 175, NS 137, NS 104, NS 811, SWARAKSHA, AMAR, BWT 1, NS 52 (BWT 2), NS 53 (BWT 3), VT 1, VT 4, ANANYA (Improved), UTTAM, (NS 8014), NS 719, NS 54 (NP 4, NS 55 (NP 5), NS 603, NS 877, NS 412, NS 428, NS 310, NS 314, NS 336, NS 157, NS 614, NS 858, NS 884, NS 644, NS 888, NS 141, NS 535, NS 716, NS 863, NS 64, NS 68, NS 73, NS 77, MS 88, NS 195

Watermelon Hybrids

Hybrid	Rind pattern	Fruit size (kg)	Fruit shape	Flesh colour	Sweetness TSS (%)
OVAL TO OBLONG TYPE					
NS 295	light green rind with dark green stripes	9-10	oval to oblong	deep crimson red	12-13
NS 296	light green rind with dark green stripes	8-10	oval to oblong	deep crimson	11-12
NS 450	light green rind with dark green stripes	8-10	oval to oblong	deep crimson red	12-13
MAYUR	light green rind with distinctive dark green broad stripes	9-12	oblong	deep crimson	11-12
NS 91	light green rind with distinctive dark green broad stripes	10-12	oblong	deep crimson	11-12
NS 92	light green rind with deep green broad stripes	10-12	oblong	deep crimson	12
NS 93	light green rind with dark green broad stripes	10-12	oblong	deep crimson	12-13
NS 116	light green rind with deep green stripes	10-12	oval to oblong	crimson red	12
NS 161	dark green stripes on green rind	9-10	oval to oblong	crimson red	12
NS 168	green rind with dark green wider stripes	9-10	oval	deep crimson	12

Contd...

Contd...

NS 246	dark green rind with bluish green stripes	8-9	oval to oblong	crimson red	12-13
NS 270	dark bluish green stripes on dark green rind	8-9	oval	deep red	12
NS 26	dark bluish green broad stripes on dark green rind	8-10	oval	deep red	12
NS 20	black head type hybrid	8-9	oblong blocky	deep red	12.5-13%
ROUND TO OVAL TYPE					
NS 16	dark green rind with with bluish green stripes	8-10	round oval	deep red	12
NS 1004	light green rind with dark green stripes	7-8	round	deep crimson	12
NS 188	light green rind with dark green medium stripes	9-12	round to oval	deep crimson	12-13
NS 170	dark green wider stripes on green skin	9-10	round to oval	deep crimson	12-13
TEJAS	dark bluish green rind	6-8	round	deep crimson	12
NS 200	dark green bluish rind with visible stripes	8-10	round to oval	deep red	12
NS 217	dark green bluish rind	8-10	round to oval	deep red	12
NS 423	dark green rind with lacing	8 9	round to oval	crimson red	12

Muskmelon Hybrids

Muskmelon Hybrids - Comparative Table

Hybrid	Fruit size (kg)	Fruit shape	Netting on fruit	Flesh colour
CANTALOUPE TYPE				
NS 7455	1.5-2.0	oval	good	deep salmon
Abhijeet	1.25-1.5	round to oval	good	deep salmon
NS 904	1.25-1.5	round to oval	good	deep salmon
NS 910	1.5-2.0	oval	good	deep salmon
NS 89	1.5-2.0	oval	medium	deep salmon
GALIA TYPE				
NS 923	1.25-1.5	round	good	light green
NS 929	0.75-1.0	round	good	light green
NS 931	0.75-1.25	round	good	light green
HONEY DEW TYPE				
NS 934	1.5-2.0	round	good	light green
NS 972	1.5	round	good	light green

Cauliflower

Early Cauliflowers

ATI SHIGHRA, NS 1, NS 60N, BASANTH

Mid Season Cauliflowers

NS 531, NS 60

Mid -Late Season Cauliflowers

NS 66, NS 76

Winter Cauliflowers

NS 75, NS 037, NS 015, NS 94, NS 84, NS 90

Cabbage Hybrids

Flat Type Hybrids

NS 43, NS 173, NS 35, NS 111

Round Type Hybrids

NS 164, NS 183, NS 160, NS 163, NS 22, NS 25

Okra Hybrids

NS 502, NS 98, NS 531, NS 532

Open Pollinated

ARKA ANAMIKA, P-8, APSARA, SUNDER, NS 20, NS 17, NS 72, NS 102, NS 787, NS 797

Chilli/Hot Pepper Hybrids

NS 1101, NS 1701, NS 1072, NS 327, NS 330, Pragati, Balwant, NS 114, Akash, Goli, NS 108, NS 222, NS 7510, NS 24, NS 250, NS 254, NS 830, NS 510, NS 541

Ridge Gourd Hybrids

NS 3, NS 401, NS 403, NS 14, NS 15, NS 21

Bottle Gourd Hybrids

KAVERI, NS 421, BHOOMIKA, MRIDULA, NUTAN

Bitter Gourd Hybrids

NS 1020, NS 1018, NS 1026, NS 241, NS 1024, NS 244, NS 250, NS 280

Sponge Gourd Hybrids

NS 441, NS 442

Capsicum Hybrids

NS 436, SINDHURI, NS 627, RATAN, ARUN, GAGAN, NS 631, NS 632

Open Pollinated Varieties

NS 10, NS 33, ARKA MOHINI

Carrot Varieties

NS 16, NS 280, NAMDHARI KURODA

Cucumber Varieties

NS 45, NS 46, HIMANGI

Coriander

SurabhI

(6LB)

This variety is very vigorous and ready for harvest in 35 days. It is suitable for multiple cuts. The leaves are very attractive large, shiny, succulent and have good aroma. It has excellent yield potential. It is late-bolting and can be grown round the year.

Broccoli

NS 50

It is a mid maturity hybrid. Heads are compact, uniform and deep dome shaped. It holds shape very well even when harvesting is delayed. Heads are uniform with no cat eye. Plants are strong leafed and highly tolerant to downy mildew and black rot. It is suitable for processing, fresh market and home garden.

R and D Centre

The main R&D research centre of Namdhari Seeds is located in Bangalore. The ever changing demands for vegetable hybrids to suit different market segments and agro-climatic zones have led to strengthening of Namdhari Research and Development wing. The recent development in this respect is the realization of a new operational research building with most modern and sophisticated facilities for Plant Breeding Division as well as different laboratory disciplines.

This company had made a modest beginning with a dedicated team of breeders whose efforts have culminated in release of several commercially successful hybrids. The team has since been expanded with specialization into crops. The major crops handled are tomato, watermelon, melons, chillies, okra, brinjal and gourds. New crops like cucumber, onion and cauliflower have been added. The breeding efforts of the R&D are enhanced via shuttle breeding through different research stations strategically located all over India.

In tomato, a major crop of subtropics, the company offers a wide range of high yielding hybrids to suit varied grower and market specificities. These include processing, dual purpose, fresh market (bush and stake culture), indeterminate (outdoor and greenhouse culture). An array of hybrids resistant to bacterial wilt, TYLCV, TMV, nematodes and a host of other temperate foliar and root diseases have also been developed and released. Watermelon and melons are major crops where concerted efforts have led to many commercially viable hybrids. These hybrids differ in rind pattern, fruit size, fruit shape and excellent quality parameters to suit different agro-regions of India and potential countries in South East Asia and Middle East. Chilli, a major crop of India is witnessing a shift from open pollinated varieties to hybrids. This company has kept pace with ever increasing demand of hybrids by releasing virus resistant NS 1701, NS 1101, Goli, Balwant and Akash which has been a big leap in stepping up the income of farmers. The company has been successful in exploiting hybrid vigour in okra. High yielding hybrids with resistance to YVMV and superior pod quality like the NS 500 series have made a mark in the domestic and offshore markets. The programme on brinjal (eggplant) has been intensified and the list of hybrids include a wide spectrum of fruit types inclusive. The plant pathology section has a highly sophisticated laboratory and specialized compartmentalised greenhouses for rapid disease diagnosis, multiplication and screening of fungi, bacteria and viruses. This has helped in churning out several lines which are prerequisite for disease resistance breeding. The molecular biology section equipped with the state of the art technologies is propelling the rapid advancement of the breeding programme. The successful use of molecular markers for disease resistance screening, male sterility, etc. has strengthened and hastened end product realisation. The lab also has a mandate of developing molecular markers helpful for DNA fingerprinting which ensures the purity of Namdhari commercial seed. The section of tissue culture has mainly focused in complementing plant breeding by the development of tetraploids, doubled haploids, embryo rescue in inter- specific crosses, maintenance and manipulation of germplasm. The application of molecular genetics and breeding in harmony with conventional breeding has catapulted the company into an enviable position in the Indian seed industry. Constant upgradation of skills, imbibition of leading edge technologies, innovations coupled with committed R&D has seen Namdhari Seeds emerge as a global player in the hybrid vegetable seed scenario.

Krishidhan Seeds Ltd

Krishidhan Seeds Ltd. is a research base organization having recognition by Department of Scientific and Industrial Research (DSIR), Govt. of India. The Company is expanding at a fast pace under dyanamic leadership of Sri Sushil J.

Karwa who has a vision of far reaching impact. The corporate office is at Pune and the multicrop sophisticated R and D facilities and laboratories are located at Jalna. In a very short spell, products of this company, namely, Research Hyb. Cotton Maruti - 9632, Paddy - Komal, Wheat-Balram, Cauliflower - Champion, Chilli - Tarun are well known to the farmers. The company has launched Bt Cotton Hybrids, namely, KDCHH - 9810, KDCHH - 9832 and KDCHH 441 - BG II. The seed production programme is organised throughout India and processed in High-Tech processing plants at Jalna, Indore, Ahmedabad and Kurnool.

It has best scientific pool to carry out R & D activities at a very high level as per needs of the farmers. It is poised for rapid strides and giant leap forward soon.

References

AGROW (1999). AGROW review of 1998. Richmond: PJB Publications.

Joon-hun, N. (1998, July 15). Biz insight: Seminis Sees Korea as Gateway to Asia. Korea Times.

Massieu, Y. (1998, March). ELM: A new global player in the vegetable market. Biotechnology and Development Monitor, 37, pp. 9-13.

Seminis. (2000). Seminis Annual Report—Fiscal Year 1999. Saticoy, CA: Seminis, Inc.

Seminis. (2001). Form 10K Annual Report Year Ending September 30, 2000. Saticoy, CA: Seminis, Inc.

www.mahyco.com

23

Field Standards in Seed Production

Minimum field standards for seed production have been prescribed for each crop. The major components to be looked into for meeting the field standards are:

- Preceding crops
- Isolation distance
- Number of field inspections

The minimum proceding crop requirement has been specified to minimize genetic contamination on account of volunteer plants germinating from the left over seeds in the previous season. The disease contamination is also possible from soil borne disease organisms if the previous crop was not certified. The preceding crop requirement would mean that for a prescribed minimum period the same crop or any other contaminating crop should not have been grown on the same field where seed production is to be taken up unless the preceding crop was a seed crop and had met the field/the seed standards.

The minimum isolation distance requirements have been prescribed to minimize genetic contamination and seed borne disease contamination from neighboring fields. The isolation distance is to be maintained from the fields of other varieties or of the same variety not conforming to varietal purity as required by the seed certification agency.

The number of field inspections are specified to ensure verification of genetic purity and other quality factors such as seed health, admixture of objectionable weeds and inseparable other crop plants at a stage of crop growth at which the seed is more prone to contamination.

The General Crop Stage for Field Inspection

Generally, five stages have been specified for field inspection. These are :

- At the time of sowing (for verification of source of seed, land requirement, isolation distance, etc.)
- Pre-flowering (to rogue out off types)
- Flowering stage (to rogue out off types and seed borne diseases)
- Post flowering and pre-harvesting stage (to rogue out off types and objectionable seed borne diseases)
- At the time of harvesting to rogue out objectionable weeds and objectionable seed borne diseases and off types

Number of Field Inspections

A minimum of two and maximum of four field inspections are required for certification of seed crop. At least one inspection at flowering stage is a must to observe off type plants and objectionable weeds and diseases.

Cultivar Maintenance Through Roguing in the Field

It is always necessary to ensure that the seed produced shall be "true to type". The trueness to the type means that the plants grown from new seed generation shall not differ significantly from the cultivar description. The loss of deterioration of cultivar unique characters is sometimes referred to as "a loss of trueness to type" or "running out". The seed stocks conforming to a cultivar description are certified.

The seed crop is inspected at various stages to ensure that any undesirable plant material/off type is removed as far as possible before pollination. This process of removing "off types" is frequency referred to as roguing. The intensity of roguing or selection during seed multiplication can have a considerable effect on the genetic purity of successive generations(Loverck and Tumer, 1995). It has been demonstrated that if selection standards were relaxed in three successive open pollinated generations of a cross pollinated crop like Brussels sprouts, there could be an increase in the proportion of off types (George, 1999). The different methods of maintaining the seed stocks by eliminating the off type plants are as follows :

Positive Selection

This system is also referred to as mass selection where plants are normally selected to a very high standard while still in vegetable stage. These selected plants (normally about 10 % of the total population) are planted in isolation to maintain breeders stock's or pre basic or basic seed. If the selected plants are to remain *in situ* then the unselected plants are removed before flowering. The seeds from positive selection are harvested in bulk unless progeny testing is required.

Negative Selection

This system is frequently referred to as roguing where the plants within the limits of cultivar description are retained for seed production, usually *in situ* while the off types are rogued out. This system of negative selection provides a lower selection pressure than positive selection and is normally used in the final multiplication stage after an open pollinated seed crop. The minimum field standards as required under Indian minimum seed certification standards for self and cross/open pollinated crops are given in Table 23.1 (Agarwal, 1995) and those for hybrids in Table 23.2.

References

Agarwal, R.L. 1995. Seed Technology-2nd Edition. Oxford and IBH Publishing Company, New Delhi. pp 829.

George, R.A.T. 1999. Vegetable Seed Production. 2nd Edn. CABI Publ. pp. 327.

Loverck, G.K. and Tumer, M.R. 1995. Roguing seed crops for genetic purity : A Review: Plant Varieties and Seeds. 8 : 29 – 46.

Table 23.1. Minimum field standards

Crop	Class of seed	Minimum				Maximum permissible level (%)			Remarks
		Land requirement	Isolation meters	No. of field inspections	Off types	Inseparable other crop plants	Objectionable weed plants	Plants/heads affected by designated diseases	
Cereals and millet									
Paddy	FS	-	3	2	0.05	-	0.01	-	O.W.; wild rice
	CS	-	3	2	0.20	-	0.02	-	
Wheat	FS	-	3, 150*	2	0.05	0.01	-	0.10	* from other wheat fields having loose smut infection and excess of 0.1 and 0.5 per cent for FS and CS class respectively. OCP; barley, oats, gram SBD; loose smut
	CS	-	3, 150*	2	0.20	0.05	-	0.50	
Barley	FS	-	3, 150*	2	0.05	0.01	-	0.10	*from other barley fields having loose smut infection an excess of 0.1 and 0.5 % for FS and CS class, respectively OCP; wheat , oats, gram SBD; loose smut
	CS	-	3, 150*	2	0.20	0.05	-	0.50	
Maize (OP)	FS	-	400	2	1.00	-	-	-	-
	CS	-	200	2	1.00	-	-	-	
Sorghum (OP)	FS	-	200, 400	3	0.05** 0.10***	-	-	0.05	* from Johnson grass and other grassy sorghum spp. - differential blooming dates for modifying isolation distance are not permitted ** for sorghum *** forage sorghum and sudan grass
Forage Sorghum and Sudan grass	CS	-	100, 400*	3	0.10** 0.20***	-	-	0.10	SBD; Karnal smut, grain smut and head smut - seed fields can be certified if diseased heads are removed and burnt and upon reinspections disease infection is within permissible limits - thoroughly rogued out the plants affected by sugary disease and ergot

Contd...

Table 23.1. Contd...

Pearl millet (OP)	FS	-	400	3	0.05	-	-	0.05*, 0.02, 0.05	- differential blooming dates for modifying isolation distance are not permitted * for downy mildew/ green year, ergotted ears heads and grain smut respectively - Complete stool is considered as one inspected unit in case of downy mildew/green ear disease - Not withstanding the ergot infection percentage within standards the seeds from such fields shall be subjected to floatation treatment - seed from rejected fields on account of higher incidence of grain smut may be certified provide seed is treated with an organomercural fungicides not earlier than a month prior to its sowing
	CS	-	200	3	0.10	-	-	0.10*, 0.04, 0.10	
Smaller* millets	FS	-	3	2	0.05	-	-	-	* smaller millets: barnyard millet, common millets, finger millets, Italian millets, kodo millets and little millets
	CS	-	3	2	0.10	-	-	-	
Pulses									
Black gram, Bengal gram, Horse gram, Khesari, Lentil, Moth bean and pea	FS	-	10	2	0.10	-	-	-	-
	CS	-	5	2	0.20	-	-	-	-
Cowpea, Green gram, Indian bean French bean, and cluster bean	FS	-	10	2	0.10	-	-	10	SBD : cowpea : Ashy stem blight, Anthracnose, *Ascochyta blight* (for hill area only) and cow pea mosaic Green gram : Halo blight Cluster bean : Ascochyta blight (for hill area only) French bean : Ascochyta blight (for hill area only) and bean mosaic
	CS	-	5	2	0.20	-	-	20	
Pigeon pea	FS	-	200	2	0.10	-	-	-	-
	CS	-	100	2	0.20	-	-	-	-
Oil seeds									
Castor	FS	-	300	2	0.10	-	-	-	-
	CS	-	150	2	0.20	-	-	-	-
Groundnut	FS	Two seasons	3	2	0.10	-	-	-	-
	CS	Two seasons	3	2	0.20	-	-	-	-
Indian rapeseed and mustard	FS	-	50*, 100	3	0.10	-	0.50	-	* for self compatible and self incompatible types respectively; and from rocket salad, and any of the other species of genus brassica OW : *Argemone mexicana*
	CS	-	25*, 50	3	0.50	-	0.10	-	
Linseed	FS	-	50	2	0.05	-	-	-	-
	CS	-	25	2	0.10	-	-	-	-
Niger	FS	-	400	3	0.05	-	-	-	-
	CS	-	200	3	0.10	-	-	-	-

Contd...

Table 23.1. Contd...

Tara mira (rocket salad)	FS	-	100*	3	0.10	-	0.05	-	* and any of the species of the genus Brassica OW : *Argemone mexicana* L.
	CS	-	50*	3	0.50	-	0.10	-	
Safflower	FS	-	400*	3	0.05	-	None	-	* and wild safflower OW : wild safflower SBD : leaf spot
	CS	-	200*	3	0.10	-	None	-	
Sesame	FS	-	100	3	0.10	-	-	0.50	-
	CS	-	50	3	0.20	-	-	1.00	
Soybean	FS	-	3	2	0.10	-	-	-	-
	CS	-	3	2	0.50	-	-	-	
Sunflower	FS	One year	400	3	0.10*	-	None	0.05, None **	OW : wild *Helianthus spp.* * sterile plants of the same variety shall not be considered as off types ** For downy mildew infection and infestation of Orobanche respectively
	CS	One year	200	3	0.20*	-	None	0.50, None **	
Fibre crops									
Cotton	FS	-	50, 5*	2	0.10	-	-	-	*from field of other varieties of different species
	CS	-	30, 5*	2	0.20	-	-	-	
Jute	FS	-	50, 5*	3	0.50	-	-	1.00	* from fields of other species SBD : Jute chlorosis
	CS	-	30, 5*	3	0.10	-	-	2.00	
Forages									
Berseem, Lucerne	FS	-	400	2	0.20	-	None	-	OW : Chikori for berseem, dodder for Lucerne
	CS	-	100	2	1.00	-	0.05	-	
Setaria grass	FS	-	400**	3	0.10	-	-	-	**differential blooming dates for modifying isolation distance are not permitted
	CS	-	200**	3	1.00	-	-	-	
Indian clover (Senji)	CS	-	50	2	0.20	-	-	-	-
	FS	-	25	2	0.10	-	-	-	
Oats	FS	-	3, 150*	2	0.05	0.01	0.01	0.10	* and from field of oats with loose smut infection beyond the permissible limit in FS/CS class respectively OCP : barley, wheat, gram OW : wild oats SBD : loose smut
	CS	-	3, 150*	2	0.20	0.05	0.02	0.50	
Vegetables									
True potato seed	CS	*	50	4	0.10	-	-	3	* seed fields preferable should be infested with wart and/or cyst forming nematodes, brown rot or non cyst forming nematodes within the previous 3 years, and common scab ** between blocks of the parental line of the same hybrids, when female and male parents are planted in separate blocks *** for mild mosaic; severe mosaic leaf roll and yellows; and brown rot respectively
TPS (hybrids)	CS		5**					1.0, 3/ha***	

Contd...

Table 23.1. Contd...

Cucurbitaceous fruits									
Ash gourd, bottle gourd Indian squash, pumpkin, ridge gourd and sponge gourd	FS	-	1000*	3	0.10	-	-	-	*for pumpkin also from winter squash, summer squash and cushaw * for ridge gourd from sponge gourd * for sponge gourd from ridge gourd
	CS	-	500*	3	0.20	-	-	-	
Bitter gourd	FS	-	1000*	3	0.10	-	None	-	*and from balsam apple, bhat karela, jangli karela (objectionable weeds)
	CS	-	500*	3	0.20	-	None	-	
Cucumber, snake gourd	FS	-	1000*	3	0.10	-	None	-	* for cucumber from *Cucumis hardwickii* (objectionable weeds) * for snake gourd from *Trichosanthes palmata*, *T. lobata* and *T. cucumerina* (objectionable weeds)
	CS	-	500*	3	0.20	-	None	-	
Long melon, snap melon	FS	-	1000*	3	0.10	-	None	-	* and from muskmelon, oriental pickling melon and wild spp. (objectionable weed species and long melon from snap melon and vice-versa OW : *Cucumis prophetarum*, weed melon and other non-dessert forms of *Cucumis melo* L.
	CS	-	500*	3	0.20*	-	None	-	
Muskmelon	FS	-	1000*	3	0.10	-	None	0.10	* and from long melon, oriental pickling melon and wild species Objectionable weed species OW : *Cucumis prophetarum*, weed melon and other non-dessert forms of *Cucumis melo* L. SBD : Cucumis mosaic virus
	CS	-	500*	3	0.20	-	None	0.20	
Summer squash	FS	-	1000*	3	0.10	-	-	0.10	* and from pumpkin, cashew, and winter squash SBD : Cucumis mosaic virus, watermelon mosaic virus
	CS	-	500*	3	0.20	-	-	0.20	
Winter squash	FS	-	1000*	3	0.10	-	-	-	* and from pumpkin, cashew and summer squash
	CS	-	500*	3	0.20	-	-	-	
Water melon	FS	-	1000*	3	0.10	-	None	-	* and from wild watermelon (objectionable weed)
	CS	-	500*	3	0.20	-	None	-	
Fruit vegetables									
Brinjal	FS	-	200	3	0.10	-	-	0.10	SBD : Phomopsis blight
	CS	-	100	3	0.20	-	-	0.50	
Sweet pepper, chilli	FS	-	400*	3	0.10	-	-	0.10	* and capsicum from chilli and vice-versa SBD : Leaf blight, anthracnose
	CS	-	200*	3	0.20	-	-	0.50	
Okra	FS	-	400*	3	0.10	-	None	-	* and from wild okra (objectionable weed) OW : Wild okra
	CS	-	200*	3	0.20	-	None	-	
Tomato	FS	-	50	3	0.10	-	-	0.10	SBD : Early blight leaf spot, tobacco mosaic virus
	CS	-	25	3	0.20	-	-	0.50	
Green/leafy vegetables									
Amaranthus	FS	-	400*	2	0.10	-	0.01	-	* and from wild amaranth (objectionable weed)
Fenugreek (Methi and kasuri methi)	FS	-	10	2	0.10	-	0.01	-	-
	CS		5	2	0.20		0.02		

Contd...

Table 23.1. Contd...

Spinach, spinach beet	FS	-	1600*	2	0.10	-	-	-	* and for spinach beet from swiss chard, sugar beet, garden beet
	CS	-	1000*	2	0.20	-	-	-	
Cole vegetables									
Cabbage, cauliflower, knoll-khol, Chinese cabbage	FS	-	1600*	3	0.10	-	-	0.10	* and other brassica spp. SBD : black rot, soft rot for Chinese cabbage; black rot, black leg
	CS	-	1000*	3	0.20	-	-	0.50	
Bulb/root crops									
Onion, Stage I (Mother bulbs)	FS	-	5	2*	0.10	-	-	-	* transplanting of seedling and lifting of bulbs
	CS	-	5	2*	0.20	-	-	-	

Note : FS = foundation seed ; CS = certified seed; OP = open pollinated varieties, composites, synthetics; OW = objectionable weeds; OCP = other inseperable crop plants; SBD = seed borne diseases

Table 23.2. Minimum field standards for hybrid crops

Crop	Minimum			Maximum permissible % of		Remarks
	Land requirement	Isolation (m)	Fields inspections	Off-types (both parents)	Shedding tassels in seed parent	
Maize						
Foundation seed						
Inbreds	-	400, 600* Time** isolation	4	20***	-	*fields of any maize with different colour or texture and teosinte. ** provided 5 percent or more of the plants (seed parent) do not have receptive s*ca when more than 0.2 percent plants (foundation seed) and 0.5 percent plants
Single crosses	-	400, 600* Time** isolation	4	20***	50, 1.0a	
Certified seed						(certified seem in other maize field failing within required isolation distance are shedding pollen. **off-type plants that have shed pollen or are shedding pollen when 5 percent or more plants of seed parent have receptive silks. Sucker tassels, portion of tassels are counted as sheddin tassels when 5 cm or more of any spike has exserted anthers and are shedding pollen. a: Total of 3 inspections at flowering.
Hybrids	-	A 200, 300 Time** isolation	4	0.50***	1.0, 2.0	a) Distance less then 200 metres may be modified by planting border rows. b) Total of 3 inspections at flowering

Contd...

Crop	Class of seed	Minimum			Max. permissible at and after flowering (%)			Remarks
		Land requirement	Isolation meters	Field inspection	Off-types (both parents)	Pollen shedders in seed parent	Pollen shedders in seed parent	
Sorghum	FS	-	300, 400*	4	0.05	0.05	0.05**	*from Jhnson grass. forage sorghum with high tillering and grassy panicle ** Kernel smut Head smut **of same cross, conforming to varietal putty requirements for certification Note 1. Differential blooming dates for modifying Isolation distances are not permitted 2. If diseased ear heads are removed and burnt and on rein-spection disease infection is within permissible limits seed field can be certified. Only one reinspection is permitted. 3. Thoroughly Rogue out to remove plants affected by sugary disease/ergot.
	CS	-	200, 400*, 5***	4	0.10	0.10	0.10	
Pearl millet	FS	-	1000	4	0.05	0.05	0.05*, 0.02**	Downy mildew/green ear disease (in both the parents at any one inspection) Ergotted ear heads in seed parent at final inspection
	CS	-	200	4	0.10	0.10	1.0*, 0.04**	
Paddy	FS	-	200	4	0.05	0.05	0.01	Note: 1. Differential blooming dates for modilyin9 isolation distance are not permitted. 2. Seeds from fields reported to contain the ergot infection shall be subjected to floatation treatment despite the fact that the incidence is within standards. 3. Seeds from fields reported to contain the grain smut infection beyond permissible limits can be certified provided the seed is treated with an organomercurial fungicide not earlier than a month prior to its sowing
	CS	-	100	4	0.20	0.10	0.02	
Sunflower	FS	One year	600	4	0.20	0.50	None	OW : wild rice SBD : downy mildew *sterile plants of the same strain shall not be considered as off-types. Note The standards for plants Inferted with Orobanches are none for each class.
	CS	One year	400	4	0.50	1.00	None	

Contd...

Contd...

Crop	Class of seed	Minimum			Max. permissible at and after flowering (%)				Remarks
		Land requirement	Isolation meters	Field inspection	Off-types (both parents)	Pollen shedders in seed parent	Objectionable weeds	Seed borne diseases	
Cotton	FS	-	50	3	0.10	0.05*	-	-	*applicable only when the male startle line Is used for the Production of hybrid seed Note : Fields of other varieties of different app, and blocks of parental lines of the same hybrid shall be separated at least by 5 metres
	CS	-	30	4	0.50	0.10*	-	-	
Castor	FS	-	300	4	0.50*	Male variants** 1.0	Female*** variants 1.0	Monoecious[1] Plants ****	*off-types includes plants found to flower over the main stem. One node less than the minimum node specified for different combiners in female lines is considered within the range. *Three secondary rachis from the base and above possessing only male flowers are considered as mostly male ***In certain cases all sex variants of dominant tamales, females with Inter spersed staminate flowers, mostly female and mostly male raceme spectra involved are considered female variants. Late opening male flowers in the female removed without removing the entire raceme. ****Incase of female, at least 25- per cent moncecious plants are required for pollen source. 1. includes racemes reverted to monoecism on female plants before anthesis - Differential blooming dates for modifying Isolation distance are not permitted
	CS	-	150	4	1.0*	2.0		2.0	
Cucurbitaceous vegetables									
Bitter gourd	FS	-	1500*	4	0.01	-	None	-	OW : balsam apple, bhat k arela and jangli karela and from OW spp.
	CS	-	1000*	4	0.05	0.10	None	-	

Contd...

Contd...

Ridge gourd, Sponge gourd	FS	-	1500*	4	0.01	-	-	-	And ridge gourd from sponge gourd and vice versa
	CS	-	1000*	4	0.05	0.10	-	-	
Pumpkin, Winter squash	FS	-	1500*	4	0.01	-	-	-	And Pumpkin from winter squash, cushaw and summer squash and winter squash from pumpkin
	CS	-	1000*	4	0.05	0.10	-	-	
Summer squash	FS	-	1500*	4	0.01	-	-	0.10	Cashew and summer squash SBD : Cucumber mosaic virus, water melon mosaic virus and from pumpkin, cushaw and winter squash
	CS	-	1000*	4	0.05	0.10	-	0.50	
Water melon, seedless water melon	FS	-	1500*	4	0.01	-	None	-	And from O.W. spp. O.W. : wild water melon
	CS	-	1000*	4	0.05	0.10	None	-	
Fruit vegetables									
Brinjal	FS	-	200	3	0.01	-	-	0.10	SBD : Phomopsis blight
	CS	-	200	3	0.05	-	-	0.50	
Tomato	FS	-	200	3	0.01	0.05*	-	0.10	Fertile segregants : Pollen shedding cy mes (applicable when CMS line is used SBD : early blight, leaf spot, TMV
	CS	-	100	3	0.05	0.10*	-	0.50	
Cole vegetables									
Cole crops	FS	-	1600	3	0.01	-	-	0.10	SBD : black leg, black rot and soft rot (soft rot is not a designated disease for Chinese cabbage)
	CS	-	1600	3	0.05	-	-	0.50	

Note : Blocks of parental lines when parent and pollinators are planted in separate blocks and hand pollination is to be adopted shall be separated at least by 5 meters

Bulb and root crops											
Crop	Class of seed	Land requirement	Mother bulb production stage			Seed production stage				Seed borne disease max. (%)	Remarks
			Isolation (m)	Min. no. of inspections	Off-type (max.) (%)	Isolation (m)	Min. no. of inspecti ons	Off-type (max.) (%)	Pollen shedders (max.) (%)		
Onion	FS	-	5	2	0.01	1000	4	0.01	-	-	-
	CS	-	5	2	0.05	500	4	0.05	-	-	-
Carrot	FS	-	5	2	0.01	1000	4	0.01	-	-	-
	CS	-	5	2	0.05	800	4	0.05	-	-	-
Radish	FS	-	-	-	-	1600*	3	0.01	0.10	0.10	* and rat-tall radish
	CS	-	-	-	-	1600*	3	0.05	0.50	0.50	SBD : black leg, black rot
Turnip	FS	-	-	-	-	1600*	3	0.01	0.10	0.10	* and Brassica spp.
	CS	-	-	-	-	1600*	3	0.05	0.50	0.50	SBD : black let, black rot

Minimum field standards for vegetatively propagated crops

Crop	Class of seeds	Minimum			Maximum permissible percentage			Remarks
		Land requirement	Isolation (m)	Field inspections	Off-types	Others	Objectionable weed plants	
*Napier grass, hybrid Napier grass	FS	Free of volunteer plants of napier, hy. Napier and pearl millet	10	After 45 days	0.01	-	None	Objectionable weeds : Canada thistle, dodder, Johnson grass, quack grass, wild moring glory
	CS		10	After 100 days	0.30	-	None	1. All of types and objectionable weed plants shall be uprooted and removed from the seed field
Little gourd, pointed gourd	FS	Free of volunteer plants	20**	1. After transplanting 2. Before cutting	0.10	-	-	* Avoid swampy, shaded conditions and residue drainage form fields of the same kind
	CS		20**	1. After transplanting 2 Before cutting	0.20	-	-	** also, between blocks of female and male plants
Lesser yam	FS	Free of volunteer plants	5	After 90, 150 and 200 days	0.05	Scale insects none	-	Avoid swampy, shaded conditions 1. All off types and plants infested with scale insects shall be ro gued out along with the tubers and destroyed 2. Gaps in the seed fields shall not be more than 10 parent.
	CS	Free of volunteer plants	5	After 90, 150 and 200 days	0.10	None	-	

Crop	Class of seeds	Minimum			Maximum permissible percentage					Remarks
		Land requirement	Isolation (m)	Field inspections	Off-types	Others	Phytophthora	Scale insects and mealy bugs	Objectionable weed plants	
Taro	FS	Free of volunteer plants	5	After 60, 90 and 160 days	0.10	0.50	None	None	-	1. Standards for designated diseases and insects shall be met at each inspection. 2. All off -types, diseased and insect infested plants shall be rogued out along with corns, cormels and destroyed 3. Gaps in the seed field shall not be more than 10 per cent
	CS	Free of volunteer plants	5	After 60 90 and 160 days	0.50	1.00	None	None	-	

Contd...

Table 36.3. Contd...

Garlic multiplier Onion	FS	Free of volunteer plants	5	When plants are large enough and before lifting of bulbs	0.10	-	-	-	-	1. All off -type plans should be rogued out along with bulbs
	CS	Free of volunteer plants	5	When plants are large enough and before lifting of bulbs	20	-	-	-	-	
(Potato Onion)										
Sweet potato	Plant bed stage			Black rot wilt			Scurf			
	FS	Free of volunteer plants	Just before transplanting	None	None	None	None		-	Avoid swampy, shaded conditions and residue drainage from other field of he
	CS	Free of volunteer plants	Just before transplanting	None	None	None	None		-	
	Seed bed stage									
	FS	Free of volunteer plants	Just after transplanting	10	-	None	-	0.5	-	1. All off -types and diseased plants should be rogued out alongwith roots 2. Standards to be met at each inspection
	CS	Free of volunteer plants	Just after transplanting	20	-	None	-	0.10	-	

Minimum Field Standards for Potato

		Maximum permissible limits (%)			Remarks
		Foundation I	Foundation II	Certified seed	
1.	Land requirement	The land shall be free from the insection of brown rot, non -cyst forming nematodes (for last three years); and also from common scab, wart and cyst forming nematodes			
2.	Isolation (meters	5	5	5	-
3.	(a) First and second inspection (hills : after 45 and 60 – 65/70-75 days Plains : after 35 and 60 – 65 days)				* of the two inspections, the higher virus percentage is considered for the purpose of the specified limits of tolerance
	- Mild mosaic	0.05	0.05	0.10	**The presence is permitted only in the areas known to be infected with this disease. Also, the neighboring plants one on either side shall also be rogued out alongwith its tubers
	- Severe mosaic, leaf roll and yellows	0.05	0.75	1.00	
	- Total viruses	1.0	2.0	3.0	
	- Brown rot	None	None	3/ha**	
	(b) Third inspection (just after haulm cutting)	Verify that the haulms have been cut as close to the ground as possible before the prescribed date. Failure to do so in time renders the crop liable for rejection			
	(c) Fourth Inspection (10 days after haulm cutting)				
	-Regrowth of haulms after cutting	0.5	0.5	0.5	

24

Seed Standards

Certified seed must meet certain seed standards in order to ensure an adequate level of seed purity along with the reasonable level of seed germination so that the farmer gets a good stand of the crop conforming to the varietal description. The components of the seed standard are as follows :

- The minimum percentage of pure seed
- The maximum permissible limit for inert matter
- Other crops seeds
- Weed seeds
- The maximum permissible limit for objectionable weeds
- The maximum permissible limits of seeds of other distinguishable varieties (ODV)
- Maximum permissible limits of seeds infected by seed bone diseases

These standards are ensured in the seed crop by strictly adhering to field standards and careful harvesting, threshing, seed drying and then brining the same to the seed processing /seed conditioning plants. The Indian minimum seed standards for different crops are given in Tables 24.1-24.3 (Agarwal, 1995).

Reference

Agarwal, R.L. 1995. Seed technology 2nd Edn.. Oxford and IBH Publishing Company, New Delhi. pp 829.

Table 24.1. Minimum seed standards for agricultural seeds

Crop	Class of seed	Minimum					Minimum permissible limit						Remarks
		Germination (%)	Pure seed (%)	Inert matter (%)	Other crop seeds (No./kg)	Weed seeds (no./kg)	Objectionable seeds (No./kg)	Diseased seeds	Other distinguishable variety seeds (No./kg)	Moisture: Ordinary pack (%)	Moisture: Vapour proof pack (%)	Other (as specified)	
Cereals and millets													
Paddy	FS	80	98	2	10	10	2	0.01	10	13	8	2[a]	A maximum huskless seeds (%) SBD; paddy bunt, OW; wild rice
	CS	80	98	2	20	20	5	0.05	20	13	8	2	
Wheat	FS	85	98	2	10	10	2	0.05* None**	-	12	8	-	SBD*; karnel bunt, nematode galls of earcockle and tundu OW; Hirankhuri, Phalaris minor
	CS	85	98	2	20	20	5	0.25* None**	-	12	8	-	
Barley	FS	85	98	2	10	10	-	-	10	12	8	-	-
	CS	80	98	2	20	20	-	-	20	12	8	-	-
Maize OP varieties, composites	FS	90	98	2	5	None	-	-	10	12	8	1.0*	*Maximum off type ears with off coloured karnels (%) (after harvest)
	CS	90	98	2	10	None	-	-	20	12	8	1.0*	
Inbreds, single crosses	FS	80	98	2	5	None	-	-	5	12	8	0.2*	*Maximum off type ears with off coloured karnels (%) (after harvest)
Hybrids	CS	90	98	2	10	None	-	-	10	12	8	0.5*	*Maximum off type ears with off coloured karnels (%) (after harvest)

Contd...

Table 24.1. Contd...

Sorghum	FS	75	98	2	5	5	-	0.02*	10	12	8	-	*Ergotted seeds, sclerotia
	CS	75	98	2	10	10	-	0.04*	20	12	8	-	
Pearl millet	FS	75	98	2	10	10	-	0.02*	-	12	8	-	*Ergotted seeds, sclerotia
	CS	75	98	2	20	20	-	0.04*	-	12	8	-	
Barnyard millet, common millet, finger millet, Italian millet, Kodo millet and little millet	FS	75	97	3	10	10	-	-	-	12	8	-	
	CS	75	97	3	20	20	-	-	-	12	8	-	
Pulses													
Black gram, green gram	FS	75	98	2	5	5	-	-	10	9	8	-	Germination including hard seeds
	CS	75	98	2	10	10	-	-	20	9	8	-	Germination including hard seeds
Bengal gram	FS	85	98	2	None	None		-	5	9	8	-	Germination including hard seeds
	CS	85	98	2	5	None	-	-	10	9	8	-	Germination including hard seeds
Cow pea	FS	75	98	2	5	5	-	-	10	9	8	-	Germination including hard seeds
	CS	75	98	2	10	10	-	-	20	9	8	-	Germination including hard seeds
Horse gram	FS	80	98	2	None	None	-	-	5	9	8	-	Germination including hard seeds
	CS	80	98	2	10	None	-	-	10	9	8	-	Germination including hard seeds
Indian bean	FS	75	98	2	None	None	-	-	5	9	8	-	Germination including hard seeds
	CS	75	98	2	None	None	-	-	10	9	8	-	Germination including hard seeds

Contd...

Table 24.1. Contd...

Khesari	FS	75	98	2	5	5	-	-	10	9	8	-	Germination including hard seeds
	CS	75	98	2	10	10	-	-	10	9	8	-	Germination including hard seeds
Lentil	FS	75	98	2	5	10	-	-	10	9	8	-	Germination including hard seeds
	CS	75	98	2	10	20	-	-	10	9	8	-	Germination including hard seeds
Moth bean, pigeon pea,	FS	75	98	2	10	10	-	-	10	9	8	-	Germination including hard seeds
	CS	75	98	2	None	None	-	-	20	9	8	-	Germination including hard seeds
Peas	FS	75	98	2	None	None	-	-	5	9	8	-	Germination including hard seeds
	CS	75	98	2	5	None	-	-	10	9	8	-	Germination including hard seeds
French bean	FS	75	98	2	None	None	-	-	5	9	7	-	Germination including hard seeds
	CS	75	98	2	None	10	-	-	10	9	7	-	Germination including hard seeds
Oil seeds, castor (OP)	FS	75	98	2	None	None	-	-	5	8	5	-	Germination including hard seeds
	CS	75	98	2	None	None	-	-	10	8	5	-	Germination including hard seeds
Castor hybrids	FS	70	98	2	None	None	-	-	5	8	5	95*	* Genetic purity in grow out test
	CS	70	98	2	None	None	-	-	10	8	5	85*	
Ground nut	FS	70	96	4	None	None	-	-	5	9*	5*	-	For hand shelled karnels
	CS	70	96	4	None	None	-	-	10		9*	5*	-

Contd...

Table 24.1. Contd...

Rape seed	FS	85	97	3	10	10	5*	-	10	8	7	-	*Argemone mexicana*
	CS	85	97	3	20	20	10*	-	20	8	7	-	
Mustard	FS	85	97	3	10	10	5*	-	10	8	5	-	*Argemone mexicana*
	CS	85	97	3	20	20	10*	-	20	8	5	-	
Linseed	FS	80	98	2	10	5	-	-	10	9	7		
	CS	80	98	2	20	10	-	-	10	9	7		
Niger	FS	80	98	2	10	5	-	-	10	9	5		
	CS	80	98	2	20	10	-	-	20	9	5		
Taramira	FS	85	97	3	10	10	5*	-	10	8	5		*Argemone mexicana*
	CS	85	97	3	20	20	10*	-	20	8	5		
Safflower	FS	80	98	2	None	5	None *	-	-	9	7		*Wild safflower
	CS	80	98	2	None	10	None *	-	-	9	7		
Sesame	FS	80	97	3	10	10	-	-	10	9	5		
	CS	80	97	3	20	20	-	-	20	9	5		
Soybean	FS	70	98	2	None	5	-	-	5	12	7		
	CS	70	98	2	10	10	-	-	10	12	7		
Sun flowers	FS	70	98	2	None	5	None*	None	-	9	7	2**	*Wild helianthus spp. ** Maximum huskless seeds
	CS	70	98	2	None	10	None*	None	-	9	7	2**	
Fibres													
Cotton	FS	65	98	2	5	5	-	-	-	10	6	-	
	CS	65	98	2	10	10	-	-	-	10	6	-	
Cotton hybrids	FS	65	98	2	5	5	-	-	-	10	6	-	Minimum genetic purity in growout test 90 %
	CS	65	98	2	10	10	-	-	-	10	6	-	
Jute	FS	80	97	3	10	10	-	-	10	9	7	-	
	CS	80	97	3	20	20	-	-	20	9	7	-	
Forages													
Berseem	FS	80	98	2	10	10	5*	-	-	10	7		*Germination includes hard seeds
	CS	80	98	2	20	20	10*	-	-	10	7		
Lucerne	FS	80	98	2	10	10	5*	-	-	10	7		*Chicory *Dodder
	CS	80	98	2	20	20	10	-	-	10	8		
Dinanath grass	FS	50	95	5	20	20	-	-	-	10	8		
	CS	50	95	5	40	40	-	-	-	10	8		
Forage sorghum, Sudan grass	FS	75	97	3	5	5	-	0.02*	10	12	8		*Ergotted seeds, sclerotia
	CS	75	97	3	10	10	-	0.04*	20	12	8		
Guar	FS	70	98	2	10	None	-	-	10	9	8		
	CS	70	98	2	20	None	-	-	20	9	8		
Oats	FS	85	98	2	10	10	2*	-	10	12	8		*wild oats
	CS	85	98	2	20	20	5*	-	20	12	8		
Rice bean	FS	70	98	2	None	5	-	-	10	9	8		* Germination includes hard seeds
	CS	70	98	2	5	10	-	-	20	9	8		
Setaria grass	FS	50	95	5	20	-	-	-	-	10	8		
	CS	50	95	5	40	-	-	-	-	10	8		

Contd...

Table 24.1. Contd...

Crop	Class of seed	Minimum					Maximum permissible limit				
		Germination (%)	Pure seed (%)	Inert matter (%)	Other crop seeds (no./kg)	Weed seeds (no./kg)	Objectionable weds (no./kg)	Other distinguishable variety seeds (no./kg)	Moisture: Ordinary pack (%)	Moisture: Vapour proof (%)	Minimum genetic purity % in growout test
Vegetables											
Fruit Vegetables											
Brinjal	FS	70	98	2	None	None	-	-	8	6	-
	CS	70	98	2	None	None	-	-	8	6	-
Brinjal hybrids	CS	70	98	2	None	None	-	-	8	6	90**
Capsicum, Chilli	FS	60	98	2	5	5	-	-	8	6	-
	CS	60	98	2	10	10	-	-	8	6	-
Okra	FS	65	99	1	None	None	None*	10	10	8	*wild okra
	CS	65	99	1	5	None	None*	20	10	8	
Eat-tail, radish	FS	70	98	2	5	10	-	-	6	5	-
	CS	70	98	2	10	20	-	-	6	5	-
Tomato	FS	70	98	2	5	None	-	-	8	6	-
	CS	70	98	2	10	None	-	-	8	6	
Tomato hybrids	FS	70	98	2	5	None	-	-	8	6	-
	CS	70	98	2	10	None	-	-	8	6	90**
Ashgourd, bottle gourd, Indian squash, pumpkin, summer squash and winter squash	FS	60	98	2	None	None	-	-	7	6	-
	CS	60	98	2	None	None	-	-	7	6	-
Bitter gourd, watermelon	FS	60	98	2	None	None	None	5	7	6	Seedless water melon ploidy test 95 %
	CS	60	98	2	None	None	None	10	7	6	Seedless water melon ploidy test 95 %
Cucumber, longmelon, muskmelon, snapmelon	FS	60	98	2	5	None	None	-	7	6	-
	CS	60	98	3	10	None	None	-	7	6	-
Muskmelon hybrids	FS	60	98	2	5	None	None	-	7	6	-
	CS	60	98	2	10	None	None	-	7	6	90**
Ridge gourd, sponge gourd	FS	60	98	2	None	None	-	5	7	6	-
	CS	60	98	2	None	None	-	10	7	6	-
Snake gourd	FS	60	98	2	None	None	None	-	7	6	-
	CS	60	98	2	None	None	None	-	7	6	-

Contd...

Table 24.1. Contd...

Green leafy vegetables											
Amaranth	FS	70	95	5	5	10	5	10	8	6	Wild amaranth
	CS	70	95	5	10	20	10	20	8	6	-
Asparagus	FS	70	96	4	5	5	-	-	8	6	-
	CS	70	96	4	10	10	-	-	8	6	-
Celery	FS	70	97	3	5	5	-	-	8	7	-
	CS	70	97	3	10	10	-	-	8	7	-
Fenugreek	FS	70	98	2	10	10	2*	10	8	6	* Senji
	CS	70	98	2	20	20	5*	20	8	6	
Lettuce	FS	70	98	2	5	5	2*	10	8	6	*Wild lettuce
	CS	70	98	2	10	10	5*	20	8	6	
Parsley	FS	65	97	3	5	5	-	-	8	7	
	CS	65	97	3	10	10	-	-	8	7	
Spinach and spinach beet	FS	60	96	4	5	5	-	-	9	8	
	CS	60	96	4	10	10	-	-	9	8	
Cole crops											
Cabbage,Chinese Cabbage,knol-khol	FS	70	98	2	5	5	-	-	7	5	
	CS	70	98	2	10	10	-	-	7	5	
Cabbage, cauliflower, broccoli, knol-khol, Chinese cabbage	Single crosses and hybrids	70	98	2	None	None	-	-	7	5	-
Bulb, root and tuber crops											
Onion hybrids	FS	70	98	2	5	5	-	-	8	6	-
	CS	70	98	2	10	10	-	-	8	6	-
Carrot hybrids	FS	60	95	5	5	5	-	5	8	7	-
	CS	60	95	5	10	10	-	10	8	7	-
Celeric	FS	70	97	3	5	5	-	-	8	7	-
	CS	70	97	3	10	10	-	-	8	7	-
Garden beet, sugarbeet	FS	60	96	4	5	5	-	-	9	8	-
	CS	60	96	4	10	10	-	-	9	8	-
Radish	FS	70	98	2	5	5	-	-	6	5	-
	CS	70	98	2	10	20	-	-	6	5	-
Radish, turnip	Single crosses and hybrids	70	98	2	None	None	-	-	6	5	-
Turnip	FS	70	98	2	5	5	-	-	6	5	-
	CS	70	98	2	10	10	-	-	6	5	-
Potato(true potato seed)	CS	80	98	2	-	10	-	-	8	6	90

Table 24.2. Seed standards for vegetatively propagating plant materials

Class of seed		Maximum permissible limit: Not conforming to seed size or weight specifications (%/No.)	Damaged (% by weight)	Off-type (% by No.)	Disease-infected/ insect infested (% by No.): Late blight	Wet rot	Common scab	Black scurf	Total diseased
Potato (tubers)									
Foundation									
Stage I		5.0*	1.0**	0.05	1.0	None	3.0	5.0	5.0
Stage II		5.0*	1.0**	0.05	1.0	None	3.0	5.0	5.0
Certified		5.0*	1.0**	0.10	1.0	None	5.0	5.0	5.0

* seed size specifications

	Hill seeds	Plains seeds
Seed size (mm)	30 – 60 (25 – 150 g)	30 – 55 (25 – 125 g)
Large size (mm)	> 60 (> 150 g)	> 55 (> 125 g)

Size of the potato tuber is decided on the basis of mean of 2 width of a tuber at middle and that of length or on the basis of corresponding weight of tuber.

** Damaged : Cut, bruised, unshaped, cracked, damaged by insects or worms, 1. Seed material shall be reasonably clean, healthy and firm. 2. Greenish pigmentatio n on tubers is not a disqualification. 3. Seed lots having even a single tuber infected by common scab shall be treated with approved fungicide. Seed lots having infected tubers beyond maximum permissible limits are not certifiable. Even after treatment with the approved fungicide. 4. Seed lots having black scurf beyond maximum permissible limits could be certified after the seed treatment with approved chemical/fungicide.

Class of seed		Not conforming	Damaged	Off-type	Late blight	Wet rot	Common scab	Black scurf	Total diseased
Lesser yam(tubers)	FS	5.0	1.0	0.05	None				
	CS	5.0	1.0	0.10	None				
Taro (corms)	FS	5.0	1.0	0.10	None				
	CS	5.0	1.0	0.50	None				
Sweet potato (roots)	FS	-	1.0	0.10	None	5	None	1.0	None
	CS	-	1.0	0.20	None	5	1.0	1.0	None
Garlic (bulbs)	FS	-	2.0	0.1	-	-	-	-	-
	CS	-	2.0	0.2	-	-	-	-	-
Potato onion (multiplier onion)	FS	-	2.0	0.1	-	-	-	-	-
	CS	-	2.0	0.2	-	-	-	-	-

Table 24.3. Standard for planting stocks

Crop	Minimum (% by No.)		Maximum (% by No.)	
	Class of seed	**Pure living root slips or stem cutting/planting stacks**	**Other living plants/ root slips, stem cuttings**	**Objectionable weeds**
Napier grass	FS	99.5	0.5	None
	CS	98.0	2.0	None
Pointed gourd	FS	99.5	0.5	None
	CS	98.0	2.0	None

25

Seed Certification

Seed certification is a legal programme to maintain and make available to the public high quality seeds and planting material of genetically distinct crop varieties. Under seed production programme, quality seed is produced by govt. farms research stations, progressive farmers and seed producers using careful quality control, pedigreed plant stock, field inspections during the growing season and seed inspections after the harvest. Certification is an officially recognized method for maintaining varietal identity of the seed. In India certification is carried out by State Seed Certification Agencies.

Classes of Seed Under Certification

Inherent in the certification concept is a generation system where by the pedigree of superior crop varieties is maintained through subsequent seed production. In India generally, a three generation scheme is followed and seed of each generation is identified by a special colour labeling tag. The classes of seed are as follows :

- Breeder seed
- Foundation seed
- Certified seed

Production of Breeder Seed

Breeder seed is produced under the first supervision of the plant breeder and represents the true pedigree of the variety, under seed production. This seed is produced by the State Agricultural Universities, ICAR Research Institutes and other organizations as per allocation through proforma (BSPI). The allocation is done by the crop coordinators on the basis of indent received by them from the Seed Division, Ministry of Agriculture, Government of India. The breeder seed requirement is placed by the different state governments and private sector seed companies. Besides this allocation, the producing institutions carry out additional breeder seed production programme to meet other requirements.

Breeder seed is produced usually from nucleus seed and in case nucleus seed is not available in adequate quantity, breeder seed can be produced from breeder seed also. However, only one cycle of breeder stage I to breeder stage II is allowed. After sowing of nucleus seed, proforma BSP II is submitted by the producing breeder to ICAR, Seed Division, Ministry of Agriculture, GOI containing information on sowing date, area sown, targeted quantity of seed production and expected date of monitoring. Throughout crop growing stage, the crop is regularly examined by the crop breeder and off type plants are rogued out.

The breeder seed production plot is monitored by a team consisting of the following:

- Crop breeder
- Representative of crop coordinator
- Area representative of National Seeds Corporation
- Dy. Director State Seed Certification Agency of the region

The monitoring is done at least at two critical stages, namely, during flowering and at maturity stage. After the satisfactory report by the monitoring team, the crop is harvested and seed is processed and bagged in appropriate size bags and labeled with yellow colour tag. The report by monitoring team is prepared in proforma BSP III. The final figure of the quantity of breeder seed produced is communicated to Seed Division, Ministry of Agriculture, GOI through proforma BSP IV. The breeder seed thus produced is distributed to the users as per allotment by the Seed Division, Ministry of Agriculture, GOI. The quantity of seed lifted or not lifted is communicated to Seed Division again through BSP V.

The nucleus/basic seed which is progenitor of the breeder seed is produced by the breeder following strict roguing procedure and isolation distance, etc. During nucleus seed production stage strict adherence to varietal genetic purity is a must. In cross-pollinated, open pollinated varieties, this is done through isolation planting of basic seed and ensuring continuous removal of off-types particularly before flowering. Leftover off types are again removed at harvesting stage as well. In self-pollinated crops, for nucleus seed production, individual plants conforming to varietal characters are harvested separately from any pure seed plot/breeder seed plot and are threshed separately as well i.e. the seed of each harvested plant is kept in separate packets. Next year, individual plant progenies (about 5 m length) are grown. The progenies are examined continuously and any progeny row showing deviation from the variety under consideration or even under doubtful category is marked and the entire row is pulled out of the seed plot. After this exercise, the leftover progenies are harvested in bulk and the seed thus produced is known as nucleus seed. The number of individual plants to be harvested depends upon the targeted quantity of nucleus seed to be produced. In wheat, rice, etc. usually 5000 plants are handled.

The contents of the breeder seed yellow colour tag are as under :

	Crop	:	
	Variety	:	
	Class of seed	:	Breeder seed
	Lot No.	:	
	Date of test	:	
*	Pure seed	:	%
*	Inert matter	:	%
*	Germination	:	%
*	Genetic Purity	:	%
Producing Institution		:	
(Name and address)			

* based on actual

Production of Foundation Seed

Foundation seed is the first generation seed from breeder seed and is usually produced by State Agricultural University Farms, National Seeds Corporation, State Farm Corporation of India, registered farmers, etc. Foundation seed is labeled with white certification tags.

Each grower has to register with State Seed Certification Agency by paying registration fee (usually Rs. 25/ seed block) and field inspection fee (Rs. 200/ to Rs. 250/ha). The seed crop is inspected 2 - 3 times, namely, before flowering, at flowering and at maturity to verify seed source, and to check isolation distance, genetic purity, seed borne diseases and report is prepared at maturity stage by the seed certification Inspector/Officer. The field and the seed must conform to minimum field and seed standards as described separately. The minimum number of field counts is given in Table 25.1 (Agarwal, 1995).

Table 25.1. Minimum number of field counts under certification

Area of field (ha)	Minimum number of counts
Up to 2	5
2-4	6
4-6	7
6-8	8
8-10	9
> 10	10

The number of plants/heads per count is given in Table 25.2 (Agarwal, 1995).

Table 25.2. The number of plants/heads per count during seed certification

Crop	Number of plants/heads per count
Wide spaced row crops	
Okra, brinjal, bulb crops, chilli, cole crops, cucurbits, potato, tomato, root crops, cotton, ground nut, maize, etc.	100
Medium spaced row crops	
Beans, cowpea, gram, black gram, green gram, pea, mustard, sesame, leaf crops	500
Thickly sown crops	
Berseem, lucerne, jute, soybean, mesta	1000
Wheat, barley, paddy, oats, pearl millet, sorghum	1000 heads

For taking field counts, procedures are common sense approach and are appropriately modified depending upon the crop density. For densely planted/ broadcast crops, generally a meter square frame is used. It is kept at 5 random locations and average number of plants/m^2 is counted. Based on this required number of 1 meter square frame is placed at randomly distributed locations and off type plants/plants affected with seed borne diseases/objectionable weeds are counted.

For wheat, barley, oats, soybean and similarly planted crops, one can enter the seed field from a randomly selected site and determine the average number of plants/heads in a row in one step and take the required number of steps to complete a count. While doing this, it is advisable to have 100 plants in one row and then cross in over 2 – 3 rows and complete 100 plants count in another row. For maize and okra like crops, one may count 10 plants per row and thus complete the 100 different rows.

In case of urgency and necessity, foundation seed can be allowed for one additional cycle of multiplication and in that eventuality, the produce is labeled as foundation seed stage II. In principle, foundation seed stage II is equivalent to "registered seed" in USA and other countries where the recognized classes of seed are:

- Breeder seed
- Foundation seed
- Registered seed
- Certified seed

Thus registered seed is the seed from foundation seed and is intended for the purpose of increasing seed another generation before the production of certified seed (Copeland and McDonald, 1995). Registered seed is not intended to be a commercial class of seed.

Certified Seed

Certified seed is produced from foundation seed (or registered seed = foundation seed stage II) and represents the final product of the seed certification programme. It is labeled with the blue tag. If needed under some special circumstances, certified seed can be allowed one additional cycle of multiplication and in that case the produce is labeled as certified seed stage II. The certified seed plot registration and certification process is similar as applicable in case of certified seed.

Post Harvest Inspection

In a developing certification scheme, it is a necessity to inspect the seed producer's or contractor's premises, particularly such features as ,the threshing place, seed drying and storage premises, bags used for storing seed and arrangements for sampling. Inadequate facilities and poor management in post-harvest handling of seeds may disqualify the seed lot from certification. During all these operations the inspection is primarily done to ensure that the identity of the lot is maintained and that the seed lot is being handled carefully.

If the harvesting is done by combines, the inspection may be made at harvesting time to ensure that combine, trailers, bags and storage places are clean. Combining one round around a field does not clean the combine. Machine cleaning can usually be accomplished only by taking the time to blow, brush or clean out. In some other manner each section of the combine, storage place, should be sprayed with Malathion to avoid insect damage.

Seed Inspection

Seed Sampling

Sampling of each seed lot to be certified should be done by certification inspectors as per rules. The bags for sampling should be stored, or piled in such a way that the inspector can reach any bag to get a representative sample of a lot. The inspector must be well-trained in seed sampling. Before sampling for seed analysis in a seed testing laboratory, the inspector should check that the moisture of the seed lot is not higher than the maximum standards and that the lot is acceptably uniform. The sample, properly identified and packed, should be transferred to a seed testing laboratory without delay. It must be sealed, if mailed or handled by unauthorised persons. A sampling report giving the following information should be submitted with sample.

1. Owner of the seed lot and address
2. See producer and location
3. Kind and variety
4. Lot number or marking
5. Weight of lot

6. Number and kind of units
7. Seed origin
8. Certification class
9. Date of sampling
10. Signature of inspector

Time of Sampling

The sampling of seed lots should be done at least two times. The first sampling should be done before the seed is processed to have a preliminary testing for seed purity, viability and/or germination and seed health. The purity check would involve checking for mechanical admixtures of other varieties and extent and kind of impurities. The cleaning of samples on laboratory model cleaning machines used at the processing plant would provide useful information, regarding the extent to which these impurities could be removed after cleaning of the seed lot and/or what precautions would be necessary in the cleaning of the seed lot. The seed health examination would provide useful guidance in selecting the appropriate seed treatment.

The second sampling should be done after cleaning the seed lot. A complete seed analysis (germination, purity, moisture, other crop seeds, seed health, bioassay test, etc.) should be done to ensure that the seed lot has met the requisite Seed Certification Standards.

Seed Testing

The testing of seed samples drawn by seed certification inspector is done in a seed testing laboratory as described in Chapter 26.

Seed Inspection During Seed Processing

Seed certification inspectors should make as many inspections of seed lots, records and the processing plant as may be required in order to satisfy that only the seed meeting certification requirements be labeled with certification tags and sealed.

The following are the general guidelines for inspecting a processing plant :

Walk through the entire area. Locate the seed dump and the flow of seed all the way to the bagging. Be sure you understand how the seeds are conveyed from one machine to the other or how the seed may bypass some equipment.

- Before inspection begins, have the plant manager run all equipment, elevators, and conveyors
- Begin your inspection at the dump pit. Proceed along the route the seed flows taking particular attention to cracks, corners sharp turns, or down spouts that may not have the proper slope.

- The bottom of the elevator, must have a drop plate in order to clean out the leg. There should be room around the leg to allow for proper cleaning. If the elevator is in a pit, it is essential that there is room for proper clean outs.
- Screw conveyors should be equipped with removable covers to allow for proper inspections upright conveyor, it must have a clean out plate at the bottom.
- Belt conveyors will need to be checked at the joints and underneath the belt, if it is in a trough, self cleaning belts are less likely to have seed remaining, however, the joints should be inspected.
- The building must be free of all seed or the seed must be properly bagged, tagged and stacked out of the way.
- Bulk seed storage must be approved by the inspector, if it is allowed in the plant. This may require a top over the bin, or disconnect the conveyor to the bin. The inspector must determine the best procedure to follow, if approval is granted.
- The inspector will record what machines are in the plant and what seed can be properly cleaned.
- The inspector will write a report approving the plant, or he will show reasons for disapproval and make recommendations.
- The inspector will leave a processing book for the manager to keep records of certified seed.

In conclusion inspector should be satisfied that the processor can handle certified seed properly, before he approves the plant. The entire operation should be handled in a business like manner, as is fitting to the certification programme. The best equipment and most modern plant is not enough. There must be a manager who has the knowledge and concern to do the right job. Be sure you have thoroughly discussed the certification programme with the plant manager.

Issuance of Tags and Seals

Tags and seals for certified seed shall be given by seed certification agency to seed producers as per rules and regulations for all seed lots approved by it. The contents of seed certification tag shall be as per following specifications (Agarwal, 1995),

Specification for Certification Tag

Length	15 cm
Breadth	7.5 cm

Quality : It shall be made of durable material such as thick paper, paper with cloth lining, wax coated paper, plastic coated paper, etc.

Colour : Both sides shall be white for foundation class and blue (ISI NO. 104 - Azure blue) for certified class

Tag No.

Certified seed

Kind........................... Name and address of certification agency

Class of seed ....

Date of issue of certificate ....

Date of test ....

Certificate valid upto ....

(Provided seed is stored under cool and dry environment)

Validity of certificate

Further extended up to

....

Use of the seed after expiry of the validity period by any person is entirely at his risk and the holder of the certificate shall not be responsible for any damage to the buyer of seed. No one should purchase the seed if seal or the certification tag has been tempered with.

Name and full address of the certified seed producer

....

N.B. if tag is to be affixed on a smaller container then the size of the tag may be reduced proportionately. However, length and breadth ratio and contents would remain the same.

Labeling of Breeder Seed

Breeder seed shall be supplied in sealed containers, duly stiched and sealed. A cloth lined label of 12 cm × 6 cm containing following information shall be affixed on the container. The colour of the tag is yellow.

Crop	
Variety	
Class of seed	Breeder seed
Lot No .	
Date of test	
Pure seed	%
Inert matter	%
Germination	%
Genetic purity	%
Producing Institution (name and address)	

The container should also have printed on it the kind, variety and name of institution.

Field Inspection Report

The general format is as follows :

Field No.………………….. Report No.………………………

Crop………………………. Variety…………………………..

Date of sowing…………… Date of inspection………………..

Expected date of harvest………………….. Time: From……………To……………

1. Name of producer/grower…………………..
2. Village………………… District………………… State………………
3. Location of farm…………………………………….
4. Class of seed…………………………………..
5. Source of seed……………………………
6. Total acreage under production of this seed crop……………………..
7. Acreage of the field inspected……………………………………….
8. Previous crop……………………………………………………………
9. Isolation distance in meters……………………………….
10. Stage of growth of contaminant……………………………….
11. Stage of seed crop at this inspection…………………………………
12. Field counts: Take 1000 heads per count for barley, oats, paddy and wheat. Take 1000 plants per count for berseem, jute, lucerne, mesta, soybean. Take 500 plants per count for medium spaced row crops such as beans, cowpea, garden pea, gram, leaf crops, mung, mustard, rape, sesame, and sun hemp. Take 100 plants per count for wide spaced row crops such a bhindi, brinjal, bulb crops, capsicum, castor, chilli, cole crops, cotton, cucurbits, ground nut, potato, red gram, root crops and tomato.

Count No.	No. of heads/plants			
	Off types	**Inseparable other crops**	**Objectionable weeds**	**Affected by seed borne disease**
1.				
2.				
3.				
4.				
5.				
6.				
7.				
8.				
9.				
10.				
Total				
Percentage				

References

Agarwal, R.L. 1995. Seed Technology 3rd Ed. Oxford and IBH.

Copeland, L.O. and McDonald, M.B. 1995. Principles of Seed Science and Technology, 3rd Ed. Chapman and Hall.

26

Seed Quality and Its Evaluation

Seed quality is a multiple criterion that encompasses several important attributes. A seed scientist may be concerned with each quality characteristic of an individual seed while the seed trader normally considers the quality components of a seed lot. Any individual seed may possess certain measurable seed quality components which include:

- Genetic and chemical composition
- Physical condition
- Physiological viability and vigour
- Seed size
- Seed appearance
- Presence of seed borne pathogens

The quality components of a seed lot include:

- Crop purity
- Cultivar purity
- Weed or crop contaminants
- Seed germination
- Seed vigour
- Seed uniformity
- Seed moisture content
- Seed borne pathogens

The important seed quality components and their evaluation are as follows:

Determination of Seed Purity

The purity test is done with the object of determining the composition by weight of the sample being tested and by inference. Components examined in the purity analysis are :

- Pure seed
- Seeds of other species
- Inert matter

The quality of seed lot is judged by the relative percentage of above components. The quality is considered superior if pure seed percentage is more than 98 per cent and other species seeds and inert matter percentages are as low as possible. The percentage of seeds of other species should be almost negligible or nil. The purity analysis and seed germination test complement each other.

General Principles of Purity Analysis

The working samples are separated into the three components, namely, pure seed, other seeds and inert matter and the percentage of each component is determined by weight.

Pure Seed

The pure seed means the seed of kind /species stated by the sender or found to predominate in the test. It includes all botanical varieties and variety of that kind/species.

Other Crop Seeds and Weed Seeds

The other crop seeds are the seeds of plants which are grown as crops other than main crop seeds. The weed seeds are the seeds of plants of those species recognized as weeds by laws, official regulations (objectionable weeds) or by general uses (common weeds).

Inert Matter

Inert matter includes all other matters and structures that do not fall in the category of pure seed, other crop seed or weed seeds.

Purity Separation

Place the working samples (or sub-sample) on the clean surface of a purity work-board and do the following.

- Examine the sample to determine whether it conforms to the variety/species name under which it has been submitted. The file samples (authentic samples of the same variety) maintained in the Seed Herbarium are of great value in making this decision. In case of any doubt, the sample under examination may be compared with the file sample of the same variety/species.
- Examine the sample to determine whether it contains small inert matter, which could be removed by sieving.
- Examine the sample to determine whether it contains light weight material which may be removed by blowing.

If small inert matter is present which can be removed by sieving or by preliminary blowing, use the appropriate sieve to separate it. If the light-weight material is present, blow the sample. In certain cases, as described in the definitions earlier, sieving or uniform blowing is obligatory.

Place the sample in a pile on the purity work-board (after sieving/blowing operations, where-ever undertaken) slightly to the rear and left of centre. With the aid of forceps, scalpel or spatula having straight smooth edges, draw a few seeds at a time from the pile spreading them apart by pulling them towards the front of the board. As small groups of seeds are being examined, pull the pure seed into the scoop or container at the front of the work-board and push all other material to the upper right hand of the board for further examination and separation into other crop seeds, weed seeds and inert matter. Continue this operation until the entire sample is separated.

Seed Viability and Its Evaluation

There is considerable disagreement and confusion regarding the precise meaning of seed viability. According to most seed technologist and commercial seed dealers, seed viability means that a seed is capable of germinating and producing a normal seedling. Therefore, it can be synonyms with seed germination capacity. In this sense, a given seed is either viable or non-viable depending upon whether it germinates and produces a normal seedling. However, according to some persons, viability, denotes the degree to which a seed is alive, metabolically active and possesses enzymes capable of catalyzing metabolic reactions required for germination and seedling growth. In this context a seed may contain both live and dead tissues and may or may not be able to germinate. This meaning deals with tissue viability as well as viability of the entire seed (Capeland and McDonald, 1995). There are several tests for determination of seed viability. Germination test is the most important and reliable method which is followed routinely. This is described as follows :

Germination, in a laboratory test, is the emergence and development from the seed embryo of those essential structures which, for the kind of seed being tested, indicate the ability to develop into a normal plant under favourable conditions in the soil. The ultimate object of testing for germination is to gain information with respect to field planting value of the seed and provide results which can be used to compare the value of different seed lots.

Method of Germination Testing

Preparing the Test

- *Number of Seeds:* The Rules specify that at least 400 seeds should be tested for germination.
- *Source of Seeds:* Special care must be taken to select 400 seed sub-sample as per rule. The seed shall not be given any pretreatment.

- When both purity and germination tests are required, seeds for germination test shall be taken from the pure seed component separated in purity analysis. The pure seed shall be well mixed and the 400 seeds counted without discrimination as to size or appearance into replicates of 100, 50, or 25 seeds.

- When only a germination test is required and the pure seed is estimated or determined to be at least 98 %, the pure seed for the germination test may be taken indiscriminately from a representative portion of the bulk.

- When only a germination test is required and the pure seed is found to be less than 98 %, the pure seed shall be taken indiscriminately from a pure seed separation made according to rules.

- The counting of 400 seeds without discrimination as to size or appearance, may be done by hand, by the aid of a counting board, or with the aid of a vacuum seed counter.

The methods of testing seed germination are summarized as under:

Apparatus

1. Germination Apparatus (Seed Germinator)

Any of the following types of apparatus would do the needful.

(a) The Germination Cabinet (Cabinet-Type Seed Germinator):

This type of germinator consists of double-walled cabinet suitably insulated against temperature changes by an air jacket, or layer of insulating material. It is equipped with appropriate trays preferred by individual laboratories. It has provision for both heating and cooling by electricity. In such cabinets, temperature can be regulated at any desired setting between approximately 8°C and 40°C. It is necessary to maintain the humidity of the air as close as possible to saturation to avoid drying of the substrata.

It would be desirable to have 2–3 cabinet germinators and one cabinet germinator (daylight type, that is having the arrangement for light, necessary for germinating light sensitive seeds). This would allow wide range of temperature conditions required for germination of different kinds of seed in a seed testing laboratory.

(b) The Room Germinator (Walk-in Room Germinator):

The room germinator is a modification of the cabinet. It is constructed on the same principles, but is large enough to permit workers to enter it and place the tests along either side of a central passage way.

2. Counting Equipment

(a) Counting Boards:

Counting boards are generally used for large seeds such as maize, beans, peas, etc. A counting board is approximately the size of the substratum on which the seeds are placed. The top consists of a stationary board with 50 or 100 holes of the general shape and size of the seeds to be counted. Below this board is a thin solid board which serves as a false bottom and can be slid forward and backward. In operation, the counting board is placed over the substratum, the seed scattered over the board, and the excess seed removed by tilting the board slightly. After checking to see that all holes are filled and that there is only one seed per hole, the movable bottom is withdrawn and the seeds fall in place upon the substratum.

(b) Vacuum Counter:

A vacuum counter is generally used for free flowing smooth seeds. It consists of a vacuum system, counting head and valve. A series of counting heads are required to handle various kinds of seeds.

3. Sand Boxes/Sand Benches

If only a few samples are required to be tested in sand, it would be preferable to have sand boxes (17.5 x 14.5 x 3.5 cm dimensions) made in metal (or plastic). On the other hand, if large-scale germination tests are to be made in a specially constructed sand germination room, it would be desirable to have wooden sand benches in which larger number of samples could be tested.

4. Refrigerator

It is required for pre-chilling certain kinds of seeds.

5. Stoves

For pre-drying certain kinds of seeds two small stoves should be available that meet the requirements as inside stainless steel, outside white enamelled steel, inside dimensions 28 x 25 x 21.5 cm, provided with 3 removable shelves made of perforated stainless steel; heated electrically by means of resistance cable (or tape) or by two sets of lamps (4 x 25 watts underneath the bottom, 2 x 15 watts in the top of stove, both sets powered with half the nominal voltage; temperature range 25 - 40°C, adjustable within ±1°C.

6. Soil/Sand Sterilizer

7. Disinfection Cabinet

8. Petri-Dishes/Plastic Boxes and Other Glasswares

9. Miscellaneous Equipment

For example, forceps, magnifiers, washer, hand scarifier, water sprinkler, thermometers, etc.

Germination Materials and Quality

The general specifications are as follows:

1. Filter Papers, Blotters or Towels

(a) Composition

The fiber content of the paper should be 100 percent bleached chemical wood, cotton or other purified vegetable cellulose. It must be free of fungi, bacteria and toxic substances.

(b) Texture

Open and porous nature, but texture should be such that the roots of the seedlings will grow on and not into the paper.

(c) Strength

The paper should possess sufficient strength to enable it to resist tearing when handled during the test.

(d) Moisture Capacity

The paper should have the capacity to hold sufficient water for whole of the test period.

(e) pH: 6—7.5

(f) Storage: at low R.H.

(g) Sterilization

It may be necessary to sterilize the paper to eliminate mold, which may have developed during storage.

(h) Quality Control:

Carry out a biological test for pernicious material. For this test, seeds of certain spp. known to be sensitive to toxic substances (for example, *Phleum, Agrostis, Festuca, Brassica, Allium,* etc.) are grown on the paper and the root development is compared with that grown on the paper of known quality on or before the first count is due. This is done so, because symptoms such as shortened roots, discoloured root tips, bunched root hairs, flattened and shortened coleoptiles in grasses due to toxic substances are more pronounced at an early stage of root growth.

2. *Sand*

Composition

The sand should be reasonably uniform and free from very small and large particles. It is recommended that nearly all the particles should pass through a sieve having holes of 0.8 mm diameter and be retained on a sieve having holes of 0.5 mm. The sand must be free from foreign seeds, fungi, bacteria or toxic substances which might interfere with the germination of seeds, the growth of the seedlings or their evaluation.

Moisture Capacity

When appropriate amount of water is added, the particles of the sand should have capacity to hold sufficient water to provide continuous movement of water to the seeds and seedlings and also provide sufficient pore space for aeration for optimal root and shoot growth.

pH: 6.0—7.5

Sterilization:

It may be necessary to wash and sterilize the sand before use.

Reuse

Several times, each time after washing, dried and resterilized sand which has been used for testing chemically treated samples should preferably be discarded to ensure that chemicals which may have accumulated in the sand do not cause phytotoxic symptoms.

Quality Control

Same as for paper substrates

3. *Soil*

Not recommended as primary testing substances.

Composition:

Good quality, non-caking and free from any large particles, foreign seeds, bacteria, fungi, nematodes or toxic substances.

Moisture Capacity:

When adjusted to the appropriate moisture content, soil should allow adequate aeration for germination and root growth.

pH: 6-7.5

Sterilization:

Same as for sand

Reuse:

No reuse is recommended.

4. Water

The water used to moisten the substrate should be reasonably free from organic or inorganic impurities. Distilled water may be used where usual water supply is unsatisfactory.

pH: 6-7.5

Quality Control:

An analysis of water should be obtained from time to time.

Germination Methods Using Paper

a. Top of Paper (TP)

The seeds are germinated on top of one, or more, layers of paper which are placed in any of the following ways:

1. Enclosed in Transparent Petri-Dishes, or Boxes: The appropriate quantity of water is added at the beginning of the test. Evaporation may be minimized by a tight-fitting lid or by enclosing the dishes in plastic bags.
2. Directly on Germination Trays in Cabinet or Room Type Germinators: In this method the relative humidity in the germinator, or room, must be maintained as close as possible to saturation. Moistened porous paper or absorbent cotton can be used as a base for the papers, or even as an immediate substrate.

b. Between Paper (BP)

The seeds are germinated between two layers of germination paper either by loosely covering the seeds with an additional layer of germination paper by placing the seeds into folded envelops which may be placed in a flat or upright position, or by placing the seeds in rolled towels. The rolls should be placed in an upright position. These can then be placed:

1. Directly on germination trays in cabinet, or room type germinators; or
2. In metal, plastic or glass boxes.

In method (1), the relative humidity in the germinator or room must be maintained as close as possible to saturation.

For most kind of seeds the paper should not be so wet that if pressed, a film of water forms around the finger.

Germination Methods Using Sand

Sand is used as a substrate for following methods:

S (in Sand)

Seeds are planted in a uniform layer of moist sand and then covered to a depth of 1 to 2 cm with sand, which is left loose.

TS (Top of Sand)

Seeds are pressed into the surface of the sand. The amount of water to be added to the sand will depend upon its characteristics and the size of the seed to be tested. The optimum amount should be determined for the main kinds of seed, so that a measured quantity could always be used in routine testing. Generally 50-60 % water holding capacity of the sand is utilized. After planting, the sand benches are watered daily.

Germination Method Using Soil

Soil, or artificial compost, may be used instead of sand for methods S and TS, although it is generally more difficult to standardize and is, therefore, liable to cause greater variation between results. It is for this reason that soil is never recommended as a primary testing substrate. But this substrate must be used to confirm the evaluation of seedlings in doubtful cases, and for testing samples which produce seedlings with phytotoxic symptoms when germinated on paper or sand.

Water should be added until the consistency of soil is such that the ball formed by squeezing it in the palm of the hand is easily broken when pressed between two fingers. After the soil is wet, it should be rubbed through a sieve and put into the containers for the test without packing.

Spacing Seed

The proper spacing of seeds to reduce to a minimum the contact of the seedlings with each other during germination cannot be over-emphasized. This is especially important for large seeded kinds and seed stocks infected with storage or other fungi. The standardized test committee - (AOSA) recommendations reproduced below should be followed for spacing seed in a germination test.

"Careful attention should be given to the spacing of seeds in blotters, towels or sand. The distance between seeds should be not less than 1 to 5 times the width or diameter of the seed, the basis of spacing being the size of the seed to be tested".

Provision and Maintenance of Environmental Conditions for Seed Germination

Permissible methods, temperatures, elite and special treatments have been presented in Table 26.1

Table 26.1. Testing of vegetable seeds for germination

Vegetable crop	Substrate	Temperature (°C)	First count (days)	Final count (days)	Additional directions
Broad bean	BP, S	20	4	14	Pre-chill
Cluster bean	BP	20-30	5	14	-
Cow pea	BP, S	20-30	5	8	-
Dolichos bean	BP, S	20-30	4	10	-
French bean	BP, S	20-30	5	20	-
Garden pea	BP, S	20	5	8	-
Lima bean	BP, S	20-30	5	9	-
Cabbage	TP	20-30	5	10	Pre-chill, KNO_3
Cauliflower	TP	20-30	5	10	Pre-chill, KNO_3
Chinese cabbage	TP	20-30	5	7	Pre-chill
Knol-khol	TP	20-30	5	10	Pre-chill, KNO_3
Brinjal	TP. BP	20-30	7	14	-
Chillies and pepper	TP, BP	20-30	7	14	KNO_3
Okra	BP, S	20-30	4	21	-
Tomato	BP, TP	20-30	5	14	KNO_3
Cucumber	BP,TP, S	20-30	4	8	-
Long melon	BP, S	20-30	4	8	-
Muskmelon	BP, S	20-30	4	8	-
Pumpkin	BP, S	20 -30	4	8	
Summer squash	BP, S	20 - 30	4	8	-
Indian squash	BP, TP, S	20 - 30	4	8	-
Watermelon	BP, S	20 - 30	5	14	-
Winter squash	BP, S	20 - 30	4	8	Low moisture
Pointed gourd	S	30 - 35	-	14	Dark, GA_3 (500 ppm) 24 hours, remove seed coat
Ridge gourd	BP, S	30	4	14	-
Snake gourd	S	30-35	-	14	Dark, GA_3 (500 ppm) 24 hours, remove seed coat
Sponge gourd	BP,S	20 - 30	4	14	-
Ash gourd	S	30-35	5	14	Light
Bitter gourd	BP, S	20 - 30	4	14	-
Bottle gourd	BP, S	20 - 30	4	14	-
Celery	TP	20-30	10	21	-
Lettuce	TP, BP	20	4	7	Pre-chill
Spinach beet	BP, TP	20 - 30	4	14	Pre-wash (Multigerm 2 hrs, monogerm 4 hours
Carrot	TP, BP	20-30	7	14	-
Radish	TP, BP	20-30	4	10	Pre-chill
Turnip	TP	20 - 30	5	7	KNO_3
Onion	BP, TP	15-20	6	12	-
Amaranth	TP	20-30	-	8	Light

TP = Top of paper

BP = Between paper (including rolled towel and pleated paper)

S = Sand

Evaluation of Germination

The interpretation of seedlings and un-germinated seeds in a germination test is done as per following rules:

Normal Seedlings

Normal seedlings show the potential for continued development into satisfactory plants when grown in good quality soil and under favourable conditions of moisture, temperature and light. To be classified as normal, the seedling must conform with one of the following categories:

1. Intact Seedlings:

Seedlings with all their essential structures well developed, complete in proportion and healthy are considered as intact seedlings. An intact seedling, depending upon the species being tested, shows a specific combination of some of the following essential structures:

- A well-developed root consisting of a long and slender primary root, usually covered with numerous root hairs and ending in a fine tip
- Secondary roots when produced within the prescribed test-period
- Several seminal roots instead of one primary root in certain genera including *Avena, Hordeum, Secale, Triticum*
- A well-developed shoot axis consisting of a straight and usually slender and an elongated hypocotyl in seedlings showing hypogeal germination
- Both an elongated hypocotyl and epicotyl in some genera with epigeal germination
- An elongated mesocotyl in certain genera of the Graminae
- A specific number of cotyledons, one cotyledon in monocotyledons or exceptionally in dicotyledons (it may be green and leaf-like or modified and remaining wholly or partly within the seed), two cotyledons in dicotyledons green, expanding primary leaves
- One primary leaf, sometimes preceded by few scale leaves in seedlings with alternating leaves or two primary leaves in seedlings with opposite leaves
- A terminal bud or shoot apex, the development of which varies with the species being tested
- A well-developed straight coleoptile in Graminae, containing a green leaf extending to the tip and eventually emerging through it

2. Seedlings with Slight Defects

Seedlings showing certain slight defects of their essential structures, provided they show an otherwise satisfactory and balanced development comparable to that of intact seedlings of the same test. The following defects are considered slight:

- Primary root with limited damage or slight growth retardation
- Primary root defective but with sufficient well-developed secondary roots in specific genera of Leguminosae (*Phaseolus, Pisum, Vicia*); Graminae (*Zea*) and in all genera of Cucurbitaceae and Malvaceae
- Hypocotyl, epicotyl or mesocotyl with limited damage (if half or more of the total tissue area is left functioning normally and if there is no evidence of damage or decay to the shoot apex or surrounding tissues)
- Three cotyledons instead of two
- Only one normal cotyledon in dicotyledons (if there is no evidence of damage or decay to the shoot apex or surrounding tissues)
- Primary leaves with limited damage only, one normal primary leaf e.g. in *Phaseolus*, provided there is no evidence of damage or decay to the terminal bud
- Primary leaves of *Phaseolus* which are properly formed but reduced in size, as long as they are larger than a quarter of the normal size
- Coleoptile with limited damage
- Coleoptile with a split from the tip extending downwards not more than one-third of the length

3. Seedlings with Secondary Infections

Seedlings in which it is evident that they would have conformed to 1 or 2 above, but have been affected by fungi or bacteria (seriously decayed) from sources other than the parent seed.

Abnormal Seedlings

Abnormal seedlings do not show the potential to develop into a normal plant when grown in good quality soil and under favourable conditions of moisture, temperature and light. The following seedlings are classified as abnormal:

- Damaged seedlings with any of the essential structures missing or so badly and irreparably damaged that balanced development cannot be expected
- Deformed or unbalanced seedlings with weak development or physiological disturbance or with essential structures deformed or out of proportion
- Decayed seedlings with any of their essential structure as diseased or decayed as a result of primary (i.e. from the parent seed) infection that normal development is prevented

Some other tests for determining seed viability are as follows:

Tetrazolium Test

The tetrazolium test is widely recognized as an accurate means of estimating seed viability. This method was originally developed in early 1940s in Germany by Professor George Lakon (1928). Presently the test is used throughout the world as an accepted method of estimating seed viability and is commonly referred to as a "quick test" since it can be completed only in a few hours in comparison to regular germination test which requires long duration (on a average 10 – 15 days). Tetrazolium test distinguishes between viable and dead tissues of the embryo on the basis of their relative respiration rate in the hydrated condition. This test primarily utilizes the activity of drhydrogenase enzymes as an index to the respiration rate and seed viability. The dehydrogenase enzymes react with the substrate tetrazolium and release hydrogen ions to the oxidized, colourless tetrazolium salt solution which is changed into red formazan as it is reduced by hydrogen ions. The reaction is shown as follows :

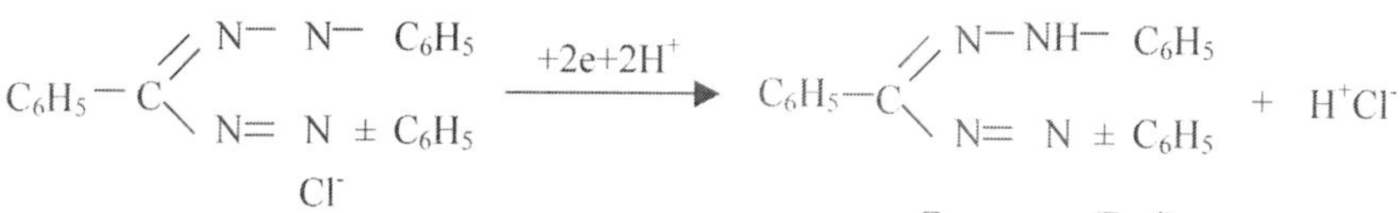

2,3,5 triphenyl tetra zolium chloride

Formazan (Red)

The seed viability is interpreted according to the topographical staining pattern of the embryo and the intensity of the colouration.

Procedure

The seeds are first imbibed on a wet substrate for complete hydration of all the tissues. For most of the species tetrazolium solution can be added to the intact seeds. After hydration seeds are placed in a tetrazolium salt solution for about two hours at 35°C. Sound embryo tissues absorb tetrazolium slowly and tend to develop a lighter colour than the embryos that are bruised or aged or disturbed.

Seed Vigour and Vigour Test

The Seed Vigour Testing Committee of the Association of Official Seed Analysts (AOSA, 1983) has compiled comprehensive review entitled "Seed Vigour Testing". According to this, "seed vigour comprises those properties which determine the potential for rapid, new form emergence and development of normal seedlings under a wide range of field conditions". Thus the seed lots having nearly identical standard germination levels may perform quite differently under poor field conditions due to differences in their vigour potential.

Factors Affecting Seed Vigour

Among the factor that influence seed vigour are :

- Genetic constitution
- Hybrid vigour
- Hard seed
- Susceptibility to mechanical damage
- Chemical composition of seed
- Environment during seed development
- Soil moisture and fertility
- Seed maturity
- Post maturation - preharvest environment
- Seed storage

Seed Vigour Test

A vigour test should possess certain essential features that can make it useful to seed producer and consumer. McDonald (1980) has described the following characteristics :

- Inexpensive
- Rapid
- Uncomplicated
- Objective
- Reproducible
- Correlated with field performance

The seed vigour tests which have attained acceptance by seed analysts and seed testing organizations are as follows (Perry, 1981, AOSA, 1983):

Cold Test

In this method seeds are placed in soil or paper towels lined with soil and exposed to cold for a specified period. Normally, the temperature is 10°C for a period of 5 - 7 days. During this process the seeds are stressed. At the end of the stressed period, the seeds are transferred to favourable temperature prescribed for the standard germination tests (AOSA, 1981).

Accelerating Ageing Test

The accelerated ageing test was originally developed to estimate longevity of seed and storage (Delouche and Baskim, 1973). Under this test the un-imbibed seeds are subjected to a high temperature of 41°C and relative humidity (RH) around 100 % for short periods (3-4 days). The seeds are then removed from stress conditions and placed under optimum germination conditions. The test is commonly conducted in either a large accelerated ageing chamber (for multiple seed samples) or a small single sample ageing chamber. A single sample procedure involves placing the seeds (40 – 45 g) in a single layer on a wiremesh screen above water inside a plastic germination box. Many of these boxes can then be placed into an incubator set at the described temperature and high RH. This test offers the advantages of being inexpensive, simple and requiring little additional training of seed analysis.

Conductivity Test

The conductivity test is a measurement of electrolytes leaching from plant tissue. Poor membrane structure is associated with higher leachate and deteriorating low vigour seed. When the seeds are soaked in water, there is greater loss of electrolytes into the water. The higher the conductivity of the soak water, the lower the seed vigour. This test has been in use to evaluate seed vigour in garden pea in Europe (Mathews and Bradnock, 1967).

Conductivity is measured by placing 25 seeds in a beaker containing 75 ml of distilled water at 20°C for 24 hrs followed by reading conductivity on a conductivity meter (AOSA, 1983). This test is rapid, precise, inexpensive and simple. However, initial seed size can affect the speed of leakage of solute. One limitation of this test is that it expresses results as an average conductivity evaluation for 25 seeds. This test would better reflect the vigour of a seed lot if they were presented on an individual seed basis. A commercial instrument is now available to monitor the electrolyte leakage of individual seeds (Copeland and McDonald, 1995).

Seedling Growth Rate Test

Seedling growth rate is based on the concept that vigorous seeds are able to efficiently synthesize new materials and rapidly transfer these to the emerging embryonic axis resulting in increased dry weight accumulation. The evaluation is based on the growth of seedling under standard germination test except that the moisture content of the paper towels is precisely controlled (AOSA, 1983). At the end of germination period, the growth of the normal seedlings is measured by :

- Length of the seedling
- Dry weight of seedling (80°C, 24 hrs) after removal of cotyledons or endosperm

Limitations of Seedling Growing Rate Test

- It is time consuming to remove and/or measure seedlings.
- It is influenced by variation in moisture and temperature.
- Differential cultivar responses can make accurate test interpretation difficult.

Seedling Vigour Classification Test

Seedling vigour classification test has been described (AOSA, 1983) and is used in some labs for soybean, field bean, cotton and ground nut. This test is an expansion of the standard germination test in the sense that the normal seedlings are further classified as "strong" and "weak". The germination test is conducted at a temperature of 25°C and a preliminary count and classification of seedlings are done 5 days after planting. First normal seedlings (free of deficiencies, viz., shortened hypocotyls, stubby primary root with no secondary roots, no epicotyl, hypocotyls and roots missing, poorly developed hypocotyls and roots) are selected and classified as :

- Strong seedlings - normal healthy seedlings without any defects
- Weak seedlings – seedlings with partially decayed epicotyl, no primary root, primary root, cracked, one primary leaf missing, split primary root, curled hypocotyls

Speed of Germination

Speed of germination is one of the oldest seed vigour concepts. The number of normal seedlings counted at the first count represents the faster germinating seeds and is a measure of seed/seedling vigour. Maguire (1962) suggested the following formula :

$$X = \frac{\text{Number of normal seedling}}{\text{days of first count}} + \ldots\ldots \frac{\text{Number of normal seedling}}{\text{days of final count}}$$

In soybean, the first count (4 days) has been used as seed vigour index component.

Osmotic Stress

When seeds are sown in the field, they are often subjected to drought stress which results in poor emergence. Such drought conditions can be simulated in lab by using standard solutions with specific osmotic potentials such as sodium chloride, glycerol, sucrose, polyethyleneglycol (PEG) and manitol. High molecular weight PEG (4000 or more) is a satisfactory compound for simulating true drought without causing toxic side effects. The rate of germination under treatment with PEG is markedly reduced and hence this method has been suggested as a vigour test (Hadas, 1977).

Seed Borne Pathogens

A large number of fungi, bacterial and virus may attack seed prior to harvest and reduce the quality of seed. Greater progress has been made to detect seed borne pathogens in seed testing labs in Europe. This has resulted in the publication of procedures for seed health testing by the International Seed Testing Association in the Official Rules for Seed Testing (ISTA, 1976) and in a Handbook for Seed Health Testing (ISTA, 1959).

Seed borne pathogens can be detected by :

- Visual examination of dry seed
- Incubation of seeds on agar plates
- Ordinary germination blotters or cellulose pads

Using agar method, seeds are usually surface sterilized in sodium hypochlorite and plated on acidified (pH 4.5) potato dextrose agar. The plates are held under fluorescent light at room temperature (22°) for 10 – 14 days and based on colony morphology, the fungal identification is done.

The blotter method combines the identification procedures commonly used by plant pathologists with the procedures used by seed analysts for germination test. The seeds are placed on moistened blotters or cellulose pads in petri dishes, plastic boxes or other suitable containers and incubated in fluorescent light for 7 days. Seeds or germinated seedlings are examined either macroscopically or with a stereomicroscope for presence of pathogen and extent of disease.

Purity Testing

Seed purity denotes the composition of a particular seed lot. It includes :

- Pure seed
- Other crop seed (OCS)
- Other distinguishable variety seed (ODV)
- Weed seed
- Inert matter

Pure seed is the portion of the true seed in the working sample representing the lot being tested. Other crop seed is the percentage of crop seeds other than the species being tested. Similarly, weed seeds are indicated. ODV is applicable in some specific cases only, for example, rice, etc. Inert matter denotes the portion of sample that is not seed. It includes :

- Chaff
- Stem
- Small stones
- Pieces of broken, damaged or immature crop or weed seeds that do not qualify as entire seeds

The criteria for this distinction are clearly defined in the Rules for Testing Seeds (AOSA, 1991). Standards for these applicable in India are given in Chapter

24 on "Seed Standards". Usually 2500 seeds are required for purity analysis. Purity separations are usually done by hand, although mechanical aids may be needed to speed up the procedure. Purity analysis equipment usually includes :

- A work board
- Adequate light
- Forceps
- Hand lens
- Stereoscopic microscope for identifying small seeds

Factors Affecting Seed Quality

- Environmental conditions
- Dates of planting
- Delayed harvesting after harvest maturity
- Alternate wetting and drying
- Fluctuations in temperature and RH in plant canopy
- Genetic factors
- Mechanical damage
- Seed borne diseases

Seed Testing Laboratory

The seed testing laboratory is the hub of the seed quality control. A seed testing laboratory can be housed as a separate building or it forms part of larger building. The entire work can be organized in a hall or in separate rooms. The space requirement for testing 10,000 samples per year as given in Table 26.2 may serve as a guide line.

Table 26.2. Space requirement of a seed laboratory

Particulars of room	Size of room (m)	Total area (m^2)
A. Rooms		
1. Sample receipt and preparations room	3 x 3.6	10.8
2. Moisture test room	3 x 3.6	10.8
3. Purity room	6 x 6	36.0
4. Germination room		
a. Walk-in-germinator	1.8x2.4	4.3
b. Cabinet-type germinator	3 x 3.6	10.8
c. Sand test room	3 x 3.6	10.8
d. Preparation room cum germination counting room	3 x 3.6	10.8

Contd...

Table 26.2. Contd...

5. Seed health test room (2 nos.)	4.5 x 6	27.0
	4.5 x 6	27.0
6. Special tests	3 x 3.6	10.8
7. Sample storage	6 x 6	36.0
8. Miscellaneous room and other office space (3 nos)	3 x 3.6	32.4
	Total	243.7
B. (i). Greenhouse (wire mesh enclosure)	4.5 x 6	27.0
(ii). Glass house	4.5 x 6	27.0
C. Grow out test plot		5000

Seed Organizations

International Organizations Associated with Seed Testing

International Seed Testing Association (ISTA)

The ISTA was established in 1924. The primary objective of ISTA is to develop, adopt and publish standard procedures for sampling and testing seeds, and to promote uniform application of them for the evaluation of seeds moving in the International Seed Trade. In addition, it also promotes research in all aspects of Seed Science and Technology, including sampling, testing, storing, processing and distribution. It encourages cultivar certification, participation in conferences and training courses aimed at furthering these objectives and establishes and maintains liaison with other organizations having common or related interests in seeds.

The technical and scientific work of the association is carried out by 15 special committees (e.g., committee on seed sampling and bulking, purity, germination, vigour, etc.). It publishes scientific and technical papers in the Association's Journal, Seed Science and Technology.

One of the foremost achievements of ISTA is the adoption of the International Rules for Seed Testing. These rules prescribe testing techniques based upon scientific evidence, which are accurate within stated statistical limits and practicable within the every day operations.

In developing the seed testing, the following objectives have served as guidelines.

- To provide methods by which the quality of seeds can be determined accurately.
- To prescribe methods by which seed analysts working in different laboratories in different countries throughout the world can obtain uniform results.
- To relate the laboratory results, in so far as is possible, to planting value.

- To complete the tests within the shortest period of time possible, commensurate with the above mentioned objectives.
- To perform the tests in the most economical manner.

The ISTA Rules for testing seeds are followed by its member countries, in carrying out seed testing work. The introduction of the International Seed Analysis Certificate, widely used in the International Seed Trade, is another important achievement.

Association of Official Seed Analysts (AOSA)

The need for standardization of seed testing methods led to the formation of an organization in 1908, then known as Association of Official Seed Analysts of North America. The phrase "of North America" was dropped from the Association's name in 1939. The basic objective of AOSA is to develop, adopt and publish rules for testing seeds, and to encourage research in seed technology. The AOSA membership currently includes official laboratories from the various states of the United States of America, U.S. federal laboratories, several research laboratories, and laboratories operated by the Canadian Department of Agriculture. The AOSA meets annually.

The significant achievements of AOSA include publication of Rules and Testing Seed. These Rules were first published in 1917 and since then have been periodically revised. Current and complete revisions of the Rules for Testing Seed (AOSA) are published every five years. In addition to Rules, the AOSA publishes the Newsletter (four issues each year) and the Journal of Seed Technology.

Indian Organizations

Central-sub- Committee on Crop Standards, Notification and Release of Varieties

This organization under the Department of Agriculture and Cooperation, Ministry of Agriculture, Government of India is responsible for deciding the crop standards for certification and notification and release of crop varieties. It is headed by Deputy Director General (Crop Sciences), Indian Council of Agricultural Research, New Delhi. Other prominent members of this committee are Deputy Director General (Horticulture), Jt. Secretary (Seed), Department of Agriculture and Cooperation, Government of India, Director (Seeds), Department of Agriculture and Cooperation, Government of India, New Delhi, Assistant Director General (F&FC), ICAR, New Delhi, Deputy Commissioner (Quality Control), Department of Agriculture and Cooperation, Ministry of Agriculture, Government of India, New Delhi and several other concerned Crop Coordinators/Directors of crop based Institutes of ICAR.

APSA (Asia Pacific Seed Association)

Mission

To improve production and trade of quality seed and planting material of agricultural and horticultural crops

Objectives

- ❒ Regional forum for formulation of recommendations on seed policy issues
- ❒ Stimulate technical and economic co-operation among seed enterprises in the region
- ❒ Systematic interchange of information on various aspects of seed (breeding, production, conditioning, quality control and marketing)
- ❒ Relationship with FAO, FIS, ISTA, UPOV and others

Contact Details

APSA Secretariat

726 and 731 (7th Floor)

Institute of Food Research and Product Development Building

Kasetsart University, Bangkok 10903, Thailand

e-mail : apsa@apsaseed.com

References

Association of Official Seed Analysts. 1978. Rules for testing seeds. Proceedings of the Association of Official Seed Analysts. 60 (2) : 39.

Association of Official Seed Analysts. 1991. Cultivar Purity Testing Handbook. *In* M.B. McDonald and R. Payhe (eds.). Handbook on Seed Testing AOSA.

Association of Official Seed Analysts. 1983. Seed Vigiour Testing- Hand Book NO. 32.

Association of Official Seed Analysts. 1981. Rules for testing seeds. J. Seed Technol. 6 : 1-26.

Copeland, L.O. and McDonald, M.B. 1995. Principles of Seed Science and Technology. 3rd edition. Chapman Hall, p 393.

Delouche, J.C. and Baskin, C.C. 1973. Accelerated aging techniques for predicting the relative storability of seed lots. Seed Sci. and Technol. 1 : 427 – 452.

Hadas, A. 1977. A suggested method for testing seed vigour under water stress in simulated arid conditions. Seed Sci. and Technol. 5 : 519 – 525.

International Seed Testing Association 1959. Handbook for Seed Health Testing, ISTA, Zurich, Switzerland.

International Seed Testing Association 1976. International rules for seed testing 1976. Seed Sci. and Technol. 4 : 1 – 177.

Lakon, G. 1928. Ist die Bestimmong der Keimfanigkeit der samen ohne Keimversuch moglich. Angewandte Botanik (Zeitschrift der vereinigung fur angewan dte Botank) 10 : 470.

Maguire, J.D. 1962. Speed of germination aids in selection and evaluation for seedling emergence and vigour. Crop Sci.2 : 176 – 177.

Mathews, S. and Brandnock, W.T. 1967. The detection of seed samples of wrinkle – seed peas (*Pisum sativum* L.) of potentially low planting value. Proc. Int. Seed Test. Assoc. 32 : 553 – 563.

McDonald, M.B. Jr. 1980. Assessment of seed quality. Horticultural Science 15 : 784 – 788.

Perry, D.A. (ed.). 1981. Handbook of Vigour Test. Methods. International Seed Testing Association, Zurich.

27

Seed Conditioning or Processing

After harvesting and threshing, the seed usually arrives at the cleaning plant with trash, green leaves, weeds, other crop seeds and insects, etc. If the seed contains leaves, etc. and other high moisture materials, it cannot be accurately cleaned and safely stored until most of the foreign material has been removed. The process of removing the unwanted materials from a seed lot is known as seed conditioning. Seed conditioning is an essential component of seed technology to ensure that the farmers get the seed of high quality.

Basic Principles of Seed Conditioning

While running a seed conditioning or processing plant, the following 5 objectives are kept in mind (Copland and McDonald, 1995).

- Complete separation and removal of all contaminants
- Minimization of seed loss during removal of contaminants
- Upgrading/improvement of seed quality through removal of decayed, cracked, broken, insect damaged, injured or low quality seed
- Highest output/capacity of the processing plant consistent with effectiveness of separation
- Minimization of labour to reduce direct operating cost

During seed conditioning, seed quality is improved in 2 ways :

- Separation of contaminating seeds of other crops, weeds and inert matter
- Upgrading through elimination of poor quality seed

The seed conditioning is done with the ultimate objective of obtaining the maximum percentage of pure seed with maximum germination potential. This concept is better stated in terms of "pure live seed" percentage which is calculated as follows (Copeland and McDonald, 1995).

Pure seed (%)	=	95
Germination (%)	=	93
Pure live seed (%)	=	$0.95 \times 0.93 \times 100$
	=	88.35

Pure live seed content is a better measure of actual quality of a given seed lot than either the seed purity or germination alone. Seed conditioners use mechanical or electrical processes to separate the quality seeds from the seeds of other crops and weeds as well as inert matter. Inert matter such as chaff, stems and stones or dust are easiest to remove than other crop seeds and weed seeds specially when these are similar in appearance and physical characteristics. Physical characteristics used to separate seeds are :

- Size
- Length
- Width
- Thickness
- Weight (specific gravity)
- Surface texture
- Colour
- Affinity for liquids
- Electrical properties

Basic Steps in Seed Conditioning

- Seed receiving
- Need base bulk storage
- Pre-cleaning
- Cleaning
- Separating and upgrading
- Treating and bagging
- Storage
- Transport

Pre-cleaning

Precleaning refers to operations such as shelling, debearding, etc. It also includes removal of trash, stones, clods, etc. larger in size than desirable crop seed size. Some pre-cleaners in addition to removing larger size particles, also remove particles that are lighter in weight and smaller in size than the crop seed. The most common equipment used in pre-cleaning operations are scalpers, debearers heavier-scarifier, buck horn machine and maize sheller.

Scalper

The most commonly used pre-cleaning machine is a scalper. This machine is used to rough clean various kinds of trash from the seed lot. The most common scalper consists of a vibrating or rotating screen (or sieve) through which the small seeds pass readily while the larger seed and inert matter are "scalped off" and removed. This separation is normally accompanied by an air flow that blows away the chaff, stems and other lighter materials. To remove the lighter contamination from the seed without a sized separation, some scalpers utilize only the air flow of an aspirator through which the seed falls (Copeland and McDonald, 1995). Scalping is the most important pre-cleaning operation.

Basic Seed Cleaning

Basic seed cleaning refers to actual cleaning and grading of seeds. Unlike pre-cleaning/pre-conditioning, which may or may not be required, basic seed cleaning is an essential process in the seed cleaning operations. Basic seed cleaning is done over an air screen machine commonly referred to as air screen cleaner. This equipment is a basic equipment in the seed processing plants (Agarwal, 1995). The air screen machine works in 3 different ways. The seed first enters the machine by gravity from a seed hopper. In some machine, it falls directly on to a scalping screen which allows the good seed to fall through while separating out the larger seeded species and foreign material. In other machine the seed first falls through an aspirating air stream which blows of the lighter material from the seed mass. Last operation grades the seed by allowing the good seed to ride over screen perforations while the smaller particles drop through. All air screen machines use these three principles though their sequence may differ. Screens using an air screen machine are constructed of either perforated sheet metal or woven wire mesh on a wooden frame. Metal screen openings may be either round, oblong or triangular while openings in wire mesh screen are either square or rectangular. The perforated metal screens are measured in 64th of an inch (hole size) and the wire mesh screens are measured in number of openings per inch. Most air screens used in conditioning seed use three or four screens. These screens can be easily changed and almost hundred sizes are available (Copeland and McDonald, 1995).

Selection of Screens While Using Air Screen Machine

The screens are selected according to the shape of crop seed being cleaned. These are as follows :

- Round seeds : top screen – round hole – bottom screen – slotted
- Oblong seeds : top screen – oblong – bottom screen - oblong
- Lens shaped seeds : top screen – oblong/ rectangular – bottom screen – round hole

Gravity Seperator

The specific gravity separator is perhaps the second most commonly used piece of seed cleaning equipment. It is used to separate undesirable seed and inert contaminates that are so similar in size, shape and seed coat characteristics to the crop seeds that they can not be removed in any other way. This is also the best machine for upgrading seed quality. The deteriorated, moldy or decayed seed which are usually similar in size and shape to good seed but have lower specific gravity can be removed by gravity separator. Insect damaged seeds, that have low specific gravity can also be removed by this machine.

Length Separators

Length separators are designed to separate seed different in length. These machines are indent disc and indent cylinder separators. The indent disc separator uses a series of indented discs that are rotated inside the cylinder through which the seed moves. Discs may be used with indents that will lift out shorter contaminating seeds from the longer crop seeds. The indent cylinder works as the indent disc but indentations are in the inside cylinder housing which lifts out undesirable seeds as it rotates on an almost horizontal axes.

Width and Thickness Separators

These are commonly called graders or sizers by the seed trade companies. There are three types of graders used:

- Ribbed horizontal flat screen type
- vertical ribbed screen type
- Cylindrical screen type

Besides these commonly used seed cleaning equipment, there are some special types of seed cleaning machines which include surface texture separators, spiral separators, colour separators, electrostatic separators, vibrator separators and affinity for liquid separators. The details are available in Copeland and McDonald (1995).

Seed Treatment

Seed treatment is the process of applying chemical substances to seeds to reduce, control or repel seed borne, soil borne or airborne pathogens. Chemical treatment of seed is usually a routine part of seed conditioning.

Advantages of Seed Treatment

- Prevention of spread of plant diseases
- Protection of seed from seed rot and seedlings blight
- Improvement in seed germination
- Protection from storage insects
- Control of soil insects

The Ideal Seed Treatment Chemicals

The ideal seed treatment chemicals should be :

- Highly effective against pathogenic organisms
- Relatively non-toxic to plants
- Harmless to humans and livestock even if misused
- Stable for relatively longer periods of time during seed storage
- Easy to use
- Economically competitive

However, it can be stated that none of the chemicals currently available meets all these requirements.

Treatment Formulation and Equipment

The types of pathogenic organisms controlled by seed treatment include:

- Fungi and bacteria causing seed rots, seedlings blight and smuts
- Soil insects such as seed corn maggot and wire worm
- Storage insects including weevils, moths, beetles

Systematic Protection

Chemicals in this group penetrate the seed and kill or suppress the activity of pest or pathogen. They may also extend into and protect the resulting plant. Seed treatment chemicals are normally combined with other materials that enhance or maintain their ability. Many formulations include several inert ingredients to act as carriers, binders, wetting agents, sticking agents, emulsifiers, suspending agents and dyes. These inert materials are not listed on the label as they are added to the formulation to improve appearance, increase coverage and adherence, present dusting off or make the formulation easily recognizable.

Seed treatment chemicals may be applied as slurry (a mixture of wettable powder and water), a liquid or a dust. Slurry method is the most common. Special equipment is available for applying each type of formulation in the seed conditioning plants. In case of non-availability of seed treatment equipments, the routinely available home methods may be applied taking care that the chemical is applied uniformly to the seed.

Design of Seed Conditioning Plant

The design and layout of seed conditioning processing plant should be planned in such a way so that :

- The seed receives the necessary conditioning and the proper sequence
- There are no bottlenecks

- Operating costs are kept to a minimum
- The seeds are not injured from excessive handling
- Facilities are completely cleanable
- The chance of contamination is kept to a minimum
- If possible seed should be elevated only once to overhead storage bins where the seed could be held until conditioning
- From the overhead storage bins the seed should be allowed to fall by gravity between the various pieces of conditioning equipment

A floor plan for seed conditioning plant as suggested by Copeland and McDonald (1995) is shown as follows :

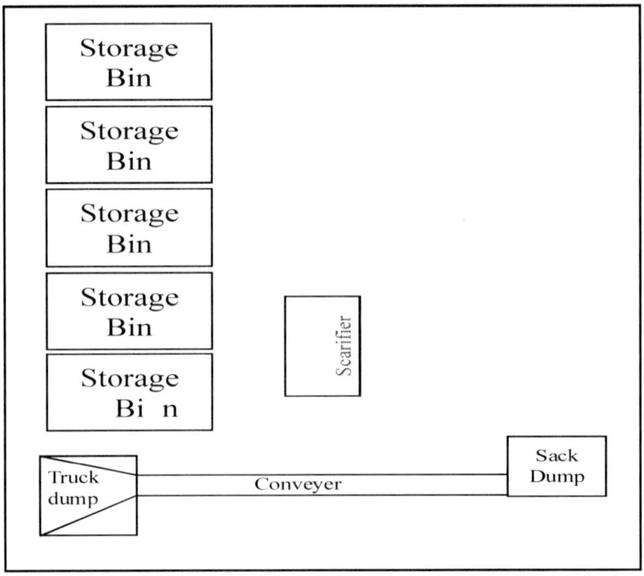

References

Agrawal, R.L. 1995. Seed Technology, Oxford and IBH.

Copeland, O.C. and McDonald, M.B. 1995. Principles of Seed Science and Technology. Chapman and Hall.

28

Seed Enhancement

Seed enhancement is the process or technology to further improve seed performances under very specific regimes and with certain planting equipment. Three major seed enhancement technologies are as follows (Copeland and McDonald, 1995).

- Seed hydration
- Biological seed treatments
- Seed coatings

Seed Hydration

Seed hydration is a process whereby seeds are hydrated using various protocols and then re-hydrated to permit routine handling. This process gives increased germination rate, more uniform emergence, germination under a broader range of environments and improved seedling vigour and growth.

To accomplish these objectives the seeds are hydrated in some way at a moisture level sufficient to initiate the early events of germination (phase II of imbibitions) but not sufficient to permit radicle protrusion (phase III). Three general approaches to hydration have been developed. These are :

- Pre-hydration
- Priming
- Solid metrics priming

Pre-hydration

According to Bradford (1986) soaking seeds in water prior to sowing enhances germination and seedling growth by controlling the imbibition conditions and reducing the vagaries of adverse weather and soil conditions. Seed soaking is usually done in water but seeds can often be pre-germinated in gels at 20°C and then planted. This process is known as fluid drilling. The gel protocols allow the germinated seeds to maintain seed moisture. Taylor and Harman (1990) have suggested the use of hydroxyl ethyl cellulose, magnesium silicate and poly-acrylamide. In some cases nutrients and pesticides are also added to improve seedlings performances.

Priming

The terms priming and osmo-conditioning have been used to describe the soaking of seeds in aerated low water potential osmotic compounds such as polyethylene glycol (PEG), KNO_3, K_3PO_4 KH_2PO_4, $MgSO_4$, NaCl, glycol and mannitol. These seeds are helpful in supplying the seeds with nitrogen and other nutrients required for protein synthesis during seed germination. However, some of these chemicals may prove to be toxic to the germinating seedlings. Presently, the most preferred osmotic compound is PEG. This is high molecular weight inert compound and because of large molecular size does not enter the seed and thus does not cause any toxic side effects which are associated with use of salts. Large quantities of seeds cannot be treated using this technique. Priming has been successful for crops like tomato, carrot, onion, pepper, celery, parsley, wheat, barley, sorghum and ryegrass. After priming the seeds are dried back to enable normal handling, storage and planting of the seeds.

Solid Matrix Priming

In this method the seed hydration is accomplished by the use of solid carriers with low metric potentials. Some of the ideal characteristics of these carriers are:

- Low metric potential
- Negligible water solubility
- High water holding capacity
- High surface area
- Non-toxicity to the seeds
- Ability to adhere to the seed surface

Natural substances with these characteristics are vermiculite, peat-moss and sand. Commercially available substances include elite and micro-cel (Copeland and McDonald, 1995). In this method seeds are generally placed into the carrier at matrix potential ranging from -0.4 to -1.5 MPa at 15° for 7 - 14 days and brought to moisture equilibrium. After that period the solid material is sieved away from the seed. This technique is found useful in enhancing seed performance of a number of crops including red beet, sugar-beet, onion, tomato, pepper, carrot and celery.

Biological Seed Treatments

Biological seed treatments are those treatments that include fungi or bacteria to control soil or seed pathogen. These are used in place of chemical seed treatments.

Biological seed treatments have a narrow range of specificity often killing just one specific pest. In this sense that are not broad spectrum pesticide (Cooka and Backer, 1983). In recent years this trend seems to be reversing.

Bioagents such as *Trichoderma* control several pathogens such as *Pythium, Rhozoctonia* and *Fusarium* spp. Some biological seed treatments can colonize plant roots and contribute to long as soil pathogen does not effect the population of bioagent on the roots. Bioagents have also been reported to induce growth leading to increased yield enhancement (Ahmad and Backer, 1987). Biological seed treatments have fungi such as *Trichoderma harzianumu, Pythium oligandrum* and *Chaetomium globosum* which are able to control damping off and other diseases. The effectiveness of bioagents as seed treatment depends considerably on iron content of soil. Certain *Trichoderma* species are ineffective in lower iron soil perhaps due to more competition for iron from *Pseudomonas* (Hubbard *et al.*, 1983). In some cases the biological seed treatments are not persistent in the soil, seed or the plant under natural conditions (Copeland and McDonald, 1995).

The application of the bioconrol agent is influenced by :

- Efficacy of biocontrol agent
- Number of propagules added to the seed
- Control of other microbes in the application

Seed Coatings

Under seed coating the seed placement and performance can be enhanced by altering the shape of the seed or placing chemicals on the seed coat which regulate an improved germination. These are of two types given below (Copeland and McDonald, 1995).

Seed Pelleting

A seed pellet is applied to a seed to improve its planting ability and performance particularly in those cases like some vegetable seeds which are not uniformly round. This irregular seed shape creates problem in precision planting. In some cases seeds are small and light because of which their accurate placement in the soil is difficult. To overcome these problems some seed companies provide seeds with coating of a material that changes the shape and size of the seed making the seeds heavier and rounder. A seed pellet is characterized by its quality to totally obscure the shape of the seed on which the seed pellet has been applied. Aamalgam of fillers (clays, lime stone, calcium carbonate, talc, vermiculites) and cementing additives (gum, gelatin, methyl cellulose, polyvinyl alcohol, poly-oxalic ethylene, glycol, waste waxes) are used to form the pellets. The seeds are introduced into a coating drum for seed pelleting. The addition of calcium oxide and peroxides in the pelleting material is believed to release oxygen to the seed under flooded conditions minimizing anaerobic damage. At the end of the pelleting process, a binder is added to harden the outer layer of the pellet. After pelleting is complete, the pelleted seeds are dried and handled as unpelleted seeds. Certain technical problems need to be carefully tackled during the pelleting process. These are :

- In compatibility of pelleting material with the seed
- Excessive wetting during pelleting leading to inadvertent seed hydration and reduced seed quality
- Too hard pelleting material after drying creating difficulty for radicle emergence through the pelleted material
- Pelleting around multiple seeds
- Pelleting around inert matters or undesirable plant material

Despite the technical problems, pelleting of seed is recognized as an important addition to the precision planting in several vegetables such as onion, lettuce, carrot and many flowers. Addition of beneficial chemicals such as plant hormone, micronutrients, microbes and fungicides to the pellet further improves performance of the seed in the field.

Seed Coating

Seed coating is one of the most economical approaches to improve seed performance. In seed coating, substances such as fungicides, micronutrients, etc. are directly applied to the seed. In seed coating unlike seed pelleting the shape of the seed remains unchanged. The ideal requirements of the seed coating polymer are as follows (Rushing, 1998):

- It should be water based polymer
- Should have low viscosity range
- Should have high concentration of solids
- Should form a hard film upon drawing

The application of seed coatings is similar to slurry seed treatments. The advantage of such treatment is that the substances are placed directly on the seed. Because of this less chemical is required in comparison to broadcast or furrow application. Another report to seed coating is film coating of seeds where additives dissolved in a dried solution of a sticky solution are applied. In this approach the seeds are dipped or sprayed with the dissolved polymer followed by immediate drying.

References

Ahmad, T.S. and Baker, R. 1987. Competitive saprophytic ability and cellulolytic activity of rihzosphere – competent mutants of *Triochoderma harzianum*. Canadian Journal of Microbiology 34 : 229 – 234.

Bradford, K.J. 1986. Manipulation of seed water relations via osmotic priming to improve germination under stress conditions. Horticultural Science 21 : 1105 – 1112.

Cook, R.J. and Baker, K.F. 1983. The nature and practice of biological control of plant pathogens. St. Paul. Minn. American Phytopathological Society.

Copeland, L.O. and McDonald, M.B. 1995. Principles of Seed Science and Technology. Third Edn. Chapman and Hall.

Hubbard, J.P., Harman, G.E. and Hadar, Y. 1983. Effect of soil borne *Pseudomonas* spp. as the biological control agent on pea seeds. Photopathology 73 : 655 – 659.

Rushing, K.W. 1988. Additives to enhance seed quality and seed performance. Proceedings of Mississippi State University, Mississippi State, MS State Seed Short Course Vol. 30.

Taylor, A.G. and Harman, G.E. 1990. Concepts of technologies of selected seed treatments. Annual Review of Phytopathology 28 : 321 – 339.

29

Seed Storage

The purpose of seed storage is to preserve planting seed from one season to the next. Further, the objective of seed storage is to maintain seed quality for the longest duration possible.

Relative Humidity and Temperature Influence Seed Storability

The two most important factors that influence the storability of seed are relative humidity and temperature. The effects of relative humidity and its subsequent effect on seed moisture and the temperature of the storage environment are highly inter- dependent. Harrington (1963) gave two useful thumb rules for relative humidity and temperature in reference to storage life of the seed. These rules are as follows :

- For each one percent reduction in moisture content, the storage life of the seed is doubled.
- For each five degree centigrade lowering of the storage temperature, the storage life of the seed is doubled.

Most crop seed loose their viability at relative humilities approaching 80 % and temperatures of 25 - 30°C but can be kept 10 years or longer at relative humidities of 50 % or less at the temperature of 5°C or lower (Toole, 1950). According to Harrington (1973), the sum of the percentage of relative humidity plus the temperature in F degrees should not exceed 100 for safe storage of seed. According to Bass (1967), this combined total may be as high as 120 as long as the temperature contributes no more than half the total. Toole (1957) suggested that relative humidity should be no high than 60% for seeds at 21°C and no higher than 70 % for seeds at 4 - 10°. However, at 5°C and 45 -50 % RH, many crop seed can be safely stored for 10 years or longer.

Seed Moisture

Harrrington (1972) recognized that there must be some qualifications to the rules of thumb regarding role of relative humidity and temperature on storage life of seed. According to him, the rule regarding seed moisture does not apply above 14 % or below 5 % seed moisture. Seeds stored at moisture content above 14 % begin to exhibit increased respiration, heating and fungal invasion and accordingly seed viability is destroyed more rapidly than indicated by the first rule of thumb. Below 5 % seed moisture a breakdown of moisture structure hastens seed deterioration. Thus, storage of most seed between 5 and 6 % seed moisture appears to be ideal for maximum longevity.

Genetic Factors

Seeds of some species have longer inherent storability than other seeds under similar conditions. Most long lived seeds belong to species having hard, impermeable seed coat. Seeds of *Canna, Lotus and Lupinus* have been reported to be viable even after 500 years (Copeland and McDonald, 1995). Seeds of lettuce, onion, parsnip and rye are short lived. Wheat has intermediate storability. Generally seed species with high oil content have poor storability in comparison to low oil content seed species. The genetic factors affecting seed storability have been well recognized and species have been divided into 3 broad categories according to their relative storability (Table 29.1, Copeland and McDonald, 1995).

Genetic differences in seed storability are not limited to the seeds of different species. Seed storability differences may also be observed among cultivars of the same species. Such differences have been reported in several crops including soybean, where the small seeded Indian local cultivars like yellow seeded Type 49 and black seeded Kalitur have shown better seed storability than the introduced cultivars like Bragg, Hardy, Jupiter, Clark 63 Monetta, etc. Kueneman (1983) demonstrated by means of reciprocal soybean crosses that the maternal plant affects the storability of F_1 seeds perhaps by influencing the permeability of seed coat. Thus, the genotype of the seed plays a distinct role in affecting seed longevity and accordingly deliberate breeding programme should be initiated to improve the seed storability. However, it should be kept in mind that the environment strongly affects the genetic potential of seed longevity.

Table 29.1. Relative storability index (1 = 50 % of the seed are expected to germinate after 1 to 2 years storage, 2 = 50 % of the seeds are expected to germinate after 3 to 5 years storage, 3 = 50 % of the seeds are expected to germinate after 5 or more years) of important crop seeds.

Crop	Relative storability Index	Crop	Relative Storability Index
Alfalfa	3	Bachiagrass	2
Barley	2	Field bean	2
Sugarbeet	3	Creeping bent grass	3
Velvet grass	3	Bermuda grass	1
Kentucky bluegrass	2	Big bluestem	2

Contd...

Table 29.1. Contd...

Mountain brome	2	Smooth brome	2
Buck wheat	2	Buffalo grass	3
Reed canary grass	1	Carpet grass	3
Alsike clover	3	Berseem clover	2
Crimson clover	2	Subterranean clover	2
Field corn	1	Cotton	1
Cowpea	1	Dallis grass	1
Chewings fescue	2	Tall fescue	2
Flax	2	Blue grass	2
Harding grass	1	Korean lespedeza	1
Yellow lupine	1	Foxtail millet	1
Pearl millet	1	Oats	2
Orchard grass	1	Peanut	1
Rape	2	Redstop	1
Rice	2	Rye	1
Perennial rye grass	2	Sorghum	1
Soybean	1	Sunflower	1
White sweet clover	3	Timothy	2
Birds foot trefoil	2	Hairy vetch	3
Wheat	2	Broccoli	2
Cabbage	2	Carrot	2
Sweet corn	2	Cucumber	2
Lettuce	1	Onion	1
Parsley	1	Parsnip	1
Pumpkin	2	New Zealand, Spinach	2
Tomato	3	Turnip	2

Seed Storage Systems / Approaches

Four principal approaches recommended for seed storage as given by Copeland and McDonald (1995) are :

- Conditioned storage
- Cryogenic storage
- Hermetic storage
- Containerized storage

Conditioned Storage

Under this approach seed can be stored for several years by careful control of temperature and relative humidity. This storage involves placing the seeds in a dry and cool environment. Seeds of most grain crops for example maize, wheat, barley, sorghum and oats will maintain satisfactory germination and vigour for about one year at a seed moisture content of 12 – 13 % under normal warehouse temperature. For longer storage, seed moisture content should be less than 11 % and temperature should not exceed 20°. Seeds of most temperate zone grasses and legumes can be stored safely for one year at moisture content of 10 - 11 %, under normal warehouse temperature. For longer storage period of these seeds, the seed moisture content should be less than 10 % and temperature should be close to or less than 20°C. Seeds of other species like soybean and peanut which are rich in oil, loose viability even over one year if kept at 11 - 12 %, moisture content at the temperature of 20°C or less. Although carryover of soybean seeds is generally not recommended, the seeds may be kept safely for 18 – 20 months in conditioned storage facilities of 20°C and 50 % relative humidity. The storage chambers to meet the above temperature and relative humidity require refrigeration and dehumidification equipment.

Cryogenic Storage

In this method the seeds are placed into liquid nitrogen at the temperature of -196°C. The advantage of system is that the seeds are placed at a temperature where little physiological activity occurs and thus the storage life of the seeds is prolonged. It is not practical for the most commercial seeds because of storage space limitation for bulk seed storage. However, it is useful for maintaining valuable seed germplasm over prolong periods. According to Stanwood (1985) most agronomic crops can be successfully stored in liquid nitrogen. Under such circumstances there is almost no metabolism and seeds are successfully stored for prolong periods.

Hermetic Storage

In recent years, packaging seed in moisture resistant or hermetically sealed containers for storage and marketing has been on increase. Under such conditions, seeds are able to maintain their moisture at a safe level. Ordinary paper and cloth containers are least effective in maintaining seed moisture. Laminated and poly-vinyl materials are moderately effective. Metal cans are completely effective in maintaining seed moisture at initial 5 % level. Such completely moisture proof containers hermetically seal the seed and are effective for long term seed storage up to 10 years or more. The moisture content of seed under airtight storage must be 2 - 3 % lower than that at which seeds are normally packaged in non-sealed storage.

Harrington (1973) has observed that starchy seeds above 12 % moisture and oily seeds above 9 % moisture deteriorate faster in sealed storage than non-sealed storage. It is believed that the atmosphere in a moisture proof container holding seeds at 13 % moisture will equilibrate at a relative humidity of 65 %.

The most extreme case of hermetic storage is to store seeds in vacuum at low moisture content. Some reports have indicated that this approach extends seed storage life but majority of the reports suggest no benefit from vacuum storage as reviewed by Copeland and McDonald (1995). There have been studies where the ambient air of the sealed container has been replaced with pure CO_2 and inert gases like nitrogen, organ and helium but most reports show little or no advantages of this approach.

Containerized Seed Storage

Humidity control in seed storage areas requires considerable investment in specifically constructed rooms treated with refrigeration unit and dehumidifying equipment. High value seed stocks and germplasm are maintained in such facilities. The National Seed Storage Laboratory operated by United State Department of Agriculture Fort Collins as a permanent repository for valuable germplasm of all kinds of plant, provides temperature and relative humidity conditions satisfactory for long term storage of all kinds of seeds. Such facilities are available in India at National Bureau of Plant Genetic Resources, New Delhi and some selected crop based ICAR institutes and State Agricultural Universities, Pantnagar Agricultural University operates a medium term storage module (MTS) where relative humidity is maintained at 30 % and temperature close to 4°C. Where such elaborate facility is not available, humidity can be closely regulated in a closed container by use of chemical desiccants that have no moisture equilibrium values.

Saturated salt solutions (barium chloride, calcium chloride, calcium nitrate, potassium chloride, potassium nitrate, potassium chloride, magnesium chloride, magnesium nitrate, sodium chloride, etc.) or acid solutions (sulphuric acid, nitric acid, hydrochloric acid) may be used. Care must be taken to keep the seed separate from the liquid desiccant by using metal platforms, cheese cloth or wire containers for holding the seed. Depending upon the size of container, small fans may be used to have uniformly distributed relative humidity.

Warehouse Seed Storage

The seeds for next year sowing are usually stored in warehouses without the requirements of refrigerators and humidity control. However, carryover seeds should not be stored for planting in the warehouses. While constructing warehouses certain measures need to be adopted for safe storage of the seeds till next planting season.

General Features of Seed Warehouse

A good warehouse should have following constructional features (Agarwal, 1995).

- Warehouse should have no windows and have only one door constructed of metal which can be sealed properly and locked.
- The material used for construction may be concrete, brick, metal or wood.
- Regardless of the material, the foundation should be made of stone or concrete and should extend 90 cm above the ground level.

- A base around the building at the 90 cm height extending out 15 cm should also be constructed. Such construction makes entrance by rodents virtually impossible, as long as the foundation remains un-cracked.
- The floor must be paved and any cracks that may develop must be repaired.
- Construction of the floor, walls and ceiling of the storage should be such that no cracks exist which can harbour insects. All cracks around openings e.g. electric conduits, ventilation openings and doors should be thoroughly sealed.
- Ventilation openings should be screened against birds and insects.

Measures for Pest and Disease Control in Warehouse Storage

Sanitation

Good sanitation in the seed store is necessary for protection from insects and rodents. The torn seed bags should either be immediately repaired or replaced with new bags to avoid spillage in warehouse. All spilled seeds or floor sweepings should be immediately removed. Discarded seed and cleanings should be carried away. In addition to cleaning, the floor and walls should be sprayed with an appropriate insecticide as often as required. In a well maintained store, spraying once a year may be quite satisfactory. The recommended materials and rates are DDT – (50 per cent WP) ½ to 1 lb/gal/1000 ft^2 ($1.25\ g/m^2$), Malathion (25 per cent WP) 1 lb/gal/1000 ft^2 ($1.25\ g/m^2$). Malathion should not be used on such surfaces as brick, cement and concrete.

Seed Treatment

An insecticide combined with a fungicide may be applied as a protectant. The most commonly used insecticide is DDT. DDT also has the advantage of long duration.

Fumigation

Once the seed storage is free of insects, the most serious source of re-infestation is infested seed which is brought in. Such infestation is controlled by fumigation. Rather than fumigating the whole storage, it is better to have a fumigation room, or to fumigate the seed on a concrete floor under a tarpaulin before it enters the main storage room. The fumigation room should have its own door to the outside and only after fumigation the seed should be brought into the storage area.

Fumigation is effective only in gas-tight storage. Numerous effective fumigants are available. However, there is a small safety margin between the dose that is toxic to insects and a dose that will cause loss of germination or vigour of seed. Reasonably safe fumigants at temperatures below 30°C and, seed moisture below 12 per cent are the following (Agarwal, 1995).

Dosage	Stored period
Methyl bromide 16 to 32 oz per 1000 ft^3 (16 to 32 mg per cubic metre)	24 hours
Hydrogen cyanide 32 to 64 ox per 1000 ft^3 (32 to 64 mg per cubic metre)	24 hours
Hydrogen phosphide 5 to 10 tablets per metric ton of seed (Phostoxin, phosphine)	3 to 7 days

It must be borne in mind that fumigation, particularly repeated fumigations, may seriously reduce the vigour and even the germination capacity of seeds. This is particularly true of seeds with a high moisture content. Seeds with moisture contents greater than 14 per cent should be dried to below this value before fumigation. High temperature also increases damage to seeds by fumigants. Hence, fumigation should be used only with entering seeds, and all other measures should be used to maintain insect control in the stored seed. Prophylactic biweekly or monthly fumigations of seed storages can lead to serious germination problems. It is far better to build an insect-proof storage and make certain that seeds, bags or anything else placed in the storage are insect-free. Of course, if a storage containing such seeds does become infested, then fumigation is required.

Temperature

As mentioned earlier, temperature is one of the most important environmental factors which influences seed viability and vigour during storage. The lower the temperature, the longer the seeds maintain germination capacity. Thus, temperature control is an important consideration in building a seed storage.

Temperature control may be achieved in one of the following ways :

1. Ventilation
2. Insulation
3. Refrigeration

These methods are not mutually exclusive and are normally used to supplement each other.

Ventilation

Ventilation could be used to reduce seed temperature and seed moisture content, if used judiciously. In addition, it also helps prevent hot spots from developing, the formation of convection air currents and maintenance of uniform seed moisture content and temperature.

Time of Ventilation

Whenever the outside temperature of air and relative humidity are low enough to benefit the seeds, either by reducing seed temperature or seed moisture content, the ventilating fans (exhaust fans) can be turned on.

Precautions to be taken during ventilation :

1. The moisture content of the seed should not be allowed to increase to a value in equilibrium with air relative humidity above 65 per cent and
2. the seed temperature should not increase above 30°C for more than few hours

Insulation

Insulation of seed storage is done to reduce the flow of heat from the warmer exterior, through the walls, roof, and floor of the storage, to the cooler air and seeds in storage.

Foam has the advantage of being light in weight. Because the air bubbles are enclosed, no convection currents are possible. It has considerable resistance to moisture penetration. It is easy to install, has mechanical strength and dimensional stability. However, it is inflammable. It is eaten by rats and cockroaches. It can become water- logged, losing its insulation value if the storage develops leaks. Certain fumigants may affect it. It does deteriorate slowly with age.

Glass wool is rat-proof, but insects can harbour in it. It is odourless. It has a low water absorption. It is fire proof. It is unaffected by fumigants. However, it allows a free flow of moisture, so an additional moisture barrier is imperative. People can be allergic to it. Convection currents are possible particularly in the lighter grades. It has less dimensional stability and mechanical strength than rigid insulation material, but is more easily worked around the curves and corners.

Refrigeration

The basic objective of refrigeration is to keep the storage temperatures below the usual ambient temperatures. An alternative to refrigeration is storing the seeds dry, either by using dehumidification or by drying and storing in sealed containers. Refrigeration often becomes necessary for carryover seeds.

Special kinds of seeds, for example, foundation seed and nucleus seed/breeder seed are not required for normal storage for next season planting.

Seed Packaging in Relation to Seed Storage

Factors to Consider

The seed package, in reality, is a small storage container. The kind of container needed is affected by several factors including:

- The quantity of seed desired in each package
- The protection desired
- The cost of the package
- The value of the seed

- The storage conditions into which the container is to be placed
- The facilities for drying the seeds

As a matter of convenience to the farmers, and in situations where land holdings are small, the package is often designed to hold seed to plant one acre or one ha.

Packaging Materials

Cereal seeds have commonly been packaged in cotton, jute and paper bags. Cost and availability of materials have been the major considerations in determining what to use. Moisture is the major consideration in determining what to use. Moisture vapour penetrates all these containers and they offer no protection against high relative humidity. They require dry warehouses, and storage conditions that will keep the seed from acquiring moisture levels above 12 per cent until use. Where these requirements can be satisfied, as is often the case, the use of cotton, jute and paper bags is probably the most satisfactory method of seed packaging.

In high humidity locations with adequate seed storage facilities, consideration should be given to methods of packaging which can protect the seed from moisture vapour. Such moisture proof containers include aluminum polyethylene laminated pouches, polyethylene bags of over seven hundred gauge thickness, sealed tins and gasketed rigid plastic containers. The cost of many of these materials has been considered too high for the packaging of cereal seeds. Polyethylene bags have been regarded as the most attractive other possibilities because of their relatively low cost compared to other kinds of sealed containers. Heat sealing devices must be available for easy sealing. Self-sealing bags, also in use, require special filling equipment and sealing bags and a bag industry that is prepared to manufacture them. Rigid plastic containers and sealed tins offer some possibility for hybrid seed, if the quantity needed is not great.

Quality Control in Storage

Stacking

At moisture contents below 14 per cent the seed stacks could be 15 to 20 feet high, without fear of the weight of stack injuring the seeds in lower bags. However, the problem of more rapid loss in germination of the seeds in the bottom bags is real one. It does occur in seed storages owing to the increased moisture content of the bottom layers and also to convection currents and moisture penetrating through the floor.

In an un-insulated storage house, the heat from the sun will warm the air under the roof and on the south side of the storage. This sets up convection currents with the expanding warm air on the south side rising, because it becomes lighter. This pushes air to the north side which is cooled by the walls and seeds. It contracts, becomes heavier and sinks to the floor. Thus a slow, circular convection current occurs. Warming lowers the relative humidity of the air. Thus moisture moves from the upper seed to the warm air. As the air descends on the north side, it cools. This raises its relative humidity until it is above the equilibrium moisture content of the lower seed. Thus, a gradient from the air to the lower seed is established and moisture moves from the cooled air to this bottom. Hence, in un-insulated storage the air is circulating so slowly that a person may not notice it. But it is circulating and transferring moisture from the upper bags of seed to the bottom bags. Thus, although all the seeds when placed in the storage may have a 12 per cent moisture content and the average moisture content may still be 12 per cent, the seeds at the bottom may rise to 15 to 16 per cent moisture content, while those on the top are dried to 9 per cent or even 8 per cent.

Another reason for an increase in the moisture content of the bottom bags is moisture penetration through a concrete floor. There is a constant flow of moisture vapour through barrier into the seed storage. This moisture is not distributed evenly in the storage. Seeds become hygroscopic desiccants, the first or bottom seeds absorb the moisture entering through the floor and not until these bottom seeds have reached equilibrium and the rate of moisture entry may rise. Thus, the moisture content of the bottom seeds may rise to 15 to 16 per cent moisture content long before the upper seeds. Therefore, they lose germination long before the upper seeds.

Solution to Moisture Problems

To eliminate convection transfer of moisture, an operating exhaust fan to break up the convection current can be used. These fans need only to move 0.1 cu ft air per 100 kg seed per minute, e.g., in a storage 20 ft × 50 ft × 20 ft holding 200 tones of seed, the fan capacity would only have to be C.F.M. 200 to eliminate the convection currents. To reduce the temperature differential, insulation or a false ceiling, and white painting of the outside wall can help. To prevent moisture transfer through the floor in a new construction, the floor should be three feet above the ground level with sand and/or rock fill and a moisture proof barrier under the concrete. Storage of seed to prevent direct contact with a damp floor and monthly re-piling of the stacks by putting the top bags on the bottom and vice versa, will help prevent rapid deterioration of the bottom seed. However, elimination of the sources of moisture is a much better protection (Agarwal, 1995).

References

Agarwal, R.L. 1995. Seed Technology. Oxford and IBH.

Bass, L.N. 1957. Controlled atmosphere and seed storage. Seed Science and Technology 1 : 463 - 492.

Harrington, J.F. 1963. Practical advise and instruction on seed storage. Proceedings of the International Seed Testing Association 28 : 989 - 994.

Harrington, J.H. 1972. Seed storage and longevity. *In* T.T. Kozlowski (ed.). Seed Biology. pp 145 – 240. New York Academic Press.

Harrington, J.F. 1973. Biochemical basis of seed longevity. Seed Science and Technology 1 : 453 - 461.

Kueneman, E.A. 1983. Genetic control of seed longevity in soybeans. Crop Science 23 : 5 – 8.

Stanwood, P.C. 1985. Cryo-preservation of seed germplasm for genetic conservation. *In* K. Kartha (ed.). Plant Cryo-preservation. pp 199 – 225. Boca Raton, FLA, CRC Press.

Toole, T.H. 1950. Relation of seed processing and of conditions during storage on seed germination. Proceedings of the International Seed Testing Association 16 : 214 – 227.

Toole, E.H. 1957. Storage of Vegetable Seeds. U.S. Department of Agriculture -Leaflet. 220.